LABORATORY STUDIES OF VERTEBRATE AND INVERTEBRATE EMBRYOS

Guide and Atlas of Descriptive and Experimental Development

Eighth Edition

Gary C. Schoenwolf, Ph.D.

Professor
Department of Neurobiology and Anatomy
University of Utah School of Medicine
Salt Lake City, Utah

Prentice Hall
Upper Saddle River, New Jersey 07458
http://www.prenhall.com

Executive Editor: Sheri Snavely
Acquisitions Editor: Karen Horton
Special Projects Manager: Barbara A. Murray
Production Editor: James Buckley
Manufacturing Buyer: Dawn Murrin
Cover Designer: Bruce Kenselaar

Front Cover Photo: Four-day chick embryo labeled by *in situ* hybridization with the transcription factor *Lmx-1*; courtesy of Dr. Shipeng Yuan.
Back Cover Photo: Pseudocolored scanning electron micrograph of a parasagittal slice through a frog yolk plug gastrula; courtesy of Dr. Robert E. Waterman.

Printed in the United States of America

10 9 8 7 6 5 4 3

ISBN 0-13-857434-0

Prentice Hall International (UK) Limited, *London*
Prentice Hall of Australia Pty. Limited, *Sydney*
Prentice Hall Canada, Inc., *Toronto*
Prentice Hall Hispanoamericana, S.A., *Mexico*
Prentice Hall of India Private Limited, *New Delhi*
Prentice Hall of Japan, Inc., *Tokyo*
Prentice Hall Asia Pte. Ltd., *Singapore*
Editora Prentice Hall do Brasil, Ltda., *Rio de Janeiro*

Contents

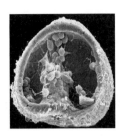

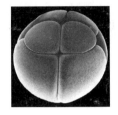

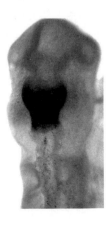

Preface

Five major changes have been made in the eighth edition. First, a new chapter (Chapter 4) has been added on the development of the mouse embryo. With the advent of modern techniques in molecular genetics, the mouse embryo has become an important model system for analyzing development. Virtually any desired gene can now be mutated in the mouse embryo, and the effects of its over expression or under expression can be readily studied. Additionally, recent advances in whole-embryo culture and techniques of experimental embryology have further increased the importance of the mouse embryo to developmental biologists. With this new chapter, we have added 69 photographs (Photos 4.1-4.69), 2 line drawings (Figs. 4.1 and 4.2), and 2 experimental exercises (Exercises 4.1 and 4.2).

Second, all images have been digitally processed. This was done to improve quality, clarity, ascetics, and ease of viewing. All illustrations have been relabeled and layout has been improved, with illustrations from each chapter grouped together within the relevant chapter rather than at the back of the book in a separate atlas. This should help the student find and learn the material quicker. Aside from the new illustrations added in Chapter 4, 28 additional illustrations (mostly scanning electron micrographs) have been included in Chapters 2 and 3, increasing the visual impact and depth of the manual. A new numbering scheme has been used for all illustrations. Both photographs and line drawings are now numbered according to chapter number (1-6) followed by a decimal point and the number of the illustration in sequence. Photographs are referred to as "Photos" and line drawings as "Figs."

Third, the layout of the text was also done digitally. In doing this, the text was reordered into 6 chapters, placed into two-column format, and updated, and the flow from topic to topic has been simplified.

Fourth, all exercises are now grouped together in Chapter 6, along with additional hands-on studies. Seven new advanced hands-on studies have been added, covering the cutting of frozen sections, in situ hybridization (plus 12 new illustrations), chick New whole-embryo culture, dye injections for tracking cell movements, BrdU labeling to study cell proliferation, TUNEL labeling to study cell death (apoptosis), and mouse whole-embryo culture. With the addition of 2 exercises on early mouse embryos, Chapter 6 now contains 22 exercises. These exercises allow students to gain some hands-on experience with, and deeper appreciation for, living embryos and the dynamic events underlying embryogenesis.

Instructors may choose to have students do all the exercises or only selected ones. Additionally, instructors may choose to do all or some of the exercises as demonstrations with various degrees of student participation. Furthermore, students may consider doing selected exercises as independent study projects. It is difficult to schedule experiments in a typical quarter or semester course format when embryos develop over hours or days, and not all experiments work every time. However, the pedagogic value of working with living embryos far outweighs the scheduling difficulties. Also, although sometimes experiments will fail, a failed experiment can often be more instructive than a successful one, leading to subsequent inquiry into what went wrong.

Fifth, a glossary has been added. This will help students find definitions quickly, facilitating their learning of important terms and concepts.

This edition has been written to allow students to start their laboratory studies with any of the organisms included, depending on the instructor's preference. However, we do recommend that students study the chick embryo prior to studying the mouse embryo; this will facilitate their understanding of the "inverted" U-shaped mouse blastoderm as compared to the flat chick blastoderm. Orientation illustrations ("orientators") are placed with photographs to indicate the levels of the sections, slices, or fractures illustrated. Unfortunately, it is not possible to indicate the level of every section shown in the photographs: sections are closely spaced and photographs are shown at a much higher magnification than are the orientators; thus, insufficient room is available to demarcate every section.

In closing, let me encourage both faculty and students to send their comments about this manual to me. Each new edition represents a stage in the manual's evolution (see History), a stage that is largely determined by the comments of users. To facilitate communication, you can contact me at the following Email address: Schoenwolf@med.utah.edu.

Dedicated to the memories of two former authors of Laboratory Studies:

Robert Milton Sweeney
and
Ray Leighton Watterson

Chapter 1

Sea Urchin Embryos

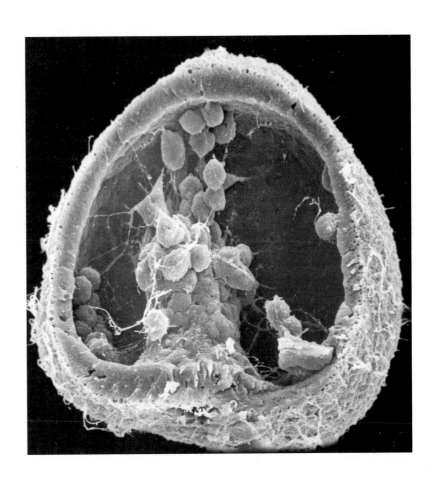

Chapter 1

Sea Urchin Embryos

A. INTRODUCTION

Sea urchin embryos have served as model systems for hundreds of descriptive and experimental studies. Of the classical studies, those of Sven Hörstadius are the most notable and lasting, in which sea urchin embryos were fate mapped with the use of vital staining. Adult sea urchins can be obtained readily and inexpensively. Simple procedures are available for inducing sexually mature adults to shed their eggs or sperm. Eggs can be fertilized artificially, and embryos develop in standard laboratory beakers containing seawater maintained at appropriate temperatures. Several species develop very quickly, and their embryos are highly translucent; thus development, including that of internal rudiments, can be easily viewed. Consequently, the sea urchin is an especially useful invertebrate for laboratory studies. For hands-on studies using sea urchin embryos, see Chapter 6 (Exercises 1.1-1.3).

This chapter is designed so that you can study sea urchin development by reading the text and examining the line drawings (Figs 1.1, 1.2) and photographs (Photos 1.1-1.44). Examination of commercially prepared slides is not necessary (commercial slides are usually of considerably lower quality than the specimens illustrated in the photographs; however, if such slides are available, they should be used in addition to the materials provided here if your instructor so directs). The intent of this chapter is to present high-quality illustrations of developing sea urchin embryos and to provide the opportunity to observe and experience living embryos. Photos 1.1-1.42 illustrate the development of a Florida Gulf-Coast sea urchin, *Lytechinus variegatus* (illustrations courtesy of Dr. Morrill). Photos 1.43 and 1.44 illustrate the development of a Japanese sea urchin, *Anthocidaris crassispina* (illustrations courtesy of Drs. Akasaka and Amemiya). A brief timetable of development of *L. variegatus* (courtesy of Dr. Morrill) at 25°C follows:

Cleavage Cycle	Potential Number of Cells Generated*	Time After Fertilization
1	2	1 hour
2	4	1 hour 30 minutes
3	8	1 hour 50 minutes
4	16	2 hours 20 minutes
5	32	2 hours 40 minutes
6	64	3 hours 15 minutes
7	128	4 hours
8	256	4 hours 30 minutes
9	512	5 hours
10	1024	5 hours 30 minutes

*The potential number of cells generated assumes that all blastomeres divide during each cleavage cycle. From the sixth cleavage cycle on, the actual number of cells generated is substantially less than the potential number (for example, only about 800 cells are present after the tenth cleavage). Thus from the sixth cleavage on, not all blastomeres divide during each cleavage cycle.

Estimates of the timing of later stages (based on the work of Dr. Morrill and co-workers) at 25°C follows:

Developmental Stages	Time After Fertilization
Hatching from fertilization envelope	7 hours 30 minutes
Primary mesenchyme formation	7 hours 30 minutes to 9 hours
Primary invagination of endoderm	11 hours
End of gastrulation	13 hours
Early prism larva	15 hours
Early pluteus larva	20 hours

B. GAMETOGENESIS AND FERTILIZATION

Eggs and **sperm** develop within the **gonads** (**ovaries** and **testes**) of adult female and male sea urchins, respectively. This process is called **gametogenesis**, specifically **oogenesis** in the female and **spermatogenesis** in the male. Sea urchins are **dioecious**; that is, there are separate male and female individuals. Because no external sexual dimorphism occurs, male and female animals appear identical to one another. **Gametes** develop within the **gonads**, which are suspended on the inner side of the **test** (the external skeleton, or "shell," of the sea urchin). Each animal contains five gonads; each gonad is covered with **coelomic peritoneum** and contains a layer called the **germinal epithelium**. Gametes are generated through the division of stem cells within the germinal epithelium. Each gonad also has **muscle layers** as well as a short duct, the **gonoduct**, which opens onto the **aboral surface** of the animal (the side opposite the **mouth**), near the **anus**. The opening of each gonoduct is called a **gonopore**.

Gametes arise in the following way. In sexually immature animals, each gonad contains stem cells, called **primordial germ cells**, within its **germinal epithelium**. As the animal approaches sexual maturity, stem cells undergo mitotic divisions, which generate millions of **oogonia** in the female and **spermatogonia** in the male. During the breeding season, oogonia and spermatogonia enlarge and form, respectively, **primary oocytes** and **primary spermatocytes**, which in turn initiate meiotic divisions. The primary oocyte is considerably larger than the primary spermatocyte. In the sea urchin, the volume ratio is estimated to be 400,000:1. The nucleus of each primary oocyte and primary spermatocyte is also enlarged and is called the **germinal vesicle**, within which one **nucleolus** is present.

In the female, each **primary oocyte** undergoes the **first meiotic division**, producing the **secondary oocyte** and the **first polar body** at the presumptive **animal pole**. The **second meiotic division** follows the first and produces the **mature egg** and the **second polar body**. Thus during **oogenesis**, one mature egg and two polar bodies are generated (note that the first polar body does not undergo the *second* mitotic division, and neither polar body plays any further role in reproduction). Only *mature* eggs can be fertilized.

In the male, each **primary spermatocyte** undergoes the **first meiotic division**, producing two **secondary spermatocytes**. Each secondary spermatocyte then undergoes the **second meiotic division**, producing two **spermatids**. Thus during spermatogenesis, four spermatids are generated. Spermatids must remain in the testes for some time to undergo a maturation process, after which they become **mature sperm** capable of fertilizing a mature egg.

During the breeding season, **eggs** and **sperm** are spawned (that is, shed) from the **ovaries** and **testes**, respectively, into the seawater by the contraction of the **muscle layers** of the gonads. The **gametes** flow during spawning from the gonad through the **gonoduct** and **gonopore** into the surrounding seawater. The presence of gametes in the seawater acts as a stimulant for the shedding of sex cells of an individual of the opposite sex, but the nature of the signal and how it is communicated is unclear. It is believed that the signal enters a buttonlike structure on the aboral surface called the **madreporite**, and from there it passes through the **water-vascular system**. Specialized **chemoreceptor cells** are believed to respond to the stimulant, thereby triggering release of the gametes.

The sea urchin **egg** is classified as **homolecithal**, which means that it contains little **yolk** and the yolk that is present is evenly distributed throughout the **cytoplasm**. At the time of spawning, the nucleus of the mature egg is called the **female** (or **egg**) **pronucleus**. The female pronucleus and its surrounding cytoplasm are enclosed within the **plasmalemma (oolemma)** of the egg; the latter is surrounded by the **vitelline envelope** (or **membrane**), which in turn is surrounded by a transparent **jelly layer**. In the West Coast sea urchin, *Strongylocentrotus purpuratus*, the egg is about 80 micrometers in diameter; its jelly coat is about 40 micrometers thick. Just beneath the plasmalemma, the cytoplasm contains numerous (about 15,000) granules called **cortical granules**.

Sea urchin **sperm** have the typical structures found in most sperm: the **head**, containing the **acrosome** and **nucleus**; the **midpiece**, containing a ring-shaped **mitochondrion** and a pair of **centrioles**; and the **tail**, containing microtubules arranged as nine outer doublets and two inner singlets. The sperm head, midpiece, and tail are all enclosed within the sperm **plasmalemma**. The acrosome contains a single, membrane-bound granule, called the **acrosomal granule**, packed with **acrosomal enzymes**. The sperm head is about 1 micrometer in diameter; its tail is about 50 micrometers long.

As gametes are shed, **fertilization** takes place in the surrounding seawater. The process is as follows. As sperm approach the jelly layer surrounding the egg, they undergo the **acrosomal reaction**. Two events occur during this reaction. First, the acrosomal membrane fuses with the sperm plasmalemma and exocytosis occurs, resulting in the release of enzymes from the acrosomal granule. These enzymes aid the sperm in penetrating the **jelly layer**. Second, a cytoskeletal protein called **actin** (located just beneath the acrosomal granule) undergoes rapid polymerization to form a filamentous rod that extends from the head of the sperm as the **acrosomal filament**. The acrosomal filament is covered with acrosomal membrane, and the two structures together constitute the **acrosomal process**. The acrosomal process establishes the initial contact between the surface of the egg and sperm after the sperm penetrates the **vitelline envelope**. Adhesion of egg and sperm during this initial contact is mediated by a species-specific **egg-binding protein** on the sperm's acrosomal membrane. Such close adherence allows the egg and sperm plasmalemmas to fuse. After the acrosomal process contacts the egg's plasmalemma, its membrane fuses with the oolemma. Typically, the egg's plasmalemma at the point of fusion with the sperm is drawn out as a cone-like projection called the **fertilization cone**. The fertilization cone is active in engulfing the sperm (actually, its nucleus, mitochondrion, centrioles, and perhaps tail microtubules) into the interior of the egg.

Fusion of the sperm with the egg initiates the **cortical reaction**. Just beneath the egg's plasmalemma is a monolayer of **cortical granules**, each approximately one-half to one micrometer in diameter. The membranes surrounding the cortical granules fuse with the egg's plasmalemma at the moment of sperm-egg fusion. Fusion of the cortical granules with the egg's plasmalemma occurs in a wavelike fashion, beginning at the point of sperm contact and radiating out circularly along the entire surface (the plasmalemma) of the egg. As fusion occurs between the cortical granules and the egg's plasmalemma, the contents of each cortical granule is expelled by exocytosis into the space between the egg's plasmalemma and the enclosing **vitelline envelope**. Several different substances are expelled. Some of the expelled substance joins with the vitelline envelope, causing it to harden and to elevate, forming the **fertilization envelope** (or **membrane**) and the **perivitelline space** (Photo 1.1). Some of the expelled substance consists of **hyaline protein**, which forms the **hyaline layer** immediately enclosing the surface of the **zygote**. Its

function is to hold blastomeres together during subsequent cleavage. The **cortical reaction** requires the entrance of **calcium ions** into the egg (see Chapter 6, Exercise 1.2). The cortical reaction changes the surface properties of the egg and provides a block to **polyspermy**, the entrance of more than one sperm into the egg. In addition to the cortical reaction, polyspermy is blocked by a much more rapid process: the **depolarization** of the egg's plasmalemma (a change in the electrical potential across the plasmalemma to a less negative state).

The sperm nucleus enlarges in the cytoplasm of the egg and is called the **male** (or **sperm**) **pronucleus**. The **male** and **female pronuclei** approach one another and undergo **syngamy**, which means that they fuse to form the **zygote nucleus** (Photo 1.1). Shortly thereafter, the first cleavage occurs. Mitotic spindles are organized for cleavage by the two **centrioles** contributed to the zygote by the midpiece of the sperm.

C. CLEAVAGE AND BLASTULATION

After fertilization, the **zygote** undergoes **cleavage**. The pattern of cleavage varies within different species of sea urchins, and new research continues to refine our knowledge of the precise patterns. Here, we will describe the pattern of cleavage as originally described by Hörstadius, with minor modification based on the pattern that occurs in the sea urchin *Lytechinus variegatus*, which is illustrated in the photographs.

The sea urchin egg, like all eggs, contains an **animal-vegetal axis**. However, the **animal pole** and the **vegetal pole** are not readily discernible (except by the fact that the **egg nucleus**, or **zygote nucleus**, resides near the animal pole) until cleavage is well under way and cell (blastomere) size can then be used for orientation. Historically, the animal pole was considered to contain "life," that is, the nucleus, and thus it was called the animal pole, whereas the opposite pole was considered to lack "life" and was therefore called the vegetative or vegetal pole. In the sea urchin, the entire egg undergoes cleavage, not just the animal pole as in some organisms. Cleavage is therefore classified as **total** (or **holoblastic**). The *first* cleavage plane passes through the animal-vegetal axis and splits the zygote into two **blastomeres** (Fig. 1.1; Photo 1.2). Experiments have shown that the two blastomeres are equivalent, and that each in isolation can form an entire, albeit small, larva. The *second* cleavage plane also passes through the animal-vegetal axis, but it is oriented perpendicularly to the first (Fig. 1.1; Photo 1.3). The four blastomeres formed after the second cleavage are also equivalent to one another and, again, each in isolation can form an entire larva. The *third* cleavage plane produces eight cells and is equatorial (it passes perpendicularly to both the first

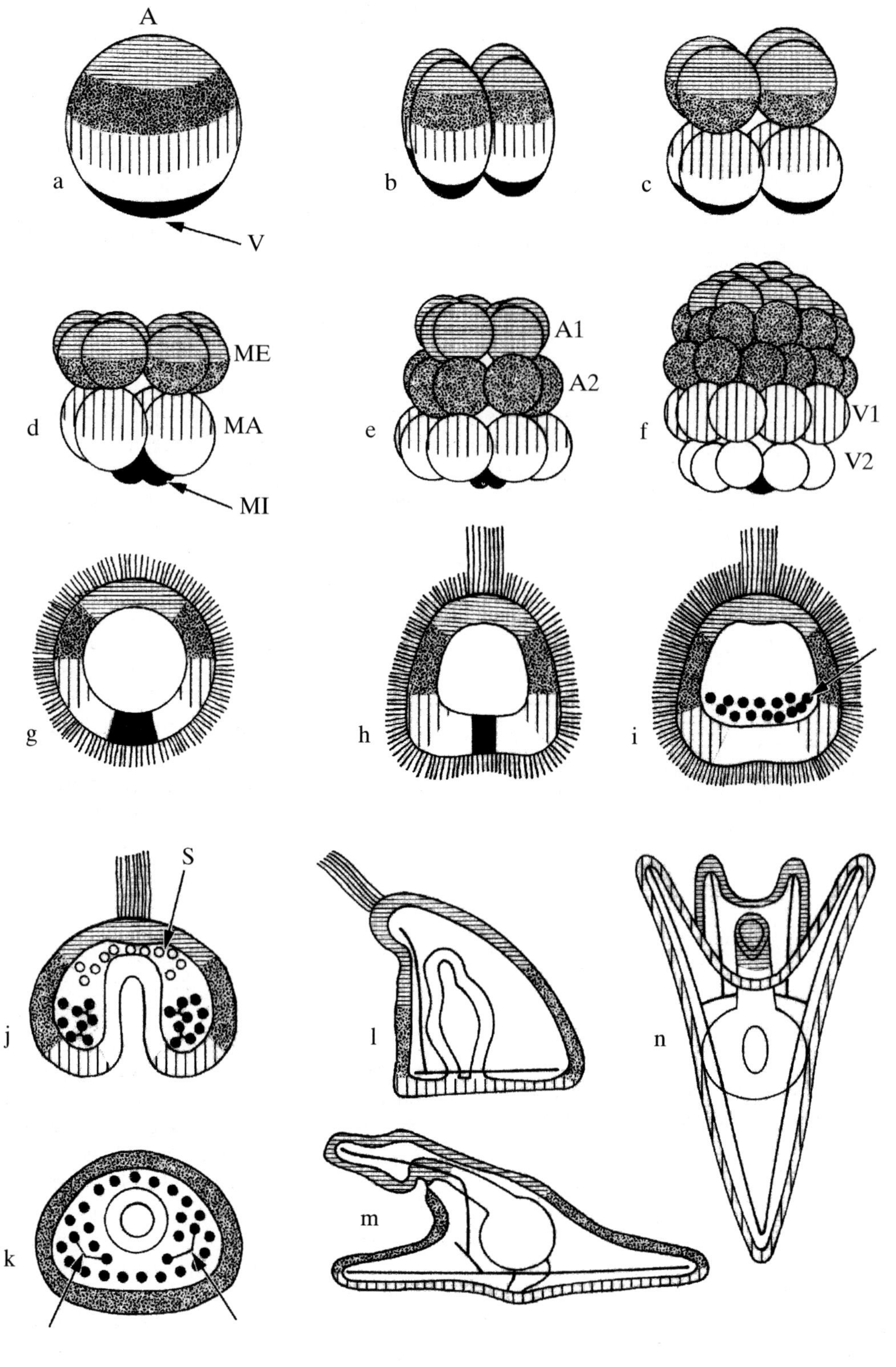

Fig. 1.1. Drawings of the development of the sea urchin from the 1-cell stage to the pluteus larva. a: fate map of the unfertilized egg. A, animal pole; V, vegetal pole. b: 4-cell stage. c: 8-cell stage. d: 16-cell stage. ME, mesomeres; MA, macromeres; MI, micromeres. e: 32-cell stage. A1, an^1 blastomeres; A2, an^2 blastomeres. f: 60-cell stage. V1, veg^1 blastomeres; V2, veg^2 blastomeres. g: early blastula stage. h: mid-blastula stage. i: late blastula stage. Arrow indicates primary mesenchyme cells. j, k: late gastrula stage; k is shown in cross section. Arrows indicate developing spicules; S, secondary mesenchyme cells. l: prism larva. m, n: pluteus larvae. l, m are shown from the side; n is shown from the bottom.

and second planes and creates two tiers of blastomeres: an animal tier consisting of four blastomeres and a vegetal tier consisting of four blastomeres [Fig. 1.1; Photo 1.4]). Experiments reveal that individual blastomeres are no longer equivalent at the eight-cell stage, and that isolated blastomeres can no longer form an entire larva, only parts of the larva.

Cleavage becomes irregular after the third division occurs. Thus in some species, there is a distinct 12-cell stage preceding the 16-cell stage. When the *fourth* cleavage is fully completed, three tiers (for a total of 16 cells) are established: a tier nearest the *animal* pole, consisting of eight blastomeres that are now called the **mesomeres** (actually, two tiers of mesomeres form at this stage, each consisting of four cells; here, Hörstadius's original description of one tier will be followed); a tier nearest the *vegetal* pole, consisting of four smaller blastomeres that are now called the **micromeres**; and a tier in between, consisting of four larger blastomeres that are now called the **macromeres** (Fig. 1.1; Photo 1.5).

The *fifth* cleavage ultimately establishes five tiers of cells, for a total of 32 cells (Fig. 1.1; Photo 1.6). The eight **mesomeres** in the animal hemisphere divide to form two tiers of eight cells each. The cells in the upper tier of eight (nearest the animal pole) are now called the **an^1 blastomeres** (that is, the first tier of animal hemisphere cells), and the cells in the lower tier of eight are now called the **an^2 blastomeres** (that is, the second tier of animal hemisphere cells). The four **macromeres** divide longitudinally to form one tier of eight blastomeres, all of which are still called **macromeres**. The four **micromeres** divide to form two tiers of four cells each. The upper tier of cells are called the **large micromeres**, and the lower tier of cells, nearest the vegetal pole, are called the **small micromeres**. The sea urchin embryo between the 32- and 64-cell stage can be called a **morula**, that is, a ball of blastomeres.

The *sixth* cleavage ultimately produces 60 cells (not the expected 64, because the four small micromeres, unlike all other blastomeres of the 32-cell stage embryo, do not divide during this cleavage cycle; Fig. 1.1). The **an^1 blastomeres** divide to form two tiers of an^1 cells for a total of 16 cells; the cells in the upper tier (that is, the tier nearest the animal pole) are called the **an^1 upper blastomeres**, and the cells in the lower tier are called the **an^1 lower blastomeres**. Similarly, the **an^2 blastomeres** divide to form two tiers of an^2 cells, for a total of 16 cells; the cells in the upper tier (that is, the tier nearest the animal pole) are called the **an^2 upper blastomeres**, and the cells in the lower tier are called the **an^2 lower blastomeres**. As shown by fate mapping studies, the an^1 upper and lower blastomeres and the an^2 upper and lower blastomeres contribute to the **ectoderm** of the embryo. The **macromeres** also divide latitudinally during the sixth cleavage to form two tiers of cells, for a total of 16 cells. The cells in the upper tier (that is, the tier nearest the animal pole) are called the **veg^1 blas-**

tomeres, and the cells in the lower tier are called the **veg^2 blastomeres**. The veg^1 cells, like the an^1 and an^2 cells, contribute to the **ectoderm** of the embryo. The veg^2 cells have a very different fate. They form the **endoderm** of the **archenteron** as well as the **secondary mesenchyme cells** (see below); thus the veg^2 cells contribute to the **endoderm** and **mesoderm** of the embryo. The large **micromeres** may also divide during the sixth cleavage to form as many as two tiers of cells, for a maximum total of 16 cells. The large micromeres form the **primary mesenchyme** (see below); thus the large micromeres, like the veg^2 cells, contribute to the **mesoderm**.

During subsequent cleavages (Fig. 1.1; Photos 1.7-1.9), blastomeres become arranged around a central cavity called the **blastocoel**, and the developing embryo at this stage is called a **blastula** (that is, with formation of the blastocoel and blastula, the process of **blastulation** occurs). **Cilia** develop on the outer surface of the blastula (one cilium per blastomere); those at the animal pole become particularly long and collectively are called the **apical tuft**. In the late blastula stage, cells begin to detach from the vegetal pole of the blastula and to move into the blastocoel; that is, these cells undergo **ingression** (Figs. 1.1, 1.2; Photos 1.10, 1.11). The ingressing cells are called the **primary mesenchyme cells**; fate mapping studies have shown that the primary mesenchyme cells are derived from the large **micromeres** and that they secrete the larval skeleton (they are therefore considered to contribute to the **mesoderm** of the embryo). The larval skeleton initially consists of a pair of triradiate **spicules**, whose orientation *overtly* defines the **bilateral symmetry** of the early **larva** (see below). Ingression of the primary mesenchyme cells involves a change in the shape of the cells composing the vegetal pole of the blastula. Such cells form a thickening of elongated cells, called the **vegetal plate**. During ingression, cells of the vegetal plate, which are derived from the large micromeres, constrict their apical side (the side away from the blastocoel and adjacent to the hyaline layer), expand their basal side (the side adjacent to the blastocoel and away from the hyaline layer), elongate apicobasally, and "slip out" of the lineup by detaching from the vegetal plate. In the process of detaching, the primary mesenchyme cells lose their affinity for the hyaline layer, retract their cilia, and break their intercellular junctions. The detached cells, which rapidly round up, constitute the **primary mesenchyme cells**. Approximately 64 primary mesenchyme cells eventually form. During the late blastula stage when primary mesenchyme cells are ingressing into the blastocoel, the blastula is often referred to as a **mesenchyme blastula**; early and late mesenchyme blastulae are identified (in Photos 1.10, 1.11).

The blastula begins to rotate within the fertilization membrane at the late blastula stage through the action of its cilia, which beat synchronously. The blastula produces an enzyme at about this time called the **hatching enzyme**. The hatching enzyme weakens the fertilization envelope, and movements of the blastula eventually rupture the fertilization envelope, allowing the ciliated blastula to "hatch." Species can differ in this respect; ingression of primary mesenchyme cells occurs after hatching in some species.

D. GASTRULATION

Gastrulation, literally formation of the **primitive gut** or **archenteron**, occurs in two distinct phases in sea urchins. (Some authors prefer to define gastrulation as the origin of the germ layers, not of the gut, as the term originally implied; with this definition, it can be said that gastrulation actually begins earlier, during ingression of the primary mesenchyme cells.) The first phase is called **primary invagination**. During primary invagination, the **vegetal plate** initiates invagination into the blastocoel to form the beginnings of the endodermal **archenteron** (Figs. 1.1, 1.2; Photo 1.12). Several processes contribute to this initial invagination. First, experiments have shown that the vegetal one-third to one-half of the embryo when isolated from more animal regions can still invaginate, suggesting that invagination is autonomous to the vegetal region. Second, during invagination at least three factors seem to act within the vegetal region: **involution** of cells over the lips of the forming blastopore (a process considered by most investigators to play only a minor role; very few cells involute); local proliferation of cells in the vegetal region (that is, **differential growth**); and **changes in cell shape** and **changes in cell adhesion**. The first factor, involution, also involves some **epiboly**, the expansion of the epithelial sheet of cells surrounding the forming blastopore.

The second phase of gastrulation is called **secondary invagination** (Figs. 1.1, 1.2; Photos 1.13-1.15). During secondary invagination, the archenteron undergoes rapid elongation or extension toward the animal pole. The ectoderm at the animal pole thickens at about this time and is called the **animal plate**. **Cilia** are particularly long at the animal plate and collectively constitute the **apical tuft**. Recent very elegant studies have shown that secondary invagination involves rapid **cell rearrangement** (Fig. 1.2).

As secondary invagination is under way, cells begin to detach from the tip of the archenteron (Figs. 1.1, 1.2; Photo 1.14). These cells, called **secondary mesenchyme cells**, extend numerous fine **filopodia** toward the overlying ectoderm. These filopodia grope at the basal side of the ectoderm (actually, at its **basal lamina**) and search for a patch of ectoderm called the **target region**. Once the target region is located, the filopodia firmly attach to it and pull the tip of the archenteron subjacent to it (Photo 1.15). Experiments have shown that in some

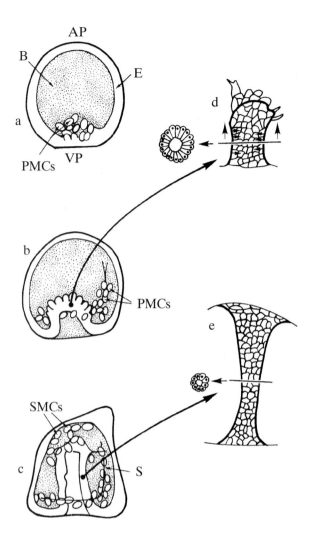

Fig. 1.2. Drawings of the development of the sea urchin from the late blastula stage through gastrulation. a: blastula stage during formation of the primary mesenchyme cells (PMCs). AP, animal pole; B, blastocoel; E, epidermis; VP, vegetal pole. b: gastrula during stage of primary invagination of the archenteron. PMCs, primary mesenchyme cells. c: gastrula during stage of secondary invagination of the archenteron. S, spicule; SMCs, secondary mesenchyme cells. d: enlargement of the cells forming the wall of the archenteron at the stage shown in b, and a cross section (left) through the archenteron; the two parallel arrows indicate the direction of elongation of the archenteron. The two columns of small arrows indicate the direction of cell rearrangement. e: enlargement of the cells forming the wall of the archenteron at the stage shown in c, and a cross section (left) through the archenteron.

species, the final one-third of archenteron invagination requires contact with the target region and filopodial traction. Through this process, the tip of the archenteron is guided toward the area of ectoderm destined to participate in formation of the **mouth**. More specifically, the ectoderm of the target region eventually becomes part of an ectodermal depression, called the **stomodeum,** on the ventral, or oral, side of the embryo. Subsequently, the ectoderm of the stomodeum and the endoderm of the tip of the archenteron rupture at the area of mutual contact, establishing the mouth opening. Thus sea urchins are classified as **deuterostomes**, that is, the *second* opening into the embryo forms the mouth. The first opening, the **blastopore**, forms the *anus* (by contrast, the blastopore forms the mouth in **protostomes**).

E. FORMATION OF THE PRISM AND PLUTEUS LARVAE

During mid- to late gastrulation, the **primary mesenchyme cells** become organized into a ring encircling the vegetal side of the invaginating archenteron (Photo 1.14). At opposite "sides" of the ring, primary mesenchyme cells cluster together, providing the first indication of **bilateral symmetry**. Each cluster of primary mesenchyme cells begins to assemble a primitive larval skeletal element, called a **spicule**, and each spicule quickly acquires a triradiate shape (Fig. 1.1; Photos 1.16, 1,17). The spicules are crystalline in nature and are composed of calcium carbonate and magnesium carbonate deposited in an organic matrix. Primary mesenchyme cells form **filopodia** that fuse together to form **syncytial cables**. Spicules are deposited within membrane-bound compartments inside of the syncytial cables.

At about this time, the embryo flattens at its **ventral (or oral) surface** and is now known as the **prism larva** (Photo 1.17). The opposite side is now called the **dorsal (or aboral) surface**. Projections of the body of the larva then begin to form. First, three projections are visible: two **anal arms** and one **oral lobe** (or **hood**; Photo 1.18). The larva at this stage is known as a **pluteus larva**. During subsequent development, two additional projections grow out from the oral lobe: the **oral arms** (Photo 1.19). Thus the pluteus larva at this stage has two anal arms and two oral arms, which rapidly lengthen (Photo 1.21). Each of these arms contains a well formed **spicule** (Photos 1.19-1.21); similarly, the body proper also contains spicules (technically, the triradiate spicules of the late gastrula grow into the larval skeleton, whose components are referred to as rods rather than spicules).

In the pluteus larva, a number of structures are developing, most of which are beyond the scope of this book. However, it should be mentioned that the archenteron in the pluteus larva is subdividing into three chambers (the **esophagus, stomach,** and **intestine**); **pigment cells**, derived from the **secondary mesenchyme cells**, have formed and dispersed throughout the ectoderm; and two **coeloms** are forming from **secondary mesenchyme cells** associated with the esophagus region of the archenteron (Photo 1.21). The pluteus larva is a free-swimming organism that feeds on plankton. **Ciliary bands** develop on its arms and along the circumference of its body proper; these beat synchronously to propel plankton toward the mouth opening and to propel the larva through the water.

F. SCANNING ELECTRON MICROSCOPY

The relative size of the **egg** and **sperm** can be readily appreciated from viewing Photo 1.22. The various regions of the sperm can be easily identified with scanning electron microscopy, including its **acrosome, head, midpiece,** and **tail** (Photo 1.23).

After fertilization, the **zygote** initiates **cleavage**. Compare the scanning electron micrographs of cleavage stages shown in Photos 1.24-1.30 with Figure 1.1. Note, as they appear, the **micromeres** (the **large** and **small micromeres**), **macromeres** (including the veg^1 blastomeres and veg^2 blastomeres), and **mesomeres** (including the an^1 upper blastomeres, an^1 lower blastomeres, an^2 upper blastomeres, and an^2 lower blastomeres). Identify the **blastocoel** (Photos 1.27, 1.30, 1.32, 1.35, 137-1.39, 1.43, 1.44).

Beginning in the late blastula stage, identify the **primary mesenchyme cells** (Photos 1.33-1.35). Note their positions at different stages (Photos 1.37-1.39, 1.43).

In the gastrula stage, identify the **blastopore, archenteron,** and **secondary mesenchyme cells** (and their **filopodia**) (Photos 1.36-1.39, 1.43, 1.44). Note the changes in appearance of the archenteron over time.

Examine scanning electron micrographs of larval stages. At the prism larva stage, identify the **ventral (or oral) surface**, the **blastopore**, the future site of formation of the **stomodeum**, and the beginnings of the **anal arms** and **oral lobe** (Photo 1.40). At the pluteus larva stage, identify the **oral lobe** and **anal arms** (Photos 1.41, 1.42).

Scanning electron microscopy reveals in exquisite detail the shapes of cells composing the sea urchin embryo. Note in particular the shapes of the **primary** and **secondary mesenchyme cells** and their **filopodia** (Photos 1.34, 1.35, 1.38, 1.43, 1.44).

G. TERMS TO KNOW

You should know the meaning of the following terms, which appeared in boldface in the preceding discussion of sea urchin embryos.

aboral surface

acrosomal enzymes

acrosomal filament

acrosomal granule

acrosomal process

acrosomal reaction

acrosome

actin

an1 blastomeres

   an1 lower blastomeres

   an1 upper blastomeres

an2 blastomeres

   an2 lower blastomeres

   an2 upper blastomeres

anal arms

animal plate

animal pole

animal-vegetal axis

anus

apical tuft

archenteron

basal lamina

bilateral symmetry

blastocoel

blastomeres

blastopore

blastula

blastulation

calcium ions

cell adhesion

cell rearrangement

cell shape

centrioles

chemoreceptor cells

cilia

ciliary bands

cleavage

coelomic peritoneum

coeloms

cortex of the egg

cortical granules

cortical reaction

cytoplasm of egg

depolarization

deuterostomes

differential growth

dioecious

dorsal surface

ectoderm

egg binding protein

egg nucleus

eggs

endoderm

epiboly

epithelial-to-mesenchymal
   transformation

esophagus

exogastrulation

female pronucleus

fertilization

fertilization cone

fertilization envelope

fertilization membrane

filopodia

first meiotic division

first polar body

gametes

gametogenesis

gastrulation

germinal epithelium

germinal vesicle

gonads

gonoduct

gonopore

hatching enzyme

head of sperm

holoblastic cleavage

homolecithal

hyaline layer

hyaline protein

ingression

intestine

invagination

involution

jelly layer

large micromeres

larva

macromeres

madreporite

male pronucleus

mature egg

mature sperm

mesenchyme blastula

mesoderm

mesomeres

micromeres

microtubules

midpiece of sperm

mitochondrion of sperm

morula

mouth

muscle layers of gonads

nucleolus

nucleus

oogenesis

oogonia

oolemma

oral arms

oral hood

oral lobe

oral surface

ovaries

perivitelline space

pigment cells

plasmalemma

pluteus larva

polyspermy

primary invagination of the
   archenteron

primary mesenchyme
   cells

primary oocytes

primary spermatocytes

primitive gut

primitive streak

primordial germ cells

prism larva

pronucleus

prospective fate

prospective fate maps

protostomes

radial symmetry

radialization

second meiotic division

second polar body

secondary invagination of the
   archenteron

secondary mesenchyme cells

secondary oocytes

secondary spermatocytes

small micromeres

sperm

spermatids

spermatogenesis

spermatogonia

spicules

stomach

stomodeum

syncytial cables

syngamy

tail of sperm

target region

test

testes

total cleavage

veg^1 blastomeres

veg^2 blastomeres

vegetal plate

vegetal pole

vegetalization

ventral surface

vitelline envelope

vitelline membrane

water-vascular system

yolk

zygote

zygote nucleus

H. COMPARISON OF GASTRULATION IN VERTEBRATES AND INVERTEBRATES

Gastrulation in vertebrates and invertebrates may at first glance appear very dissimilar. Indeed, gastrulation among vertebrates even as diverse as amphibians, birds, and mammals may seem to be very different. This section focuses on the similarities that occur in gastrulation in the invertebrate and vertebrate embryos studied in this guide. Its purpose is to solidify your knowledge of morphogenetic movements underlying gastrulation. If you have started your laboratory study with sea urchin embryos, return to this section after you have studied gastrulation in vertebrate embryos (Chapters 2-4).

First, compare Fig. 1.1, which shows the prospective fate of different areas of the sea urchin embryo, with Figs. 2.4, 3.7, and 4.2, which show the prospective fate maps of amphibian, avian, and mammalian embryos, respectively. After you have acquired a sufficient understanding of the locations of various prospective regions in the four types of embryos, consider the following types of morphogenetic movements that occur during gastrulation: epiboly, involution, ingression, and invagination. Epiboly occurs in all four types of embryos, although the amount of epiboly is far less in the sea urchin embryo than it is in the other embryos. Epiboly allows areas on the surface of the embryo to replace other areas formerly on the surface, as they move into the interior of the embryo during involution, ingression, or invagination. Involution plays a major role in amphibian gastrulation. Cells composing an epithelial sheet roll over the lips of the blastopore as a unit (that is, they involute) to move into the interior. Ingression plays a major role in avian and mammalian gastrulation. Cells move into the primitive streak where they line up craniocaudally. They then undergo an epithelial-to-mesenchymal transformation to move into the interior as individual mesenchymal cells (that is, they ingress). Both invagination and ingression play major roles in sea urchin gastrulation. Cells composing the vegetal plate bend inward (that is, invaginate) to form the archenteron, which subsequently elongates. Primary mesenchyme cells, prior to the beginning of invagination of the archenteron, and secondary mesenchyme cells, as invagination is under way, leave the epithelial sheet (that is, they undergo an epithelial-to-mesenchymal transformation) to move into the blastocoel (that is, they ingress).

Finally, realize that although the three types of embryos share similar morphogenetic movements, many subtleties exist in how gastrulation occurs. The similarities give us hope that common cellular and molecular mechanisms underlie gastrulation in a wide range of organisms, whereas the differences emphasize the complexity and difficulty of the problem and the need for continued research.

I. PHOTOS 1.1-1.44: SEA URCHIN EMBRYOS

Photos 1.1-1.44 depict the stages of sea urchin embryos discussed in Chapter 1. These photos and their accompanying legends begin on the following page.

12   Chapter 1

Photos 1.1-1.9
Sea Urchin Embryos
Legend

1. Fertilization envelope
2. Zygote nucleus
3. Blastomere nuclei
4. First cleavage furrow
5. Macromeres
6. Mesomeres (out of focus)
7. Micromeres
8. Large micromeres
9. Small micromeres
10. Blastocoel
11. Animal pole
12. Vegetal pole
13. Perivitelline space
14. Hyaline layer

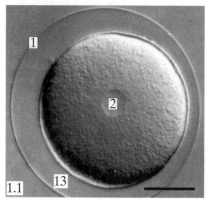

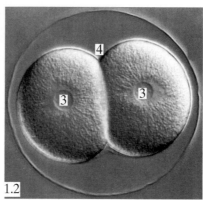

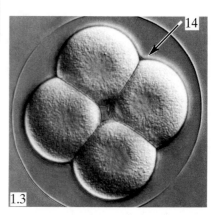

Photo 1.1. Fertilized sea urchin egg (zygote at the one-cell stage; viewed from the animal pole). The bar indicates 40 micrometers.

Photo 1.2. Sea urchin two-cell stage (viewed from the animal pole).

Photo 1.3. Sea urchin four-cell stage (viewed from the animal pole).

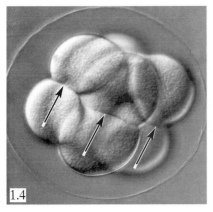

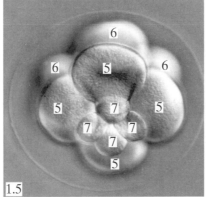

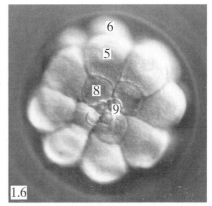

Photo 1.4. Sea urchin eight-cell stage (viewed from the side). The unlabeled arrows indicate the third cleavage furrows.

Photo 1.5. Sea urchin twelve- to sixteen cell stage (viewed from the vegetal pole).

Photo 1.6. Sea urchin thirty-two cell stage (viewed from the vegetal pole).

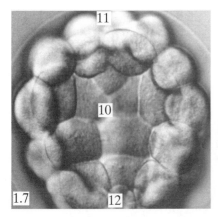

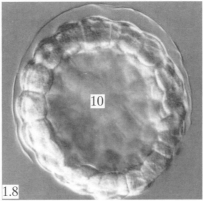

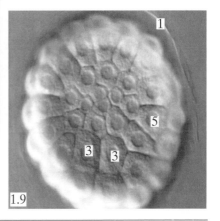

Photo 1.7. Sea urchin mid-blastula stage (sixty-cell stage; viewed from the side but focused through the cells composing the side wall of the blastula).

Photos 1.8, 1.9. Sea urchin late blastula stage (one hundred and eight-cell stage; Photo 1.8 is viewed from the side but focused through the cells composing the side wall of the blastula; Photo 1.9 is viewed from the vegetal pole).

Photos 1.10-1.18

Sea Urchin Embryos

Legend

1. Vegetal plate
2. Blastocoel
3. Primary mesenchyme cells
4. Blastopore

5. Archenteron (endodermal wall of)
6. Secondary mesenchyme cells
7. Animal plate

8. Triradiate spicules
9. Apical tuft
10. Anal arms
11. Oral lobe

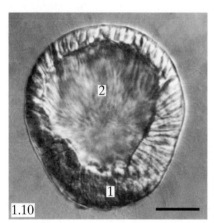

1.10

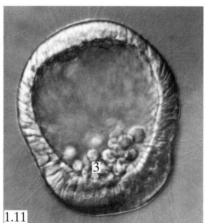

1.11

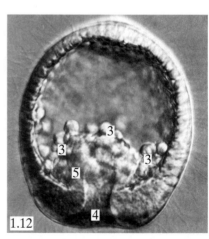

1.12

Photo 1.10. Sea urchin early mesenchyme blastula (viewed from the side but focused through the cells composing the side wall of the blastula). The bar indicates 40 micrometers.

Photo 1.11. Sea urchin late mesenchyme blastula (viewed from the side but focused through the cells composing the side wall of the blastula).

Photo 1.12. Sea urchin gastrula. Primary invagination of endoderm (viewed from the side but focused through the cells composing the side wall of the gastrula).

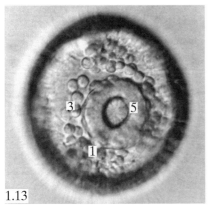

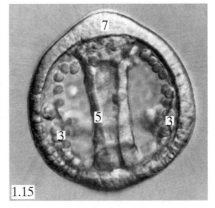

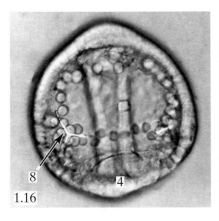

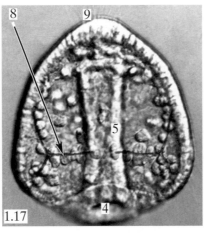

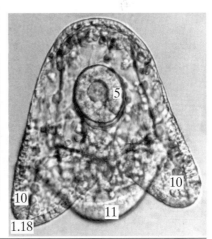

Photo 1.13. Sea urchin gastrula. Cross-sectional view during invagination of the archenteron (that is, viewed from the animal pole but focused through the cells composing the animal wall of the gastrula).

Photo 1.14. Sea urchin gastrula. Early secondary invagination of the archenteron (viewed from the side but focused through the cells composing the side wall of the gastrula).

Photo 1.15. Sea urchin gastrula. End of secondary invagination of the archenteron (viewed from the side but focused through the cells composing the side wall of the gastrula).

Photo 1.16. Sea urchin gastrula. End of secondary invagination of the archenteron (viewed from the side and illuminated with polarized light to show the triradiate spicules).

Photo 1.17. Sea urchin early prism larva (viewed from ventral or oral side).

Photo 1.18. Sea urchin early pluteus larva (viewed from the dorsal or anal side).

Photos 1.19-1.27

Sea Urchin Embryos

Legend

1. Oral arms
2. Anal arms
3. Body proper of larva
4. Archenteron
5. Spicules
6. Acrosome
7. Head of sperm
8. Midpiece of sperm
9. Tail of sperm
10. Surface of blastomeres (visible after partial dissection of the hyaline layer)
11. Depression in hyaline layer marking the position of the second cleavage furrow
12. Depression in hyaline layer marking the position of the third cleavage furrow
13. Animal pole
14. Vegetal pole
15. Micromeres
16. Macromeres
17. An^2 blastomeres
18. Hole in blastomere plasmalemma
19. An^1 blastomeres
20. Blastocoel

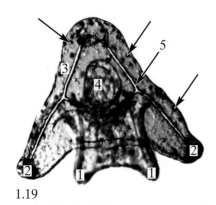

1.19

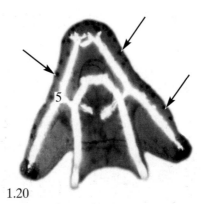

1.20

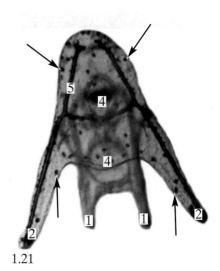

1.21

Photos 1.19-1.21. Sea urchin pluteus larvae. Photo 1.20 was illuminated with polarized light. The unlabeled arrows indicate pigment cells.

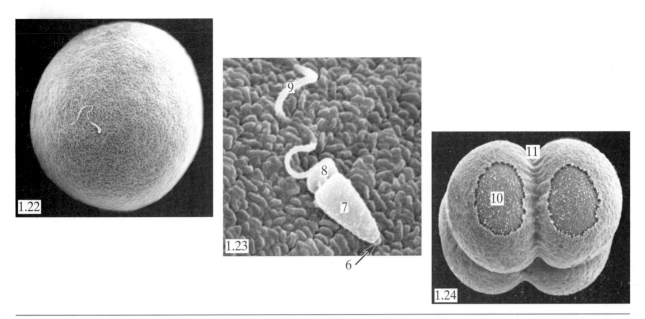

Photo 1.22. Scanning electron micrograph of an unfertilized sea urchin egg with one sperm attached to the surface of the vitelline envelope (the egg was dejellied). The egg is about 110 micrometers in diameter.

Photo 1.23. Scanning electron micrograph of a sea urchin sperm on the surface of a dejellied sea urchin egg.

Photo 1.24. Scanning electron micrograph of a sea urchin four-cell stage embryo (viewed from the side).

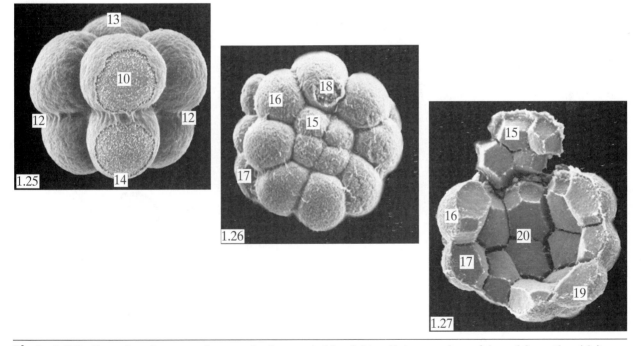

Photo 1.25. Scanning electron micrograph of a sea urchin eight-cell stage embryo (viewed from the side).

Photo 1.26. Scanning electron micrograph of a sea urchin twenty-eight-cell stage embryo (viewed from the vegetal pole).

Photo 1.27. Scanning electron micrograph of a fractured sea urchin twenty-eight-cell stage embryo.

Photos 1.28-1.36

Sea Urchin Embryos

Legend

1. Small micromeres
2. Large micromeres
3. Macromeres
4. An^2 blastomeres
5. Veg^2 blastomeres

6. Veg^1 blastomeres
7. Remnants of the fertilization envelope
8. Wall of blastula
9. Blastocoel

10. Animal pole
11. Vegetal pole
12. Vegetal plate
13. Primary mesenchyme cells
14. Blastopore

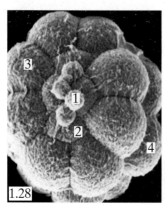

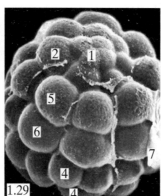

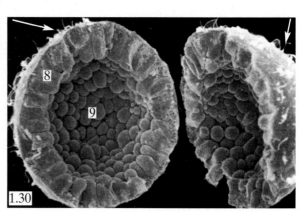

Photo 1.28.   Scanning electron micrograph of a sea urchin thirty-two-cell stage embryo (viewed from the vegetal pole).

Photo 1.29.   Scanning electron micrograph of a sea urchin fifty-six- to sixty-cell stage embryo (viewed from the side).

Photo 1.30.   Scanning electron micrograph of a sea urchin mid-blastula stage embryo sliced into two halves. The unlabeled arrows indicate cilia.

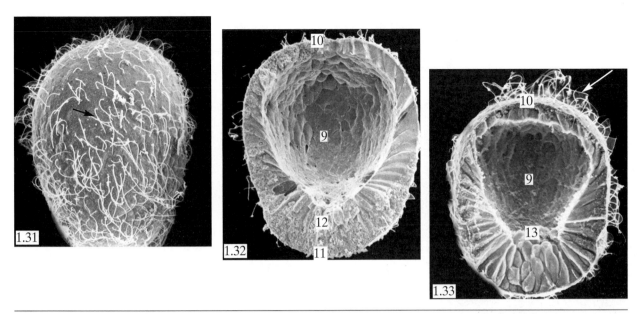

Photo 1.31.   Scanning electron micrograph of a sea urchin mid-blastula stage embryo (viewed from the side). The unlabeled arrow indicates cilia.

Photo 1.32.   Scanning electron micrograph of a slice through a sea urchin late blastula stage embryo.

Photo 1.33.   Scanning electron micrograph of a slice through a sea urchin early mesenchyme blastula stage embryo during initial ingression of the primary mesenchyme cells. The unlabeled arrow indicates cilia.

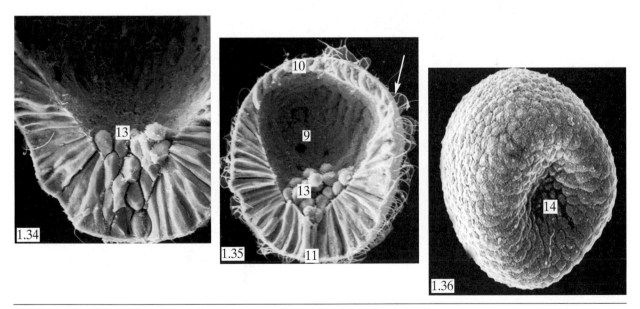

Photo 1.34.   Scanning electron micrograph of an enlargement of a slice through a sea urchin early mesenchyme blastula stage embryo.

Photo 1.35.   Scanning electron micrograph of a slice through a sea urchin late mesenchyme blastula stage embryo. The unlabeled arrow indicates cilia.

Photo 1.36.   Scanning electron micrograph of a sea urchin gastrula stage embryo (deciliated) at the stage of primary invagination of the archenteron (viewed from the blastopore side).

Photos 1.37-1.44

Sea Urchin Embryos

Legend

1. Animal pole
2. Vegetal pole
3. Blastocoel
4. Archenteron
5. Secondary mesenchyme cells (note numerous filopodia)

6. Primary mesenchyme cells
7. Lumen of cross-sectioned archenteron
8. Blastopore
9. Future site of the stomodeum

10. Beginnings of anal arms
11. Body of pluteus larva
12. Anal arms
13. Oral lobe
14. Filopodia on secondary mesenchyme cells

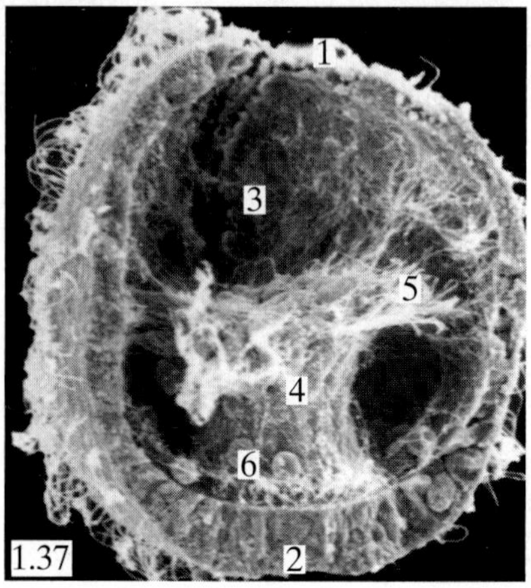

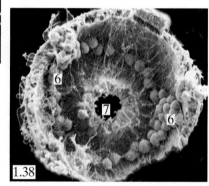

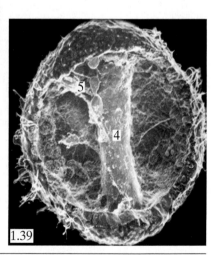

Photo 1.37. Scanning electron micrograph of a slice through a sea urchin mid-gastrula stage embryo.

Photo 1.38. Scanning electron micrograph of a cross-sectional slice through a sea urchin late-gastrula stage embryo.

Photo 1.39. Scanning electron micrograph of a slice through a sea urchin late-gastrula stage embryo.

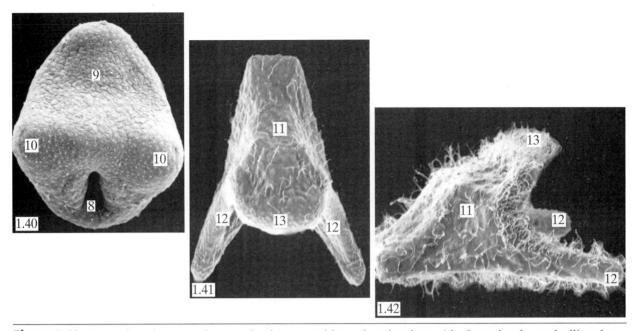

Photo 1.40. Scanning electron micrograph of a sea urchin early prism larva (the larva has been deciliated and is viewed from its oral or ventral side).

Photo 1.41. Scanning electron micrograph of a sea urchin early pluteus larva (viewed from the aboral or dorsal side).

Photo 1.42. Scanning electron micrograph of a sea urchin early pluteus larva (viewed from the lateral side).

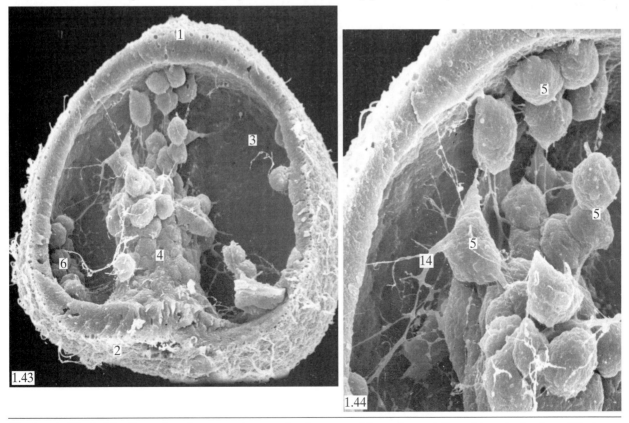

Photos 1.43, 1.44. Scanning electron micrograph of a sea urchin embryo at the mid-gastrula stage (Photo 1.44 is an enlargement of Photo 1.43).

Chapter 2

Frog Embryos

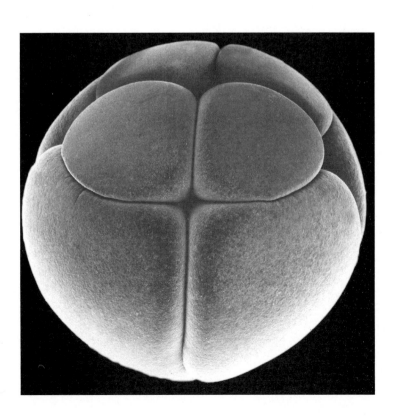

Chapter 2

Frog Embryos

A. INTRODUCTION

Amphibian eggs are relatively large and can be readily obtained. Many outstanding experimental embryologists (including Hans Spemann, Ross Harrison, and their several students) took advantage of these facts when they began to experiment on developing vertebrate embryos by removing parts, adding parts, and recombining parts by microsurgery. Because so much of classical experimental embryology involved experiments on early amphibian embryos, it is essential that you have an understanding of the structure of these embryos and how this structure originated in order to appreciate experimental analyses. The study of early amphibian development, when compared to that of other organisms covered in this laboratory guide and atlas, will also demonstrate the basic similarities in developmental events and processes among different vertebrates such as amphibians, birds, and mammals, and between these vertebrates and invertebrates such as sea urchins. For hands-on studies using frog embryos, see Chapter 6 (Exercises 2.1-2.4).

B. HOW TO USE SERIAL SECTIONS

Laboratory work in introductory courses in developmental biology and embryology usually includes studies of *developmental anatomy*. Suppose you obtained some preserved frog embryos at the stage illustrated by Fig. 2.1a and wanted to study their anatomy. With a microscope, you could identify a few poorly defined external features as well as the body axes (*cranial-caudal, dorsal-ventral*, and *right-left*). To study internal features in detail, you could slice (section) the entire embryo into thin sections of a given thickness; sections cut perpendicularly to the cranial-caudal axis of the embryo are called *transverse* (*cross*) sections. A slide or a collection of slides containing every transverse section from the first one (the most *anterior* one) to the last one (the most *posterior* one) is called a *set of serial transverse sections*. The most anterior section of this set (#1) is mounted at the

upper left-hand corner of the slide, and successive sections are mounted in the following way (the numbers shown are for illustrative purposes only; the actual numbers of sections in each row, and the number of rows per slide, vary):

1	2	3	4	5	6	7	8	9	10
11	12	13	14	15	16	17	18	19	20
21	22	23	24	25	26	27	28	29	30

If there are too many sections to mount on one slide, the more posterior ones are mounted on slide #2 of the series in the following way:

31	32	33	34	35	36	37	38	39	40
41	42	43	44	45	46	47	48	49	50
and so forth.									

One other type of serial section will be encountered extensively in your laboratory studies: the *sagittal section*. A *set of serial sagittal sections* cuts parallel to the long axis of the embryo. A *midsagittal section* is cut exactly down the midline of the embryo, whereas a *parasagittal section* is cut either to the right or left of the midline. Thus sagittal sections pass through the dorsal-ventral extent of the embryo, either to the right or left of the midline or on the midline.

Fig. 2.1b illustrates four representative *transverse sections*. Fig. 2.1c illustrates these same sections after they were transferred to a glass slide and mounted from left to right in the order in which they were cut. Exactly how much of the anatomy of the embryo can one expect to see in any one representative section? Suppose that your set of serial sections contained a total of 100 sections. The sixth section of the set might cut through the level of the developing *eyes*; in this section you could determine the relationship of the eyes to other structures (Figs. 2.1b, 2.1c). You might then examine a more posterior section, such as #15 through the *heart*, or still more posterior sections (#50, #80). Unfortunately, the study of individual sections provides only a two-dimensional picture of the embryo. To understand the anatomy of the embryo in *three* dimensions you must visualize each section as again part of the whole embryo. For example, the *notochord* can be identified in sections #15, #50, and #80. By connecting the section of the notochord at the level of section #15 with the section of the notochord at the next level (section #50) and those at successive levels, you get an accurate picture of the craniocaudal extent of the notochord, as well as its relationship to other structures. In the same way, you can get an accurate picture of the craniocaudal extent of the *neural tube* (*brain* and *spinal cord*) and *digestive tube,* as well as their relationship to each other and to other structures. *The most difficult task facing the beginning student is to learn to visualize relationships of parts of an embryo to one another in three dimensions.* We have attempted to help you with this difficult task by providing three types of visual aids in this laboratory manual: (1) line drawings (called "Figs."); (2) section "orientators," placed on most photo legends (these orientators show the exact levels where embryos were sectioned, sliced, or cryofractured); and (3) scanning electron micrographs, which portray a more three-dimensional image than do flat, two-dimensional serial sections.

Methods have been included in Chapter 6 (Advanced Hands-on Studies) to help you understand how embryos are prepared for light microscopy, scanning electron microscopy, immunocytochemistry, and in situ hybridization, as well as to help you understand why different processing procedures result in different types of images.

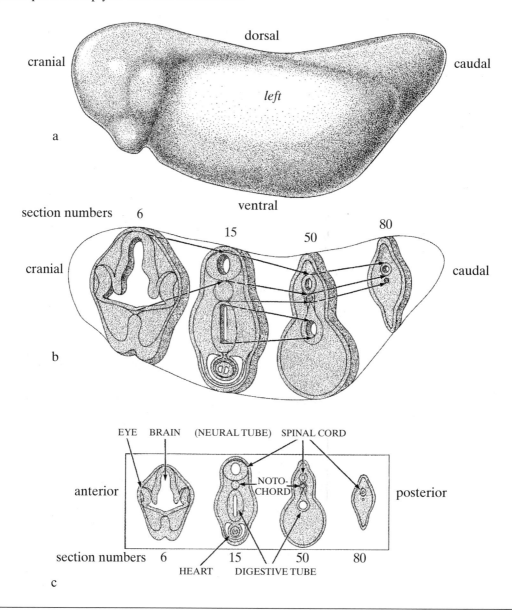

Fig. 2.1. Drawings illustrating the relationships between a preserved 4-mm frog embryo viewed from the left side (Fig. 2.1a) and four representative transverse sections through this embryo (Figs. 2.1b, 2.1c). Section numbers are for example only.

C. OOGENESIS AND FERTILIZATION

Oogenesis (development of the ovum) begins in the paired **ovaries** of the mature female frog. During the breeding season (which begins in the spring), each ovary consists of a sac containing a cluster of spherical structures called **follicles** (Fig. 2.2). Each follicle consists of a large central cell containing a lot of **yolk** in its cytoplasm, the **primary oocyte**, surrounded by a layer of much smaller, flattened cells called **follicle cells**. The primary oocyte contains a large nucleus, the **germinal vesicle**. The **vitelline membrane** lies between the follicle cells and the **plasmalemma** of the primary oocyte. A thin sheath of connective tissue, the **theca folliculi externa**, forms the surface layer of the ovary. Another sheath, the **theca folliculi interna**, partially surrounds each follicle but is lacking in the region where the follicle contacts the theca folliculi externa; at this region **ovulation** (the rupture of the follicle and release of its contained oocyte) occurs.

Each ovary also contains cells called **oogonia**. These cells undergo rapid mitotic divisions, increasing in number. After the breeding season is completed (that is, in the autumn), a few thousand oogonia within each ovary lose the ability to divide mitotically. Each enlarges slightly as a **primary oocyte** and becomes surrounded by a single layer of **follicle cells**, forming an ovarian **follicle**, which slowly enlarges due to the accumulation of yolk. These primary oocytes enter the **prophase** stage of the **first meiotic division** but remain there until the following spring.

Fully grown primary oocytes undergo **ovulation** in response to hormones secreted by the **anterior pituitary gland** (**adenohypophysis**). Each oocyte is slowly squeezed through the follicular wall at the region where the theca folliculi interna is lacking (Fig. 2.2) and enters the body cavity (coelom) of the female. Many oocytes (2,000-20,000, depending upon the species) are ovulated by a single female each breeding season. Primary oocytes complete the **first meiotic division** during ovulation, with each forming a **first polar body** and **secondary oocyte**. Both of these structures are contained within the vitelline membrane formed earlier, while the primary oocyte was in the ovary. Cilia on the lining of the coelom beat toward the **ostium** of the **oviduct** and propel the secondary oocytes into this opening. The **second meiotic division** is *initiated* by each secondary oocyte at about the time that it enters the oviduct, but it then arrests in the **metaphase** stage. As the secondary oocytes pass through the oviduct, a multilayered, gelatinous **egg capsule** is secreted outside of the vitelline membrane by the cells lining the oviduct.

Fertilization occurs externally as the **secondary oocytes** are spawned (shed) by the female into the water. The **second meiotic division** is completed as a **sperm** contacts and penetrates each secondary oocyte, resulting in the formation of a **second polar body** and a **mature ovum** containing the **female pronucleus**. The nucleus of the penetrating sperm enlarges within the ovum as the **male pronucleus,** and the male and female pronuclei unite to complete the process of fertilization.

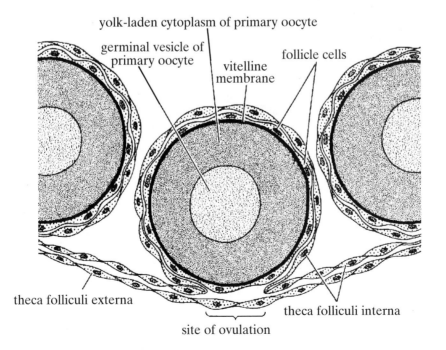

Fig. 2.2. Schematic drawing of a section through three ovarian follicles of a mature female frog.

D. FORMATION OF THE GRAY CRESCENT

The outer portion of the egg (the **cortex**) contains a distinct pattern of pigmentation at the time of ovulation (Fig. 2.3a). About two-thirds of the cortex is heavily pigmented; the remainder contains almost no pigment. The uppermost part of the pigmented portion is the **animal pole**. This pole corresponds to the *cranial* end of the future embryo. The half of the egg that contains the animal pole is the **animal hemisphere**. The **vegetal pole** lies directly opposite the animal pole. This pole corresponds to the *caudal* end of the future embryo. The half of the egg that contains the vegetal pole is the **vegetal hemisphere**.

Following contact by and entrance of the **sperm**, the pigmented cortex shifts relative to the less pigmented deeper portion of the egg, toward and past the site of sperm entrance and away from the side of the egg opposite the sperm entrance point (Fig. 2.3b). This reduces the pigmentation of a crescent-shaped area opposite the point of sperm entrance. This crescent-shaped area between the heavily pigmented cortex above and the essentially nonpigmented cortex below constitutes the **gray crescent** (Figs. 2.3b, 2.3c). A plane passing through the animal and vegetal poles and through the center of the gray crescent corresponds to the *midsagittal plane* of the future embryo (Fig. 2.3c).

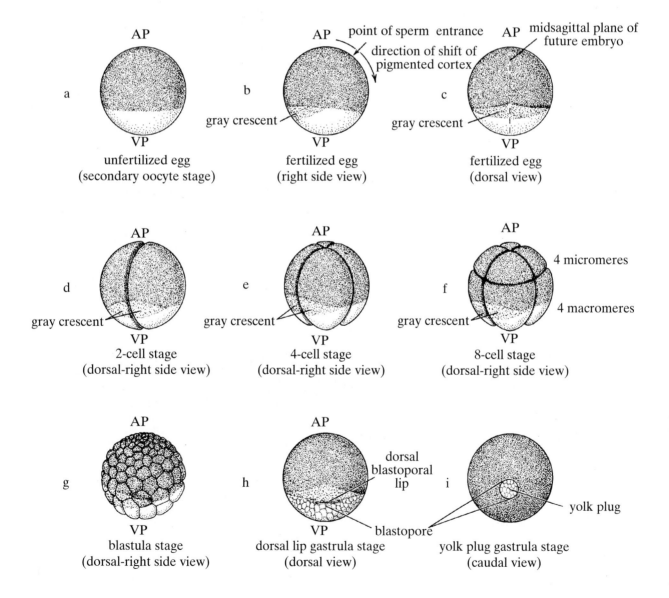

Fig. 2.3. Drawings of early developmental stages of the frog. AP, animal pole; D, dorsal side; L, left side; R, right side; V, ventral side; VP, vegetal pole.

E. CLEAVAGE AND BLASTULATION

Cleavage consists of a series of rapid mitotic divisions that result in **blastulation** (that is, the formation of a **blastula,** which consists of a group of cells called **blastomeres** surrounding a cavity, the **blastocoel**). **Cleavage furrows** pass through the entire egg, so cleavage is classified as **total** (**holoblastic**). However, furrows pass through the vegetal hemisphere much more slowly than through the animal hemisphere, presumably because the former contains far more yolk than the latter. A rough timetable of the first three cleavages for the frog, *Rana pipiens*, is as follows (Figs. 2.3d-2.3f): (1) first cleavage: meridional (that is, passes through both the animal and vegetal poles), usually bisecting the gray crescent (that is, passes through the midsagittal plane of the future embryo), 2.5 hours postfertilization; (2) second cleavage: meridional, at right angles to the first, 3.5 hours postfertilization; and (3) third cleavage: horizontal, displaced toward the animal pole due to the yolk content of the vegetal hemisphere, producing four smaller animal hemisphere cells, the **micromeres**, and four larger vegetal hemisphere cells, the **macromeres**, 4 hours postfertilization. The egg reaches the **blastula** stage near the end of cleavage (approximately 16 hours postfertilization) (Fig. 2.3g). Numerous small blastomeres occupy the animal hemisphere, whereas the vegetal hemisphere consists of a lesser number of larger blastomeres. The blastula is only slightly larger at the end of cleavage than is the newly fertilized egg.

Examine models, or preferably living eggs (see Chapter 6, Exercise 2.1), showing early cleavage stages. Note the position of early cleavage furrows (that is, whether meridional or horizontal), the difference in size of the micromeres and macromeres, and the difference in pigmentation of animal and vegetal hemisphere cells. Try to identify the **gray crescent**; the region of the gray crescent that is broadest will become the *dorsal* surface of the embryo, and the opposite side of the egg will become the *ventral* surface.

Examine a section of the frog blastula that closely resembles Photo 2.1. Identify the main cavity, the **blastocoel**. Note that it is displaced toward one side of the blastula, the **animal pole**; thus, it is contained within the **animal hemisphere**. The blastocoel is filled with a fluid that may have coagulated during preparation of your slide. Note that the wall of the blastula is composed of distinct cells, the **blastomeres**. Identify their **nuclei**. Spaces between blastomeres in the **vegetal hemisphere** are shrinkage spaces produced during preparation of your slides.

Examine the **animal hemisphere**. It has the following characteristics: (1) there is a heavily pigmented cortex, with pigmentation being most intense at the animal pole and grading off progressively toward the vegetal pole;

(2) it is four or five cells thick, with smaller cells and fewer cell layers at the animal pole and with a progressive increase in cell size and number of cell layers toward the vegetal pole; and (3) the blastomeres contain very little yolk.

Examine the **vegetal hemisphere**. Its characteristics are exactly the opposite: (1) a pigmented cortex, if present at all, is much less evident than in the animal hemisphere; (2) the blastomeres are very large and few in number, indicative of less frequent cleavage; and (3) the blastomeres are packed with yolk.

Try to identify the **gray crescent**. In your sections it usually lies either to the left or right side of the blastocoel and also slightly ventral to it. It has the following characteristics (compare sides indicated by letters *D* and *V* in Photo 2.1): (1) the pigmented cortex is thinner in the gray crescent than on the opposite side; and (2) the blastocoel lies nearer the surface on the gray crescent side than on the opposite side (that is, the wall of the blastula is thinner on the gray crescent side than on the opposite side).

The blastula consists of a mosaic of cellular areas, each of which will normally produce a certain structure during subsequent development. In other words, each area of cells has a certain **prospective fate** that will be realized during normal development. In blastulae of some chordates (Urochordata or tunicates) the outlines of these cellular areas can be determined directly because the cytoplasm of cells within certain areas is colored differently. But in most cases it is necessary to determine the prospective fate of each cellular area indirectly by marking experiments. **Vital dyes** have been used most frequently for this purpose in amphibians. Several areas of the blastula are stained with a vital dye, and the structure or structures that are formed from each stained area are observed (see Chapter 6, Exercise 2.2). Another technique has been used more recently. A **cell marker** (for example, the enzyme **horseradish peroxidase**, which can be demonstrated histochemically by incubating tissue containing the enzyme with the appropriate substrate; or **fluorescein-** or **rhodamine-labeled dextran**, which can be demonstrated with a fluorescence microscope after illumination with the proper wavelength of light) is injected into a single cell or groups of cells at the blastula stage. As injected cells cleave, the marker is passed to their descendants. A **prospective fate map** is constructed with the aid of the information gained by these techniques. The amphibian fate map should be carefully compared with the ones for the chick (Fig. 3.7), mouse (Fig. 4.2), and sea urchin (Fig. 1.1). A prospective fate map indicates the location of specific groups of cells prior to the onset of **gastrulation**. These groups of cells are shifted in an orderly way into appropriate positions during gastrulation, which will enable them to cooperate and interact in formation of tissues and organs. The blastocoel appears to be essential in many species to provide a space into which certain groups of cells can move either *en masse* or individually during gastrulation.

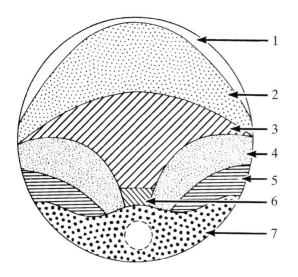

Fig. 2.4. Prospective fate map of the frog blastula. The approximate site of blastopore formation is indicated by a circular dashed line.

1. Prospective ectoderm (epidermis)
2. Prospective neural plate
3. Prospective notochord
4. Prospective segmental plate mesoderm
5. Prospective lateral plate mesoderm
6. Prospective head mesenchyme
7. Prospective endoderm

F. GASTRULATION

During gastrulation some of the cells originally located on the surface of the blastula turn inward or undergo **involution** to move into the interior. These cells will give rise to two **germ layers**: the **endoderm**, or innermost layer, and the **mesoderm**, or middle layer; cells that remain on the surface form the outermost germ layer, the **ectoderm**. A depression, the **blastopore**, begins to form just below the **gray crescent** as cells initiate involution (Fig. 2.3h). Simultaneously, a liplike structure, the **dorsal blastoporal lip**, forms just above the blastopore. With formation of the blastopore and dorsal blastoporal lip, the blastula is transformed into a **gastrula**. Cells continue to involute over the dorsal blastoporal lip with further development, and involution progressively occurs laterally and ultimately ventrally as well. This results in formation of a circular blastopore containing a mass of yolk-filled endodermal cells called the **yolk plug** (Fig. 2.3i). The circular blastopore is surrounded by continuous **dorsal**, **lateral**, and **ventral blastoporal lips**. The directions of gastrulation movements can be altered experimentally, resulting in **exogastrulation,** a process during which surface cells move but fail to involute over

the blastoporal lips (see Chapter 6, Exercise 2.3).

The locations of several areas (designated arbitrarily as areas 1-10 and 25-27) before and during their involution over the **dorsal** and **ventral blastoporal lips** are shown in Fig. 2.5. All of these areas are located on the surface at the blastula stage. Areas 1-5 have undergone involution over the *dorsal* blastoporal lip by the dorsal lip gastrula stage. Similarly, areas 1-8 have undergone involution over the *dorsal* blastoporal lip by the yolk plug gastrula stage, and areas 27 and 26 have undergone involution over the *ventral* blastoporal lip. The remaining numbered areas will undergo involution during subsequent development. Other cellular areas undergo involution over the *lateral* blastoporal lips in a similar manner.

Fig. 2.4 shows the locations of several prospective areas, some of which undergo involution during gastrulation. Some of the cells of the **prospective endoderm** and all the cells of the **prospective head mesenchyme** and **prospective notochord** involute over the *dorsal* blastoporal lip. Similarly, some of the cells of the **prospective endoderm**, some of the cells of the **prospective lateral plate mesoderm**, and all the cells of the **prospective segmental plate mesoderm** involute over the *lateral* lips of the blastopore. The remaining cells of the **prospective lateral plate mesoderm**, as well as some of the cells of the **prospective endoderm**, involute over the *ventral* blastoporal lip. Only a relatively small number of prospective endodermal cells undergo involution over the blastoporal lips. Most prospective endodermal cells remain relatively stationary during gastrulation stages forming the **yolk plug** and floor of the **archenteron** (Fig. 2.5). As areas involute, the **prospective neural plate** and **prospective ectoderm** (**epidermis**) undergo spreading, or **epiboly**, toward the blastopore to replace areas that have moved into the interior of the gastrula.

Obtain a slide containing sagittal sections of the dorsal lip gastrula and select a section closely resembling Photo 2.2. Identify the **blastopore, dorsal blastoporal lip**, and **blastocoel**. The blastopore represents the future *caudal* end. The future *cranial* end lies directly opposite. The blastopore opens into a narrow cavity, the primitive gut, or **archenteron**. The floor of the archenteron is formed by yolk-filled endodermal cells. The cranial end of the archenteron roof is formed from endoderm because the first cells to involute over the dorsal and lateral blastoporal lips, and thus to contribute to the wall of the archenteron, are the cells of the **prospective endoderm** (Fig. 2.4). The remainder of the archenteron roof at the gastrula stage is formed by mesoderm, which involutes over the dorsal and lateral blastoporal lips following involution of prospective endoderm. The mesoderm of the archenteron roof will *later* be covered by endoderm, which migrates upward over the inner surface of the mesoderm (see below). Thus the archenteron is ultimately lined entirely by endoderm. Note that the ectoderm in the frog consists of two layers: an

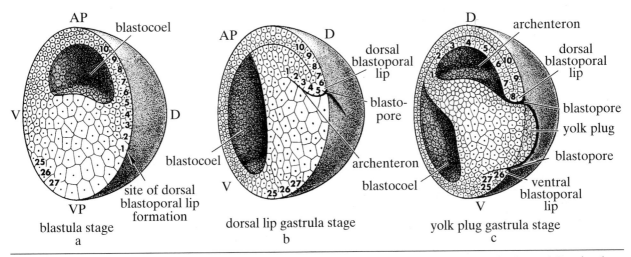

Fig. 2.5. Drawings of the cut surfaces of right halves of blastula and gastrula stages of the frog. AP, animal pole; D, dorsal side; V, ventral side; VP, vegetal pole; numbers 1-10 indicate specific areas of the blastula that undergo involution over the dorsal blastoporal lip; numbers 25-27 indicate specific areas that undergo involution over the ventral blastoporal lip. The numbers are arbitrary and do not indicate particular groups of cells.

outer (superficial) ectodermal layer and an **inner (deep) ectodermal layer**.

Obtain a slide containing sagittal sections of a yolk plug gastrula and select a section closely resembling Photo 2.3. Identify the **yolk plug**, a protruding mass of yolk-filled endodermal cells located between the very prominent **dorsal blastoporal lip** above and the less prominent **ventral blastoporal lip** below. Also identify the **archenteron** located above the mass of poorly defined endodermal cells forming its floor. The yolk plug fills the entrance to the archenteron, the **blastopore**. Beneath the archenteron and separated from the latter by endodermal cells is the irregularly shaped **blastocoel**. This cavity is later squeezed out of existence. Note that the pigmented cortex now covers all cells except those of the yolk plug. The more rapidly dividing cells of the **prospective ectoderm (epidermis)** and **prospective neural plate** undergo epiboly between the blastula stage and this advanced gastrula stage, growing down over the yolk-filled cells of the vegetal hemisphere. Meanwhile the large endodermal cells of the vegetal hemisphere have sunken inward to some extent and become rearranged to form the floor and ventrolateral walls of the archenteron. Most of the archenteron roof still consists of mesoderm at the yolk plug gastrula stage.

For a comparison of gastrulation in vertebrates and invertebrates, see Chapter 1, Section H.

G. NEURULATION

The ectodermal cells directly overlying the archenteron roof at the yolk plug gastrula stage constitute the **neural ectoderm**. The neural ectoderm is *induced* to thicken as the **neural plate**. Induction of the neural plate begins with prospective mesodermal cells within the

dorsal lip of the blastopore, which send their induction signal through the horizontal plane of the ectoderm. The initial signal is later reinforced by mesodermal cells within the roof of the archenteron, which send their induction signal vertically to the overlying ectoderm. The gastrula is transformed into the **neurula** with formation of the neural plate.

Obtain a slide containing transverse sections of a neurula at the neural plate stage. Select a section closely resembling Photos 2.4, 2.5 and identify the **neural plate**. The roof of the **archenteron** at the neurula stage is formed by *endoderm*. Between the gastrula and neurula stages, endoderm migrated upward over the inner surface of the mesoderm that previously formed the archenteron roof. Note that the dorsal mesoderm is now organized into a midline rod of cells, the **notochord**, flanked by two bands of cells, the **segmental plate mesoderm**. This latter mesoderm is the source of the **somites**. The segmental plate mesoderm gradually merges laterally with the **lateral plate mesoderm**. The **coelom** forms within the lateral plate mesoderm.

Obtain a slide containing transverse sections of a neurula at the neural fold stage and select a section closely resembling Photos 2.6, 2.7. Identify the paired **neural folds**, which have formed at the lateral margins of the neural plate, and the **neural groove**. The **notochord, segmental plate mesoderm**, and **lateral plate mesoderm** are more clearly defined at this stage, and the *endoderm* of the roof of the **archenteron** exists as a more distinct layer. During subsequent development the neural folds will fuse in the dorsal midline to close the neural groove and thus establish the **neural tube**. **Neural crest cells** will *later* form from the roof of the neural tube and give rise to a multitude of structures, including **pigment cells** (see Chapter 6, Exercise 2.4).

H. 4-MM FROG EMBRYOS

1. Introduction

Fig. 2.1a shows the general shape of the body at this stage, and Fig. 2.6 shows the structures visible in a midsagittal section. Familiarize yourself with the spatial relationships of these structures before examining serial transverse sections.

2. Serial transverse sections

Position your slide on the microscope stage so that when viewed through the microscope, each section is oriented as in Photo 2.8. *Do not place microscope stage clips (if present) on slides to hold them in place as this could fragment sections.* Examine slides in anteroposterior sequence unless directed otherwise (see Chapter 2, Section B).

a. Ectodermal derivatives

The first few sections cut through the tip of the head. Identify the large **prosencephalon** (future **telencephalon** and **diencephalon** of the brain), which is enclosed by **skin (surface) ectoderm** (Photo 2.8). The skin ectoderm consists of two layers (not readily distinguishable from one another): the **outer ectodermal layer** is heavily pigmented; the **inner ectodermal layer** contains far less pigment. Try to identify a region of thickened skin ectoderm on each side ventral to the prosencephalon. These thickenings are the **nasal (olfactory) placodes**. The nasal placodes eventually form the linings of the **nasal cavities**, and they are derived from only the *inner* ectodermal layer. The increased density of pigment granules at the periphery of these placodes is characteristic of invaginating ectodermal derivatives. Presumably, the peripheral ends of invaginating cells become narrowed, thus concentrating their pigment content, whereas the inner ends of the cells enlarge. **Young neurons** *later* originate from these placodes and produce the **axons** of the **olfactory (I) cranial nerves**, as well as the **dendrites** that function as **olfactory receptors** (receptors for the sense of smell). Nasal placodes may not have yet developed in some embryos. Identify the **pineal gland (epiphysis)**, which originated as a dorsal evagination from the prosencephalon. Continue to trace sections posteriorly and identify the **optic cups** (Photo 2.9). Each optic cup is derived from a lateral evagination from the prosencephalon, which secondarily invaginated at its blind end forming a double-layered optic cup. The thickened layer of the optic cup is the **sensory retina**; the thin layer is the **pigmented retina**. The optic cups are connected to the prosencephalon by the **optic stalks**. Note that the *inner* ectodermal layer adjacent to the sensory retina has thickened to form the **lens placode** (Photo 2.10). The *outer* pigmented layer of the ectoderm will *later* be in-

duced by the lens to form the transparent **corneal epithelium**. At the level of the optic cups, sections cut across the prosencephalon (ventrally) continuous with the **mesencephalon** (dorsally).

The brain constricts into two separate parts as sections are traced posteriorly (Photo 2.10). The ventral part is the **infundibulum**, the source of the **posterior pituitary gland (neurohypophysis)**. The dorsal part is the **rhombencephalon** (the future **metencephalon** and **myelencephalon** of the brain). Identify a solid ectodermal rod at about this level, just ventral to the infundibulum. This is the **rudiment of the anterior pituitary gland**. Reverse direction and trace sections *anteriorly*. The rudiment of the anterior pituitary gland is continuous with an ectodermal invagination, the **stomodeum**, from which it originated as an outgrowth. (See Photos 2.10, 2.9 for the location of the stomodeum. The rudiment of the anterior pituitary gland and the stomodeum are continuous at a section level between those illustrated by Photos 2.9, 2.8). Reverse direction and trace sections *posteriorly*. Identify prominent ectodermal thickenings (Photo 2.10), the **adhesive glands (ventral suckers)**, to either side of the stomodeum and extending through many sections.

Return to the level where the infundibulum first appears (Photo 2.11). At about this level, or slightly more posteriorly, as the optic cups fade from view, identify an accumulation of ectodermal cells (**neural crest cells**) just above each of them. These are the **semilunar ganglia** of the **trigeminal (V) cranial nerves** (Photo 2.11). (*Later* cells derived from a thickening of the *inner* ectodermal layer, on each side, an **epibranchial placode**, will also contribute to these ganglia.) Continue to trace sections posteriorly and identify the paired **auditory vesicles** lying ventrolateral to the rhombencephalon (Photo 2.12). These vesicles originated from thickenings of the *inner* ectodermal layer, which subsequently invaginated. The auditory vesicles later differentiate into the **inner ears**.

Continue to trace sections posteriorly and note that the neural tube gradually narrows, indicating the level of the **spinal cord** (Photo 2.14). Examine the spinal cord with high magnification. Note its characteristic thin **roof** and **floor plates** and its thick lateral walls. At about this level, identify pigmented or nonpigmented cells lying dorsal to the spinal cord and beneath the skin ectoderm; these are **neural crest cells** (Photo 2.14). As you approach the final sections in your set, identify the **dorsal fin** above the spinal cord (Photos 2.15-2.18). The loose cells forming its core are **neural crest cells**, which *induce* formation of this structure. Note that the body narrows progressively at caudal levels; identify the **ventral fin** at the caudal end of the body (Photo 2.18). Identify the **proctodeum**, a ventral invagination of skin ectoderm that is a few sections *anterior* to the ventral fin (Photo 2.17).

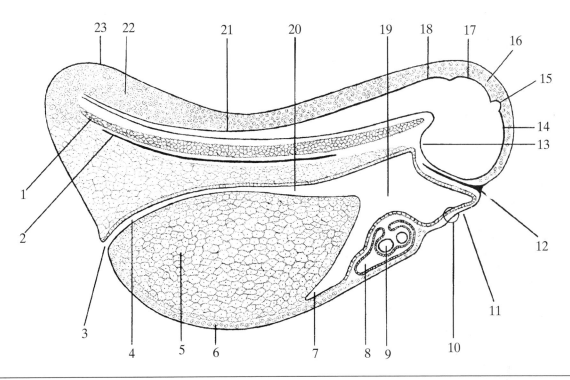

Fig. 2.6. Schematic drawing of a midsagittal section of a 4-mm frog embryo.

1. Notochord
2. Subnotochordal rod
3. Anus
4. Hindgut
5. Yolk-filled endodermal cells
6. Lateral plate mesoderm
7. Liver rudiment
8. Pericardial cavity
9. Heart (ventricle)
10. Oral membrane
11. Stomodeum
12. Rudiment of the anterior pituitary gland
13. Infundibulum
14. Prosencephalon
15. Pineal gland
16. Head mesenchyme
17. Mesencephalon
18. Rhombencephalon
19. Pharynx
20. Midgut
21. Spinal cord
22. Neural crest cells
23. Dorsal fin

b. Endodermal derivatives

Return to the level of the nasal placodes (Photo 2.8) and trace sections posteriorly. Identify a small cavity lying beneath the prosencephalon (Photo 2.10). This is the cranial end of the **foregut**; its walls are formed from endoderm. The **stomodeum** has invaginated toward the foregut, and the stomodeal *ectoderm* and the foregut *endoderm* are in contact as the **oral membrane**. This membrane *later* ruptures to form the **mouth opening**. The outer part of the mouth is therefore lined by stomodeal ectoderm, and the inner part by foregut endoderm.

The foregut expands as the **pharynx** as sections are traced posteriorly (Photo 2.12). (In a few embryos the lateral walls of the pharynx may contact the skin ectoderm, which invaginates slightly to meet them. Such localized pharyngeal expansions are the **pharyngeal pouches**.) Continue tracing sections posteriorly. The foregut narrows and then forms a prominent ventral evagination, the **liver rudiment** (Photo 2.13). The level of the foregut that is continuous ventrally with the liver rudiment is the **duodenum**. The liver rudiment separates from the duodenum and then fades out a few sec-

tions more posteriorly. The remaining portion of the gut is the **midgut** (Photos 2.14, 2.15). It contains a small cavity bounded dorsally by a thin layer of endodermal cells and ventrally by a large mass of yolk-filled endodermal cells. Continue to trace sections posteriorly following the midgut. It gradually moves ventrally and enlarges somewhat as the **hindgut** (Photo 2.16). (In some embryos the *endoderm* of the hindgut fuses with the *ectoderm* of the proctodeum in more posterior sections to form the **cloacal membrane**. The cloacal membrane ultimately ruptures to form the **anus**.)

c. Mesodermal derivatives

Return to the level where the infundibulum first appears (Photo 2.11) and trace sections posteriorly. Identify a group of mesodermal cells, the **notochord**, lying beneath the rhombencephalon (Photo 2.12). The vacuolated condition of the notochord is apparently responsible for its rigidity, allowing this structure to serve as a longitudinal supporting rod in young embryos. Quickly trace sections posteriorly, noting the changes that the notochord undergoes. It is smaller at caudal levels than at more cranial levels, and vacuolization is progressively reduced toward the caudal end (compare

Photos 2.14-1.18). The caudal end of the notochord is less developed at this stage because the notochord develops in craniocaudal sequence. In posterior sections identify the **somites** (Photos 2.15, 2.16), paired blocks of mesoderm lying ventrolateral to the spinal cord. Also identify a small cluster of mesodermal cells lying beneath the notochord in caudal regions, the **subnotochordal rod (hypochord)** (Photos 2.15-2.17). Its developmental significance is unknown.

Return to the level at which the liver rudiment is continuous with the duodenum (Photo 2.13) and trace sections posteriorly. The mesoderm ventrolateral to the somites is usually in the form of more or less distinct epithelial vesicles (Photo 2.14). The round vesicle on each side farthest from the spinal cord is the **pronephric duct**. Just above the latter, and often continuous with it, are the **pronephric tubules**. The pronephric duct and tubules on each side constitute the **pronephric kidney**, which is functional in amphibian larvae. The pronephric kidney when well formed causes the body to bulge laterad as the **pronephric ridge**.

The **lateral plate mesoderm** (often difficult to identify as a distinct area) lies ventral to each pronephric kidney, between the skin ectoderm and the endoderm. This mesoderm is in the process of splitting into an *outer* layer of **somatic mesoderm**, adjacent to the skin ectoderm, and an *inner* layer of **splanchnic mesoderm**, adjacent to the endoderm. If somatic and splanchnic mesoderm have formed, identify a space (or a series of small spaces) between them, the **coelom**.

Return to the level at which the foregut first appears (Photo 2.10) and trace sections posteriorly. Identify the **conotruncus (bulbus cordis)** region of the developing heart lying beneath the pharynx (Photo 2.12). The heart enlarges as the **ventricle** in more posterior sections (Photo 2.13). Note that the heart consists of an *inner* layer, the **endocardium**, surrounded by an *outer*, thicker layer, the **myocardium**. Both these layers are derived from *splanchnic* mesoderm. (*Splanchnic* mesoderm will *later* contribute to a third layer of the heart: its outermost covering.) Identify a thin layer of cells enclosing the heart, the **parietal pericardium**, derived from *somatic* mesoderm. The parietal pericardium is usually separated from the skin ectoderm by a shrinkage space. The space between the heart and the parietal pericardium is the **pericardial cavity**. The heart is suspended within the pericardial cavity by a dorsal bridge of *splanchnic* mesoderm, the **dorsal mesocardium**.

The major blood vessels are in early stages of formation at this time, and they are thus difficult to identify with certainty in most embryos. Two major blood vessels can sometimes be identified. The **first aortic arches** mainly lie ventrolateral to the pharynx (Photo 2.12). (They will soon establish connections with the conotrun-

cus via a pair of blood vessels that will lie ventral to the pharynx, the **ventral aortae**. These latter vessels can sometimes be identified at this stage.) The dorsal end of each first aortic arch is continuous with a blood vessel lying dorsolateral to the pharynx, the **dorsal aorta** (Photo 2.12). The dorsal aortae extend caudad and fade out at about the level of the midgut (Photo 2.14).

3. Summary of the contributions of the germ layers to structures present in the 4-mm frog embryo

Ectoderm

adhesive glands

auditory vesicles

corneal epithelium

dorsal fin

infundibulum

lens placodes

mesencephalon

nasal placodes

neural crest cells

optic cups

optic stalks

pineal gland

proctodeum

prosencephalon

rhombencephalon

rudiments of the anterior pituitary gland

semilunar ganglia

stomodeum

spinal cord

ventral fin

Mesoderm

conotruncus

dorsal aortae

dorsal mesocardium

first aortic arches

notochord

parietal pericardium

pronephric kidneys

somites

subnotochordal rod

ventricle

Endoderm

duodenum

foregut

hindgut

liver rudiment

midgut

pharynx

Ectoderm and endoderm

cloacal membrane

oral membrane

I. SCANNING ELECTRON MICROSCOPY

Examine scanning electron micrographs (Photos 2.19-2.32) to help you visualize the shapes of developing frog eggs and embryos three-dimensionally. Note at cleavage stages (Photos 2.19-2.24) the positions and orientations of **cleavage furrows** separating **blastomeres** and the differences in sizes of the **micromeres** and **macromeres**. Boundaries between blastomeres become less well defined as cleavage advances (compare Photos 2.20 and 2.24).

The lips of the blastopore form during gastrula stages (Photos 2.25-2.27). First the **dorsal lip** forms, followed by the **lateral lips**, and finally the **ventral lip**. The **yolk plug** continues to occupy the **blastopore** throughout gastrulation.

The neural plate forms and rolls up into the **neural tube** between the gastrula stage and early embryo stage (Photos 2.28-2.30). Concomitant with formation of the neural tube, the embryo begins to elongate craniocaudally. This elongation is an obvious feature of developing frog embryos (Photos 2.31-2.32). The neural tube extends throughout the length of the embryo. It bulges laterad at its cranial end to form the developing **eyes** in conjunction with the overlying **skin ectoderm**. Note that the skin ectoderm in the embryos illustrated in Photos 2.31, 2.32 is covered with **ciliary tufts**. These structures establish currents around the embryos as they beat, circulating fluids. These structures also function in primitive locomotory movements before swimming begins. In 4-mm embryos the skin ectoderm at the cranial end of the embryo has invaginated in the midline forming the **stomodeum**. Caudally, the skin ectoderm is attenuated, demarcating the **dorsal** and **ventral fins** (Photo 2.32).

J. TERMS TO KNOW

You should know the meaning of the following terms, which appeared in boldface in the preceding discussion of frog embryos.

adenohypophysis	cloacal membrane	epiboly
adhesive glands	cloacal valves	epibranchial placode
animal hemisphere	coelom	epidermis
animal pole	conotruncus	epiphysis
anterior pituitary gland	corneal epithelium	exogastrulation
anus	cortex of egg	eyes
archenteron	deep ectodermal layer	female pronucleus
auditory vesicles	dendrites	fertilization
axons	diencephalon	first aortic arches
blastocoel	dorsal aorta	first meiotic division
blastomeres	dorsal blastoporal lip	first polar body
blastopore	dorsal fin	floor plate
blastula	dorsal lymph sac	fluorescein-labeled dextran
blastulation	dorsal mesocardium	follicles
cell marker	duodenum	follicle cells
chorionic gonadotropin	ectoderm	foregut
ciliary tufts	egg capsule	gastrula
cleavage	endocardium	gastrulation
cleavage furrows	endoderm	germ layers

germinal vesicle

gray crescent

hindgut

holoblastic cleavage

horseradish peroxidase

hypochord

infundibulum

inner ears

inner ectodermal layer

involution

lateral blastoporal lips

lateral plate mesoderm

lens placode

liver rudiment

macromeres

male pronucleus

mature ovum

mesencephalon

mesoderm

metaphase

metencephalon

micromeres

midgut

morphogenetic movements

mouth opening

myelencephalon

myocardium

nasal cavities

nasal placodes

neural crest cells

neural ectoderm

neural folds

neural groove

neural plate

neural tube

neurohypophysis

neurula

notochord

nuclei

nuptial pads

olfactory (I) cranial nerves

olfactory placodes

olfactory receptors

oogenesis

oogonia

optic cups

optic stalks

oral membrane

ostium of oviduct

outer ectodermal layer

ovaries

oviduct

ovulation

parietal pericardium

pericardial cavity

pharyngeal pouches

pharynx

pigment cells

pigmented retina

pineal gland

plasmalemma

posterior pituitary gland

primary oocyte

proctodeum

pronephric duct

pronephric kidney

pronephric ridge

pronephric tubules

prophase

prosencephalon

prospective ectoderm

prospective endoderm

prospective fate

prospective fate map

prospective head mesenchyme

prospective lateral plate
mesoderm

prospective neural plate

prospective notochord

prospective segmental plate
mesoderm

rhodamine-labeled dextran

rhombencephalon

roof plate

rudiment of the anterior
pituitary gland

second meiotic division

second polar body

secondary oocytes

segmental plate mesoderm

semilunar ganglia

sensory retina

skin

skin ectoderm

somatic mesoderm

somites

sperm

spinal cord

splanchnic mesoderm

stomodeum

subnotochordal rod

superficial ectodermal layer

surface ectoderm

telencephalon

theca folliculi externa

theca folliculi interna

total cleavage

trigeminal (V) cranial nerves

vegetal hemisphere

vegetal pole

ventral aortae

ventral blastoporal lip

ventral fins

ventral suckers

ventricle of heart

vital dyes

vitelline membrane

yolk

yolk plug

young neurons

K. PHOTOS 2.1-2.32: FROG EMBRYOS

Photos 2.1-1.32 depict the stages of frog embryos discussed in Chapter 2. These photos and their accompanying legends begin on the following page.

Photos 2.1-2.5

Frog Embryos

Legend

1. Pigmented cortex
2. Vitelline membrane
3. Blastocoel
4. Area of gray crescent
5. Shrinkage spaces
6. Nuclei
7. Blastomeres
8. Outer ectodermal layer
9. Inner ectodermal layer

10. Archenteron roof (mesoderm, except at cranial end where roof formed from endoderm)
11. Direction of epiboly
12. Dorsal blastoporal lip
13. Blastopore
14. Archenteron
15. Yolk-filled endodermal cells
16. Yolk plug
17. Ventral blastoporal lip

18. Neural plate
19. Skin ectoderm
20. Lateral plate mesoderm
21. Directions of cellular migration to form endodermal roof of archenteron
22. Segmental plate mesoderm
23. Notochord
24. Archenteron roof (endoderm)

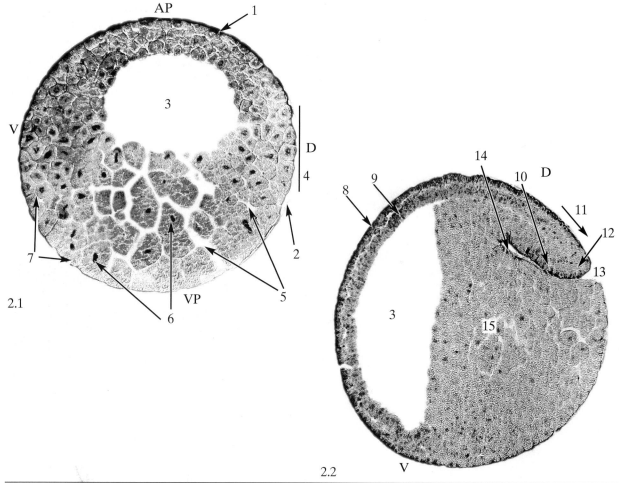

Photo 2.1. Frog blastula (sagittal section). AP, Animal pole; D, Dorsal side; V, Ventral side; VP; Vegetal pole.

Photo 2.2. Frog dorsal lip gastrula (sagittal section). D, Dorsal side; V, Ventral side.

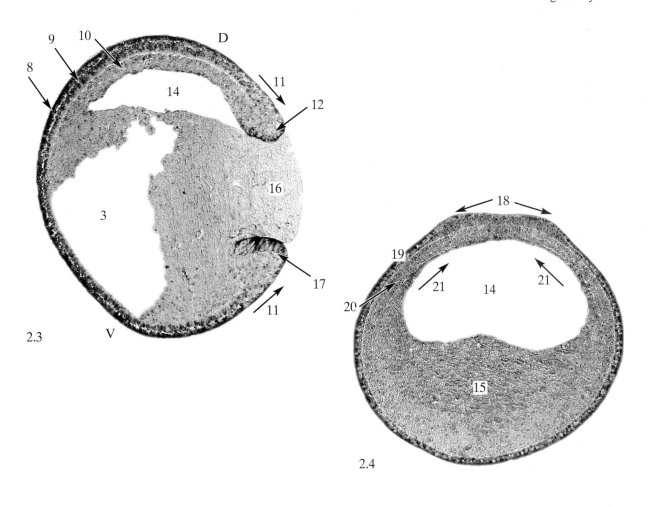

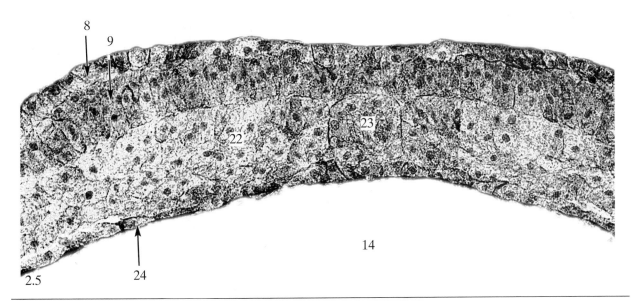

Photo 2.3. Frog yolk plug gastrula (sagittal section). D, Dorsal side; V, Ventral side.

Photos 2.4, 2.5. Frog neural plate neurula (transverse section). Photo 2.5 is an enlargement of the dorsal portion of Photo 2.4.

Photos 2.6-2.9

Frog Embryos

Legend

1. Neural groove
2. Skin ectoderm
3. Lateral plate mesoderm
4. Archenteron
5. Yolk-filled endodermal cells
6. Neural fold
7. Outer ectodermal layer
8. Inner ectodermal layer

9. Segmental plate mesoderm
10. Notochord
11. Archenteron roof (endoderm)
12. Pineal gland
13. Prosencephalon
14. Nasal placode
15. Mesencephalon

16. Pigmented retina of the optic cup
17. Sensory retina of the optic cup
18. Optic stalk
19. Rudiment of the anterior pituitary gland
20. Head mesenchyme
21. Stomodeum

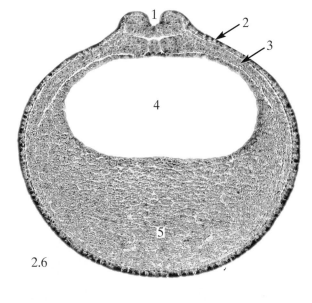

2.6

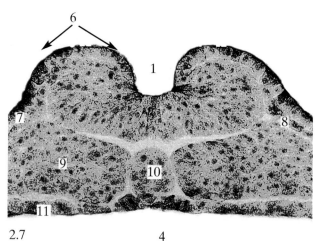

2.7

Photos 2.6, 2.7. Frog neural fold neurula (transverse section). Photo 2.7 is an enlargement of the dorsal portion of Photo 2.6.

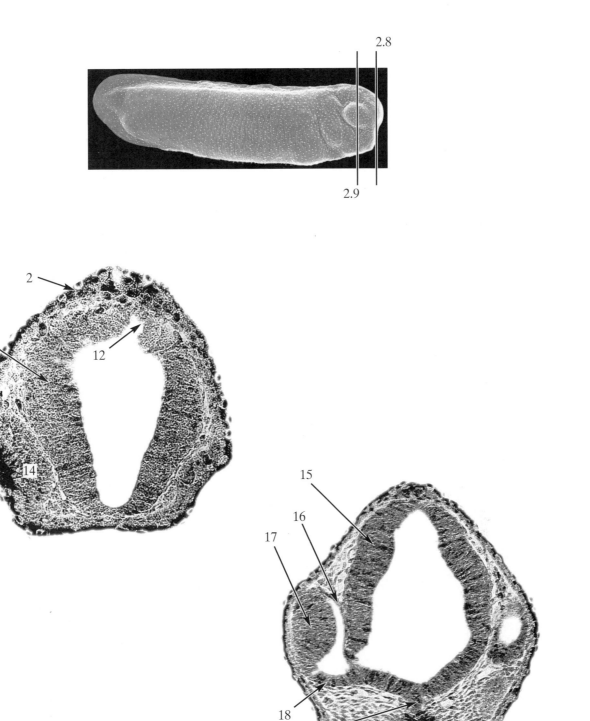

Photos 2.8, 2.9. 4-mm frog embryo serial transverse sections numbered in anterior to posterior sequence.

Photos 2.10–2.13

Frog Embryos

Legend

1. Mesencephalon
2. Future corneal epithelium
3. Lens placode
4. Rudiment of the anterior pituitary gland
5. Foregut
6. Oral membrane
7. Adhesive gland
8. Stomodeum
9. Rhombencephalon
10. Skin ectoderm
11. Semilunar ganglion
12. Infundibulum
13. Auditory vesicle
14. Notochord
15. Dorsal aorta
16. Pharynx
17. First aortic arch
18. Conotruncus
19. Duodenum
20. Liver rudiment
21. Dorsal mesocardium
22. Pericardial cavity
23. Endocardium of ventricle
24. Myocardium of ventricle
25. Parietal pericardium
26. Shrinkage space

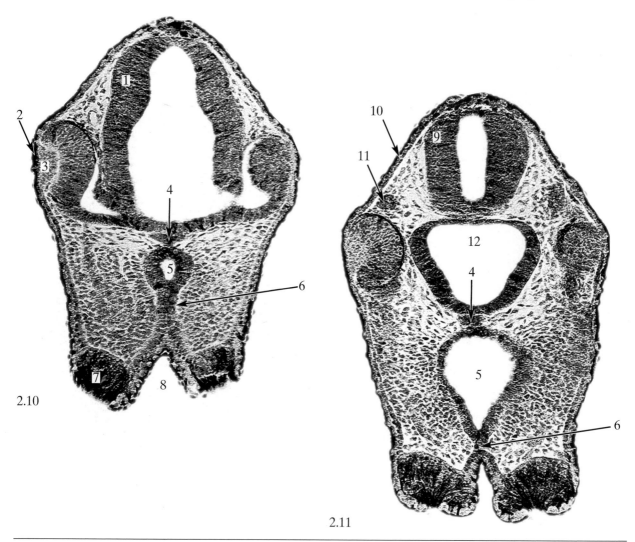

2.10

2.11

Photos 2.10, 2.11. Continuation of 4-mm frog embryo serial transverse sections numbered in anterior to posterior sequence.

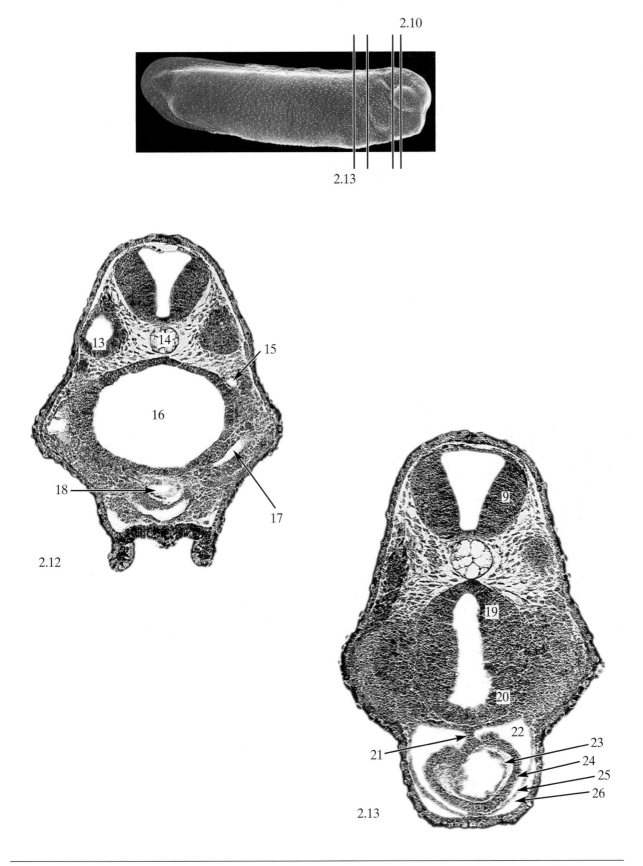

Photos 2.12, 2.13. Continuation of 4-mm frog embryo serial transverse sections numbered in anterior to posterior sequence.

Photos 2.14-2.18

Frog Embryos

Legend

1. Roof plate
2. Neural crest cells
3. Spinal cord
4. Floor plate
5. Dorsal aorta
6. Pronephric tubule
7. Pronephric duct

8. Pronephric ridge
9. Midgut
10. Yolk-filled endodermal cells
11. Skin ectoderm
12. Liver rudiment
13. Dorsal fin
14. Somite

15. Notochord
16. Subnotochordal rod
17. Hindgut
18. Proctodeum
19. Ventral fin

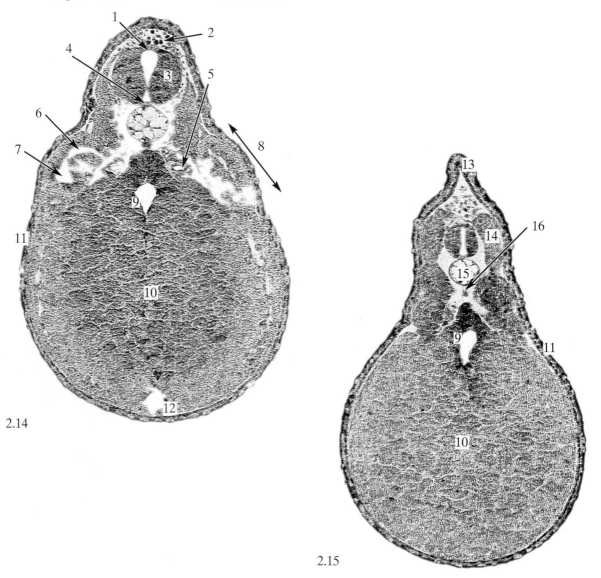

2.14

2.15

Photos 2.14, 2.15.   Continuation of 4-mm frog embryo serial transverse sections numbered in anterior to posterior sequence.

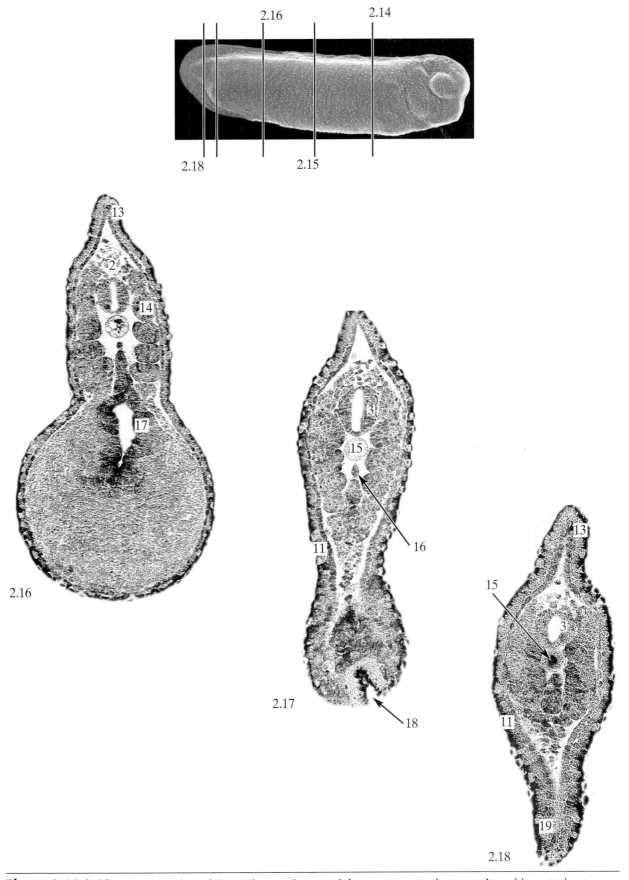

Photos 2.16-2.18.   Continuation of 4-mm frog embryo serial transverse sections numbered in anterior to posterior sequence.

Photos 2.19-2.24
Frog Embryos
Legend

1. Cleavage furrow
2. Blastomere

3. Micromere (blastomere)
4. Macromere (blastomere)

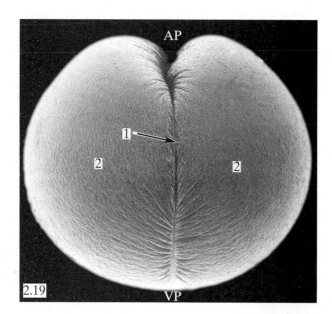

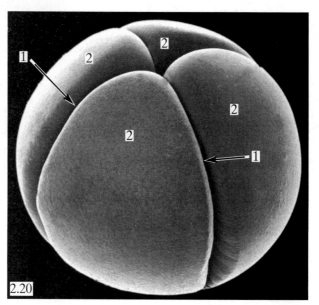

Photo 2.19. Scanning electron micrograph of a frog two-cell stage (viewed from the side). AP, Animal pole; VP, Vegetal pole.

Photo 2.20. Scanning electron micrograph of a frog four-cell stage (viewed from animal pole and side).

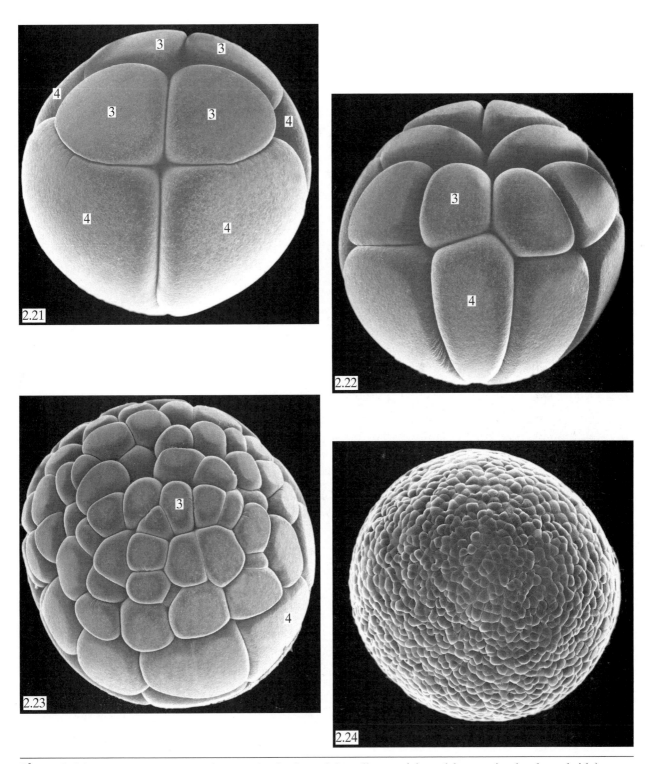

Photo 2.21. Scanning electron micrograph of a frog eight-cell stage (viewed from animal pole and side).

Photo 2.22. Scanning electron micrograph of a frog sixteen-cell stage (viewed from animal pole and side; all macromeres are not visible).

Photo 2.23. Scanning electron micrograph of a frog early blastula (viewed from animal pole and side).

Photo 2.24. Scanning electron micrograph of a frog late blastula (viewed from animal pole).

Photos 2.25-2.30

Frog Embryos

Legend

1. Yolk plug
2. Dorsal blastoporal lip
3. Lateral blastoporal lip
4. Archenteron
5. Ectoderm
6. Roof of archenteron
7. Blastocoel

8. Blastopore
9. Future brain level of neural plate
10. Future spinal cord level of neural plate
11. Neural fold
12. Skin ectoderm

13. Future brain level of neural groove
14. Future spinal cord level of neural groove
15. Future brain level of early neural tube
16. Future spinal cord level of early neural tube

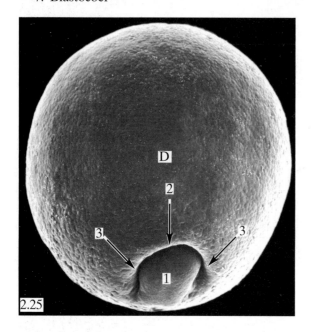

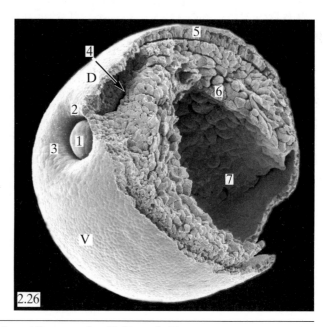

Photo 2.25. Scanning electron micrograph of a frog dorsal lip gastrula. D, Dorsal side.

Photo 2.26. Scanning electron micrograph of a parasagittal slice through a frog yolk plug gastrula. D, Dorsal side; V, Ventral side.

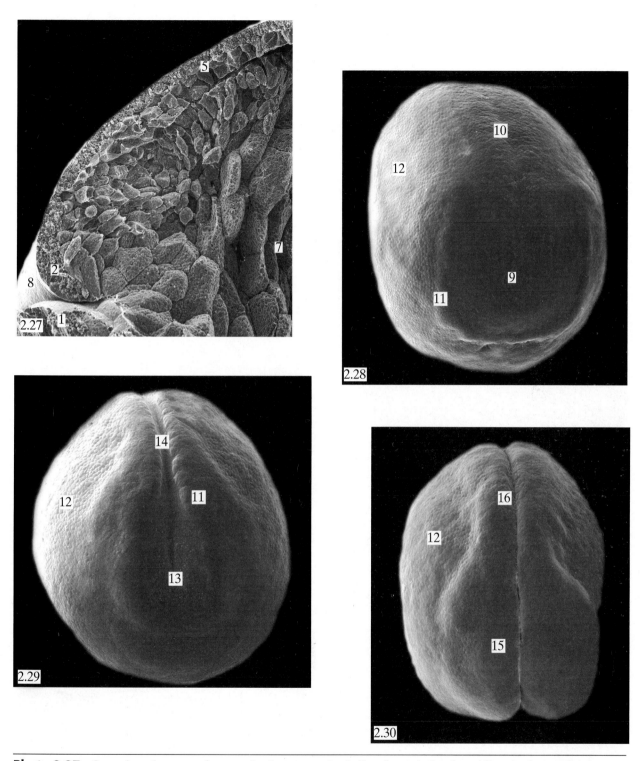

Photo 2.27. Scanning electron micrograph of a parasagittal slice through the dorsal lip of a frog yolk plug gastrula.

Photo 2.28. Scanning electron micrograph of a frog neural plate neurula.

Photo 2.29. Scanning electron micrograph of a frog neural fold neurula.

Photo 2.30. Scanning electron micrograph of a frog early neural tube neurula.

Frog Embryos

Legend

1. Neural tube
2. Ciliary tufts

3. Developing eye
4. Dorsal fin

5. Ventral fin
6. Stomodeum

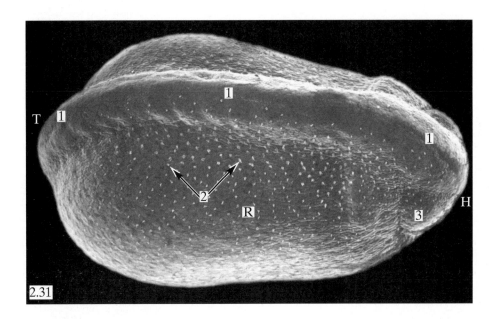

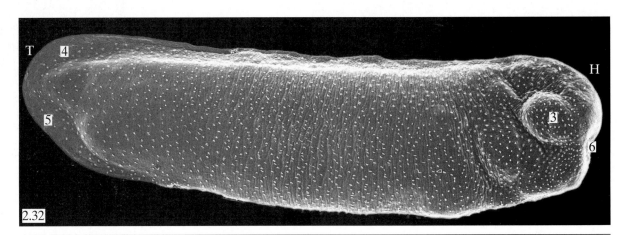

Photo 2.31.   Scanning electron micrograph of a frog early embryo (dorsolateral view). R, Right side.

Photo 2.32.   Scanning electron micrograph of a frog 4-mm embryo (lateral view). H, Head (cranial) end; T, Tail (caudal) end.

Chapter 3

Chick Embryos

Chapter 3

Chick Embryos

A. INTRODUCTION

The chick embryo has long been used for the study of early embryonic development because the stages desired can be readily obtained. Many of the developmental changes in the chick embryo are almost identical with those characteristic of other vertebrates, and particularly of mammals, whose young embryos are difficult to obtain in sufficient numbers in the stages needed for study. Consequently, the chick will be used extensively for the study of early stages of development. This is especially appropriate because many outstanding experimental embryologists (including Frank Lillie, Benjamin Willier, and Viktor Hamburger and their several students) chose the chick embryo as their model to analyze vertebrate development. The stages we will study are those that have been designated classically as 18-, 24-, 33-, 48-, and 72-hour embryos. Embryos within each classical stage vary somewhat in the degree of development attained, and thus your embryos may be slightly different than those described and illustrated here. Descriptions have been sufficiently generalized to permit study of most embryos within each classical stage.

Study of the early chick embryo begins with the 33-hour stage. You will then examine *earlier* stages in detail to learn how the structural relationships in the 33-hour embryo originated. You will next study *later* stages in detail to learn how the structural relationships in the 33-hour embryo are progressively modified during further development. Although chick embryos are initially studied "out of order," this approach has been beneficial for thousands of students (well over 100,000 have used this manual so far). For hands-on studies using chick embryos, see Chapter 6 (Exercises 3.1-3.12).

B. THE PART OF THE EGG TO BE STUDIED

From an embryological point of view, the bird's egg consists of a large spherical mass of **yolk** on which rests a small white disc of cytoplasm. If this disc has not initi-

ated **cleavage** (that is, if it is unsegmented), it is called the **blastodisc**. If this disc has initiated cleavage (that is, it is segmented into two or more cells by the process of mitosis), it is called the **blastoderm**. All parts of the embryo and its membranes originate from the blastodisc.

Developmental stages can be examined in detail after the blastoderm is removed from the yolk and then either mounted flat on a slide so it can be studied as a whole mount or cut into serial sections and then mounted. The center of the blastoderm is separated from the underlying yolk by a **subgerminal cavity**; the periphery is not. Furthermore, the periphery of the blastoderm contains cells laden with yolk, but yolk is sparse in the cells of the central region. Thus the central region of the blastoderm is more translucent than its periphery. The central region of the blastoderm is therefore named the **area pellucida**, and the peripheral region is called the **area opaca**.

C. 33-HOUR CHICK EMBRYOS

1. Whole mounts

First examine your slide with the naked eye. Note the round to oval shape of the mounted specimen. The specimen is divisible into two main areas, the peripheral **area opaca** and the central **area pellucida** (Fig. 3.1). The area opaca is further divided into the inner mottled **area vasculosa**, containing developing **blood islands**, and the outer **area vitellina**, a nonmottled area overgrowing the yolk. Try to identify the **sinus terminalis**, a circular blood vessel separating the area vasculosa from the area vitellina. This vessel may appear light or dark, depending on whether it contains blood cells. Most of the area vitellina is usually trimmed off from mounted specimens, so the sinus terminalis will probably be located just within the margin of the blastoderm. Running lengthwise within the area pellucida is a dark region, the **body** of the developing embryo. The darker and wider part of this region is the **head (cranial) end** of the embryo; the lighter part is the **tail (caudal) end**. Note a very clear area beneath and in front of the head; this area is called the

proamnion (a misnomer because it is not the source of the amnion, an extraembryonic membrane that develops later). It is the only region of the blastoderm that still lacks **mesoderm**, the middle **germ layer**.

Position your slide on the microscope stage so that when viewed through the microscope, the embryo is oriented as in Fig. 3.1 and Photos 3.1-3.4. Examine the slide under low magnification. First examine the **area vasculosa**. The **blood islands**, which give this area its mottled appearance, consist of irregularly shaped accumulations of mesodermal cells. The peripheral cells of each blood island form the **endothelium** constituting the walls of the blood vessels; central cells form **primitive blood cells**. Note that blood islands are most numerous lateral and caudal to the body of the embryo and that they are fusing together to form blood vessels. The primitive blood cells are therefore enclosed within these developing blood vessels, the **vitelline blood vessels**, which may open peripherally into the sinus terminalis and extend across the area pellucida toward the body of the embryo. Note that within the area pellucida, blood vessels are clear (empty) channels containing no blood cells and that the vitelline vessels are all about the same size. Circulation of the blood begins slightly later, although the heart at this stage contracts weakly. Initiation of the heartbeat and onset of circulation begin long before the heart is supplied with nerve fibers.

Next examine the **head end** of the embryo. Focus up and down on the head with the coarse adjustment of your microscope. Different structures and lines will become apparent to you at different levels of focus. This means that in addition to *length* and *width*, the body of the embryo has considerable *depth*; therefore, you can focus only on a part of its thickness at one time.

Notice two dense, thick bands running parallel to each other throughout the middle region of the embryo (Photos 3.1-3.3). Just lateral to these bands are paired blocks of mesoderm, the **somites**, separated by **intersomitic furrows**. These bands are the thick lateral walls of the **neural tube**. The neural tube is formed from the outermost **germ layer**, the **ectoderm**. The bands diverge in a bilaterally symmetrical fashion at the cranial end of the embryo where they constitute the lateral walls of the **brain**. Caudal to the brain these bands constitute the lateral walls of the **spinal cord**. Thus, most of the **central nervous system** (that is, the brain and spinal cord) present at this stage exists as a *hollow ectodermal tube*. More caudally the bands diverge laterally as **neural folds**. The space between the neural folds is the **neural groove**. At the caudal end of the embryo the converging bands represent the **primitive ridges (folds)**, which are the lateral margins of the primitive streak. The **primitive streak** is the site of **mesoderm** formation at this stage.

Turn your attention now to the brain region and note

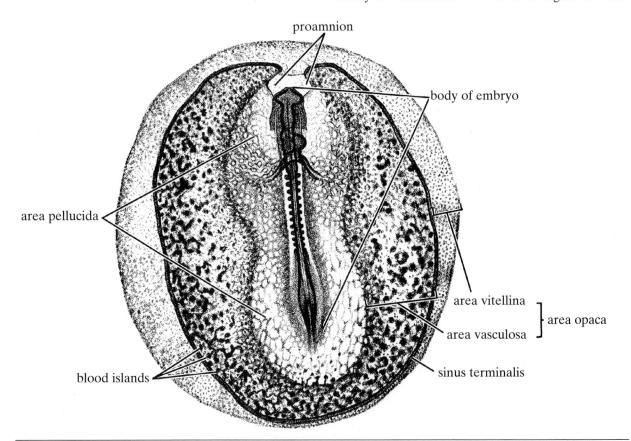

Fig. 3.1. Drawing of a 33-hour chick embryo whole mount.

the two largest lateral expansions (that is, evaginations) of the neural tube; these are the **optic vesicles**, the earliest rudiments of the **eyes** (Photos 3.3-3.5). The region of the brain from which these evaginate is the **diencephalon**. Note by carefully focusing that these vesicles are in close contact laterally with the overlying **skin (surface) ectoderm**. These regions of skin ectoderm are probably not thickened, but they will *later* thicken and transform into the **lenses** of the eyes. Note a dark, shadowy mass lying in the midline between the optic vesicles. This is the **infundibulum**, which forms as an evagination from the floor of the diencephalon (Photos 3.2, 3.4, 3.5). It will *later* form the **posterior pituitary gland (neurohypophysis)**.

The level of the brain cranial to the diencephalon is the **telencephalon** (Photo 3.4). It is poorly demarcated from the diencephalon at this stage, and the two are usually spoken of collectively as the **prosencephalon**. Focus down into the telencephalon and notice a white median notch or cleft, the remnant of the **cranial (anterior) neuropore**. Earlier, the cavity of the neural tube opened broadly to the outside through the cranial neuropore. At the 33-hour stage, the neural folds that originally flanked the cranial neuropore are approximated but not yet completely fused.

The division of the brain caudal to the diencephalon is the **mesencephalon**, and the next caudal expansion is the **metencephalon** (Photo 3.2). The most caudal region of the brain is the **myelencephalon**, which is characterized by several smaller enlargements of the neural tube, the **neuromeres**. The metencephalon and myelencephalon are spoken of collectively as the **rhombencephalon**. Note that the space between the walls of the brain and the skin ectoderm is filled with loosely packed cells, the **head mesenchyme** (Photos 3.4, 3.5).

Recall that the somites are located lateral to the neural tube throughout the middle region of the embryo. How many pairs are present in your embryo? The most cranial pair is incomplete (has no definite cranial boundaries). It lies just caudal to the last neuromere. All somites cranial to the *fifth* intersomitic furrow are later contained within the head of the chick. Thus, more than one-third of the embryonic body formed by this stage is future head. Note that lateral to each neural fold a band of tissue extends caudally from the last pair of somites. These two bands are the **segmental plates**, which will form additional somites in craniocaudal sequence (Photo 3.1). By carefully focusing down through the neural tube, you will be able to identify a narrow mesodermal rod, the **notochord** (Photos 3.2-3.5). This structure lies directly beneath the neural tube and neural groove, between the infundibulum and the cranial end of the primitive streak. The cranial tip of the notochord *induces* the infundibulum to form. Notice that the notochord is widest just cranial to the primitive streak.

Observe lines extending craniocaudally by carefully focusing just laterally to the mesencephalon and rhombencephalon. These lines represent the lateral boundaries of the **foregut**. The cranial boundary of the foregut usually lies near the caudal boundaries of the optic vesicles, but it is often very difficult to see. The foregut is formed from the innermost **germ layer**, the **endoderm**, and resembles a finger of a glove: closed cranially but open caudally. The opening is called the **cranial (anterior) intestinal portal** (Photo 3.4) and can be seen more readily if slides are viewed from the ventral surface (Photo 3.5). It lies beneath the myelencephalon, and its margin forms a distinct curvature with its concavity facing caudally.

Note that the cranial part of the head is sharply bounded and that it protrudes above the underlying blastoderm. The head is separated from the blastoderm by an ectoderm-lined space, the **subcephalic pocket**.

Focus down through the level of the rhombencephalon and observe the **heart**, usually bent somewhat toward the right (Photo 3.4). The heart can be seen more readily from the ventral surface (Photo 3.5). Two large veins, the **vitelline veins**, enter the heart from behind, immediately cranial to the cranial intestinal portal. The cranial ends of these veins are continuous with the **sinoatrial region** of the heart (*future* **sinus venosus** and **atrium**). The region of the heart that bulges slightly to the right (or the widest part of the heart, if bending has not yet occurred) is the **ventricle**. The heart narrows cranial to the ventricle, indicating the level of the **conotruncus (bulbus cordis)**.

2. Serial transverse sections

Position your slide on the microscope stage so that when viewed through the microscope, each section is oriented as in Photos 3.6-3.14 (that is, with the dorsal surface at the top). *Do not place microscope stage clips (if present) on slides to hold them in place as this could fragment sections.*

For instructions on how to use serial sections, see Chapter 2, Section B. It is recommended that you prepare a graphical reconstruction of *your* transversely serially sectioned embryo to help you visualize relationships of parts of an embryo to one another in three dimensions. A sample is given in Fig. 3.2. Directions and graph paper are given in Chapter 6 (Exercise 3.1).

Begin your slide study by examining sections in anteroposterior sequence (see Chapter 2, Section B) until you find a section resembling Photo 3.6. This section cuts through the portion of the head containing the **diencephalon**. The **proamnion** lies beneath the head, separated from it by a space called the **subcephalic pocket**. The proamnion consists of two layers: upper **ectoderm**, lower **endoderm**. To either side of it are four cell layers. The dorsal two constitute the **somatopleure** (upper **ectoderm**, lower **somatic mesoderm**); the ventral two,

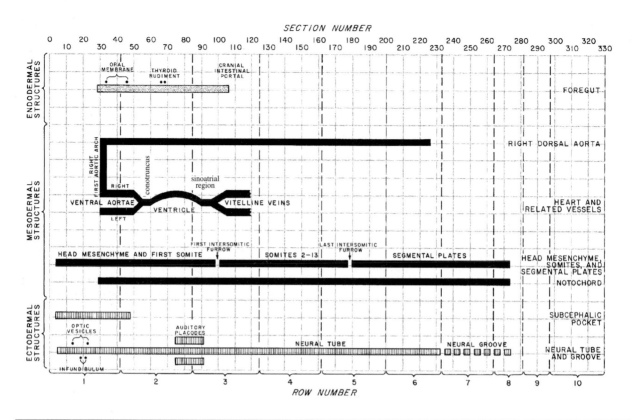

Fig. 3.2. Sample graphical reconstruction of a 33-hour chick embryo, showing relationships of parts to one another along the craniocaudal axis. See Chapter 6, Exercise 3.1 for instructions on how to prepare a similar reconstruction from your set of serial transverse sections.

the **splanchnopleure** (upper **splanchnic mesoderm**, lower **endoderm**). The space between somatic and splanchnic mesoderm is the **coelom**.

From this level trace sections *anteriorly* until the head disappears. Identify the proamnion, somatopleure, splanchnopleure, and coelom at this level. Then trace sections *posteriorly* and note the reappearance of the head. The first part of the brain seen is the **telencephalon**. [Record the first section through the neural tube on the *neural tube and groove* line of your graph.] It will usually seem divided into right and left halves by a narrow white slit, the remnant of the **cranial neuropore**.

Continue to trace sections posteriorly. The region of the brain to which the **optic vesicles** attach is the **diencephalon** (Photo 3.6). Part of each optic vesicle may lie caudal to its connection to the diencephalon and thus appear in some sections as an isolated vesicle. [Record the craniocaudal extent of the optic vesicles by blue dots above the *neural tube and groove* line. Record these boundaries for one side only if your embryo is not sectioned symmetrically.] As the optic vesicles begin to fade out, note an evagination from the floor of the diencephalon, the **infundibulum**. [Record the craniocaudal extent of the infundibulum by blue dots below the *neural tube and groove* line.] Identify the thin **skin ectoderm**, the outer layer of the head, and the **head mes-**

enchyme between the skin ectoderm and brain wall.

A few sections behind the infundibulum a small mass of cells appears beneath the neural tube, closely applied to its surface. This is the cranial tip of the **notochord**. [Record the first section through the notochord on the *notochord* line.] Continue to trace the neural tube and notochord caudally, noting that the notochord becomes more distinct and separated from the neural tube. The level of the neural tube that is oval or pear shaped is the **mesencephalon** (Photos 3.7, 3.8). It appears as the optic vesicles fade out, just caudal to the infundibulum. The neural tube usually becomes smaller in more posterior sections. In this general region the **metencephalon** can be identified as that portion of the neural tube that lies above the conotruncus region of the heart (Photo 3.9). The caudal end of the **subcephalic pocket** can also be identified at about this level. [Record the craniocaudal extent of the subcephalic pocket by a blue bar on the appropriate line under *ectodermal structures*.] In more posterior sections the **myelencephalon** can be identified as that portion of the neural tube that lies above the ventricle and sinoatrial region (Photos 3.10, 3.11). The skin ectoderm has thickened alongside part of the myelencephalon, forming the **auditory (otic) placodes**, the rudiments of the **inner ears**. [Record the craniocaudal extent of the auditory placodes by blue bars above and below the *neural tube and groove* line.]

Formation of the auditory placodes is *induced* by the myelencephalon and adjacent head mesenchyme.

Focus on the dorsal surface of the myelencephalon just cranial to or just caudal to the auditory placodes. You may be able to identify a straggling line of cells wedged between the skin ectoderm and myelencephalon. (If you cannot identify these cells at the level of the myelencephalon, look more cranially at the level of the metencephalon or mesencephalon; Photos 3.8, 3.9.) These are **neural crest cells**, which are derived from *ectoderm*. Both neural crest cells and mesoderm contribute to the **head mesenchyme**.

Note the change in shape of the neural tube as you continue caudally from the level of the myelencephalon, indicating that you have reached the level of the **spinal cord** (Photo 3.12). The lateral walls of the spinal cord are much thicker than its dorsal and ventral walls. Recall that only the thick lateral walls of the neural tube can be readily seen in whole mounts. The dorsal wall is designated as the **roof plate** and the ventral as the **floor plate**. **Skin ectoderm** overlies the roof plate.

In more posterior sections the neural tube may seem to contain no cavity. This is because the walls of certain levels of the spinal cord are normally in contact during this stage, occluding the lumen (the fact that the lumen of the spinal cord is occluded is usually not ascertainable in whole mounts). This occlusion confines **cerebrospinal fluid** to the brain region, aiding in the subsequent enlargement of the brain as fluid pressure increases (compare the size of the brain in 33-, 48-, and 72-hour embryos; Photos 3.1-3.5, 3.78-3.80, 3.112-3.114). The roof plate disappears as sections are traced posteriorly; this is the beginning of the **neural groove** (Photo 3.13). [Record the craniocaudal extent of the neural tube by a blue bar on the *neural tube and groove* line.] The neural groove is flanked by the **neural folds**. Continue following the neural groove caudally and note the progressive enlargement and flattening of the notochord. The notochord then becomes indistinct, seeming to fuse with the floor of the neural groove. [Record the craniocaudal extent of the notochord by a red bar on the *notochord* line.] The neural folds are widely separated at this level (see Photo 3.1). Slightly more caudally, a large mass of cells, which extends across the midline, gradually appears in place of the notochord. You are now near the caudal end of the neural groove in the region of the **primitive streak**. [Record the craniocaudal extent of the neural groove by a broken blue bar on the *neural tube and groove* line.] In more posterior sections, identify the smaller **primitive groove**, which is flanked by the **primitive ridges** (Photo 3.14). The primitive ridges are directly continuous with the neural folds, so the transition between the two is gradual. The primitive ridges and groove disappear in more posterior sections, indicating that you are now caudal to the primitive streak.

Return in your sections to the level of the optic vesicles and again trace sections posteriorly. A small endoderm-lined cavity, the **foregut**, appears just beneath the neural tube near the caudal ends of the optic vesicles (Photo 3.6). [Record the cranial boundary of the foregut on the *foregut* line.] In more posterior sections, the foregut widens and its endodermal floor contacts the ventral skin ectoderm of the head, forming the **oral membrane** (Photo 3.17). If a wide space exists between the two layers, it is a shrinkage space caused by section preparation. These two layers will *later* rupture to establish the **mouth opening**. [Record the craniocaudal extent of the oral membrane by yellow dots above the *foregut* line.]

Examine the floor of the foregut as you continue to trace sections caudally. Identify a midline thickening, the **thyroid rudiment**, which usually appears as a shallow depression above the ventricle (Photo 3.10). [Record the craniocaudal extent of the thyroid rudiment by yellow dots above the *foregut* line.] Continue to trace the foregut caudally and note the sudden disappearance of its floor; this is the region of the **cranial intestinal portal**. [Record the craniocaudal extent of the foregut by a yellow bar on the *foregut* line, and label the level of the cranial intestinal portal.] Note that in more posterior sections the embryo progressively flattens dorsoventrally and that the endoderm forms its lower layer (Photo 3.12). The endoderm becomes indistinguishable in the midline at the level of the primitive streak.

Return again to the level of the rhombencephalon and identify the loosely packed cells of the **head mesenchyme**. The mesoderm to either side of the neural tube just caudal to the auditory placodes forms compact masses, the **somites**. The first somite pair is difficult to identify because it has no distinct cranial boundaries. Two to three sections posteriorly, the first somite pair begins to fade out as sections cut through the first **intersomitic furrows**. [Record the approximate craniocaudal extent of the head mesenchyme and first somite pair by a red bar on the *head mesenchyme, somites, and segmental plates* line.] The second somite pair appears slightly more caudally and soon begins to fade out at the level of the second intersomitic furrows. Continue tracing the somites caudally and determine the approximate number of pairs present in your embryo. [Record the distance between the first and last intersomitic furrows by a red bar on the *head mesenchyme, somites, and segmental plates* line, and indicate how many somite pairs this bar represents.]

Examine in detail the structure of one of the best-formed somites under higher magnification. It is roughly triangular in transverse section and usually contains a small core of cells (Photo 3.12). Note the close relationship of the somites and notochord to the neural tube. Experiments have shown that the characteristic shape of the neural tube (that is, its thick lateral walls and thin roof and floor plates) depends on this relationship. At this level it might be possible to observe **neural crest cells** lying between the neural tube, skin ectoderm, and somites.

Just lateral to each somite is a slender area of mesodermal cells, the **nephrotome** (**intermediate mesoderm**). The nephrotome on each side produces a solid mass of cells called the **pronephric cord** (Photo 3.12). The cords constitute the paired rudimentary **pronephric kidneys**. (The pronephric cord on each side *later* separates from the underlying nephrotome and elongates to form a **mesonephric [pronephric] duct rudiment**.) The mesoderm lateral to the nephrotome is the **lateral plate**. It is divided into an upper layer of **somatic mesoderm** (adjacent to the ectoderm) and a lower layer of **splanchnic mesoderm** (adjacent to the endoderm). The space between the somatic and splanchnic mesoderm is the **coelom**. Note the presence of **vitelline blood vessels** in the splanchnic mesoderm.

Identify the **segmental plates** in sections posterior to the last pair of intersomitic furrows. As you trace them caudally, observe that they become continuous with the more lateral mesoderm (Photo 3.13) and that they become indistinguishable more caudally at the level of the primitive streak. [Record the craniocaudal extent of the segmental plates by a red bar on the *head mesenchyme, somites, and segmental plates* line.]

There remains for consideration only the circulatory system, a mesodermal derivative. Return to the level of the diencephalon and again trace sections posteriorly. A white space should be visible on each side of the foregut at about the level where its cranial end appears. These spaces are the **first aortic arches** (Photo 3.7). [Record the craniocaudal extent of the right first aortic arch by a red bar interconnecting the sixth and ninth centimeter lines under *mesodermal structures*.] Each is continuous with a small **ventral aorta**, lying below the foregut, and a larger **dorsal aorta**, lying above the foregut. The first aortic arch is often very small, so it may not be readily seen in your sections.

Trace sections posteriorly, watching all aortae. The two ventral aortae may be difficult to follow because they are just forming. Note that they merge with the **conotruncus** caudal to the oral membrane. [Record the craniocaudal extent of the ventral aortae by red bars on the fifth and sixth centimeter lines under *mesodermal structures*. Connect the right ventral aorta with the right first aortic arch.] The conotruncus lies in the midline and has a small diameter. It consists of an inner, very thin lining of **endocardium**, continuous with the endothelium of the ventral aortae, reinforced by an outer thick layer, the **myocardium**. The space between the endocardium and myocardium is filled with a cellular secretion called the **cardiac jelly** (usually not visible in conventional paraffin sections). (A third cellular layer, forming the outermost covering of the heart, will ap-

pear between the third and fourth days of development.) When sections are traced posteriorly, the heart usually seems to bend from the midline somewhat toward the *embryo's right*, the *apparent left* (Photo 3.10). This is the region of the **ventricle**. Even if the ventricle has not yet bent to the right in your embryo, it can be readily identified because its endocardium and myocardium are widely separated. Note that the ventricle is suspended within the coelom by a bridge of mesoderm, the **dorsal mesocardium**. The portion of the coelom surrounding the heart is the **pericardial cavity**. The endocardium and myocardium are no longer widely separated in more posterior sections, and the heart lies in the midline. This is the beginning of the **sinoatrial region**. The sinoatrial region is incompletely formed caudally, consisting of two rudiments, one on either side of the cranial intestinal portal. As you continue tracing sections posteriorly, there is a gradual transition between the sinoatrial region and **vitelline veins**. The vitelline veins shift laterad and consist only of endothelium. [Record the approximate craniocaudal extents of the conotruncus, ventricle, sinoatrial region, and vitelline veins by red bars under *mesodermal structures*.]

Shift your attention to the dorsal aortae and trace them caudally. The dorsal aortae lie just below the nephrotomes and somites throughout most of their extent (Photo 3.12). Their endothelium is so thin that it may not be visible in some sections. [Record the craniocaudal extent of the right dorsal aorta by a red bar on the *right dorsal aorta* line. Connect the right dorsal aorta with the right first aortic arch.]

3. Serial sagittal sections

Careful study of serial sagittal sections can help you immeasurably to understand relationships among developing parts of early chick embryos. These sections are cut lengthwise in the dorsoventral plane. First examine your slide with the naked eye. Only the darkest sections are cut through the body of the embryo. Position your slide on the microscope stage so that when viewed through the microscope, the neural tube is uppermost (Photo 3.15). Trace sections until you find one closely resembling Photo 3.15. Identify the various levels of the **neural tube** in midsagittal sections. Similarly, identify the **neural groove** caudal to the neural tube. The **notochord** lies beneath the neural tube and groove, but only part of the notochord's length will be cut through in any one section. Determine the craniocaudal extent of the notochord. It merges caudally with the **primitive streak**. Its cranial end lies just caudal to the **infundibulum**. Identify the endoderm-lined **foregut**, the double-layered **oral membrane**, and the **cranial intestinal portal**. Note that the head is separated from the **proamnion** by the **subcephalic pocket**.

Beneath the foregut is the portion of the **coelom** containing the heart, the **pericardial cavity**. Identify the

conotruncus, **ventricle**, and **sinoatrial region**. It may be necessary to trace sections to either side of the midline to identify these structures. By carefully tracing sections, you may also be able to identify the **dorsal aortae** above the foregut, the **ventral aortae** below the foregut, and the **first aortic arches**.

Note that as the **prosencephalon** fades out in sections to either side of the midline, the **optic vesicles** remain for several more sections. Also note that as the **myelencephalon** begins to fade out, the skin ectoderm overlying it thickens as the **auditory placodes**. Identify the **head mesenchyme**, **somites**, **intersomitic furrows**, and **segmental plates** in sections to either side of the neural tube.

4. Summary of the contributions of the germ layers to structures present in the 33-hour chick embryo

Ectoderm

> auditory placodes
> infundibulum
> neural crest cells
> neural folds and groove
> neural tube
> optic vesicles

Mesoderm

> blood islands
> blood vessels (endothelium)
> dorsal mesocardium
> heart
> lateral plates
> nephrotomes
> notochord
> primitive blood cells
> pronephric cords
> segmental plates and somites

Endoderm

> foregut
> thyroid rudiment

Ectoderm and mesoderm

> head mesenchyme

Ectoderm and Endoderm

> oral membrane
> proamnion

5. Scanning electron microscopy

Scanning electron microscopy of intact embryos provides a view that is much different from that provided by light microscopy of whole mounts (compare Photos 3.1-3.3 to Photo 3.16). Because the embryo is covered by **skin ectoderm** and scanning electron microscopy reveals only surface features (that is, we cannot "see" through the ectoderm as we can in stained and cleared whole mounts examined with light microscopy), details of the shapes of the neural tube and somites are largely obscured. Identify the **auditory placodes** near the caudal end of the head. Where the neural tube had not yet formed (and hence is not covered by skin ectoderm), the **neural groove** and **neural folds** are readily visible (Photos 3.16, 3.17). Less well defined is the **primitive streak**, consisting of the shallow **primitive groove** flanked by the paired **primitive ridges**. The most prominent feature of intact embryos at this stage occurs at their head end. The neural tube bulges bilaterad as the **optic vesicles** (Photos 3.16, 3.18). The remnant of the **cranial neuropore** lies in the midline between the optic vesicles.

Transverse slices through the region of the **optic vesicles** reveal the **infundibulum**, the remnant of the **cranial neuropore** on the wall of the **telencephalon**, and the connections of the **optic vesicles** to the **diencephalon** (Photo 3.19). A close spatial relationship exists between the elongated cells of each optic vesicle and the much shorter covering cells of the future **lens** ectoderm (Photo 3.20). More caudally, sections pass through the level of the **mesencephalon** (Photo 3.21). Identify at this level the **foregut**, the paired **dorsal aortae** lying above the foregut, the paired **ventral aortae** lying ventral to the foregut, and the double-layered **oral membrane** (consisting of **foregut endoderm** and **skin ectoderm**). The **heart** is encountered in more caudal levels (Photo 3.22-3.24). The **conotruncus** is present at the level of the **metencephalon** (Photos 3.22, 3.23). Also present at this level are the **notochord** (not readily visible in Photos 3.22, 3.23), **dorsal aortae**, and **foregut**. The space surrounding the heart is the **pericardial cavity** (lined with **somatic mesoderm** above and **splanchnic mesoderm** below), and the conotruncus is suspended in this cavity by the **dorsal mesocardium**.

The conotruncus is continuous caudally with the **ventricle** (Photos 3.22, 3.23). (Note that the ventricle joins a third region of the heart more caudally, the **sinoatrial region**; Photos 3.23, 3.24.) The two layers of the ventricle, the **endocardium** and **myocardium**, are widely separated by the **cardiac jelly** (Photos 3.24, 3.25). Note that this extracellular material has a complex structure consisting of numerous slender fibrils and tiny, spherical bodies. Many elegant investigations have concentrated on determining the chemical constituents of the cardiac jelly and their role in the morphogenesis of the heart.

The division of the brain at the level of the ventricle is the **myelencephalon** (Photo 3.24). The **skin ectoderm** lateral to the myelencephalon has thickened (forming the **auditory placodes**) and (in the embryo shown in

Photo 3.24) invaginated, forming the **auditory vesicles**. Each auditory vesicle consists of a bowl-shaped depression in surface views of intact embryos (Photo 3.26).

The **foregut** is present at the levels of the conotruncus and ventricle. The floor of the foregut is thickened and evaginated at the level of the ventricle, forming the **thyroid rudiment** (Photo 3.24). More caudally, the **cranial intestinal portal** is present; consequently, the gut has not yet formed. The embryo has a relatively simple organization in areas caudal to the foregut, consisting of three germ layers arranged in dorsoventral sequence (Photos 3.27, 3.28). The outer (dorsal) surface of the embryo consists of **skin ectoderm**. The lower (ventral) surface consists of **endoderm**. The **mesoderm** occupies most of the area between these two layers, except in the central region occupied by the ectoderm-derived **spinal cord**. The mesoderm is subdivided into the midline **notochord** (ventral to the spinal cord), the **somites** (flanking the spinal cord), the ill-defined **nephrotome**, and the **lateral plate** (subdivided into an upper layer of **somatic mesoderm** and a lower layer of **splanchnic mesoderm**; the space between these two layers is the **coelom**). The **dorsal aortae** (also derived from mesoderm) lie beneath the nephrotomes and lateral portions of the somites.

A neural tube has not yet formed at more caudal levels. Depending on the exact level, a **neural groove** (flanked by **neural folds**) or a flat **neural plate** is present. The **notochord** occupies the midline, beneath the neural groove and neural plate. The remainder of the mesoderm at these levels consists of the **segmental plates** (Photo 3.29), which merge laterally without distinct boundaries with mesoderm that will *later* form the

nephrotomes and **lateral plates**.

A short **primitive streak** is present at the caudal end of the embryo (Photo 3.30). Three well-defined germ layers are present lateral to the streak. Medially, cells are involuting through the streak to contribute to the mesodermal layer. A pair of poorly defined **primitive ridges** and a shallow **primitive groove** can be identified.

Micrographs of tissue blocks sectioned sagittally and then examined by scanning electron microscopy greatly aid beginning students in visualizing embryos in three dimensions. Photo 3.31 shows such a micrograph of an embryo slightly more developed than most 33-hour embryos. Most of the major subdivisions of the brain are clearly visible (that is, the **diencephalon**, **mesencephalon**, **metencephalon**, and **myelencephalon**), as well as associated structures such as the **infundibulum** and left **optic vesicle**. Also visible are small sections of the **notochord** and associated mesoderm, **foregut**, **oral membrane**, **cranial intestinal portal**, **sinoatrial region**, and the ectoderm-lined **subcephalic pocket**.

Photo 3.32 shows a parasagittal cryofracture through the caudal end of a 33-hour embryo. Identify the **skin ectoderm**, **somites**, **segmental plate**, and **endoderm**.

D. PHOTOS 3.1-3.32: 33-HOUR CHICK EMBRYOS

Photos 3.1-3.32 depict the 33-hour chick embryos discussed in Chapter 3. These photos and their accompanying legends begin after the following page.

Photos begin on the following page. Use the space below for notes.

Photos 3.1-3.5

Chick Embryos

Legend

1. Area vasculosa
2. Blood islands
3. Somites
4. Spinal cord region of the neural tube
5. Segmental plates
6. Neural folds
7. Region of the primitive streak
8. Infundibulum
9. Proamnion
10. Mesencephalon region of the neural tube
11. Caudal extent of the subcephalic pocket
12. Metencephalon region of the neural tube
13. Neuromeres of the myelencephalon region of the neural tube
14. 5th intersomitic furrow
15. Notochord beneath broadened neural groove
16. Optic vesicle(s)
17. Notochord beneath mesencephalon
18. First somite
19. Area pellucida
20. Continuity between a neural fold (upper arrow) and primitive ridge (lower arrow)
21. Remnant of the cranial neuropore
22. Skin ectoderm
23. Head mesenchyme
24. Foregut
25. Ventricle of heart
26. Sinoatrial region of heart
27. Cranial intestinal portal
28. Conotruncus region of heart
29. Continuity between the sinoatrial region of heart and a vitelline vein
30. Telencephalon region of the neural tube
31. Diencephalon region of the neural tube

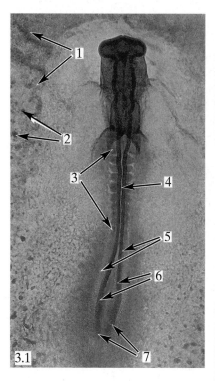
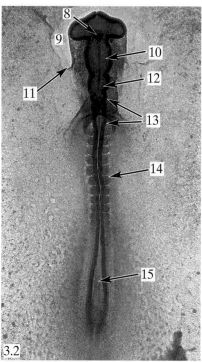
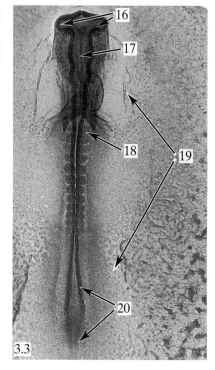

Photos 3.1-3.3.   33-hour chick embryo whole mounts (dorsal views).

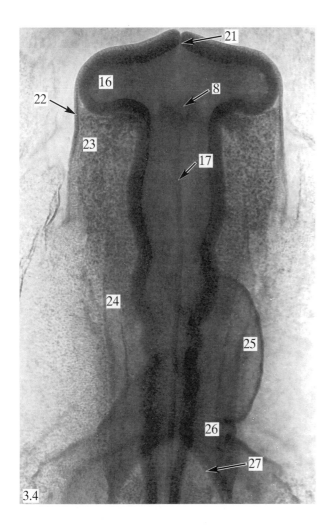

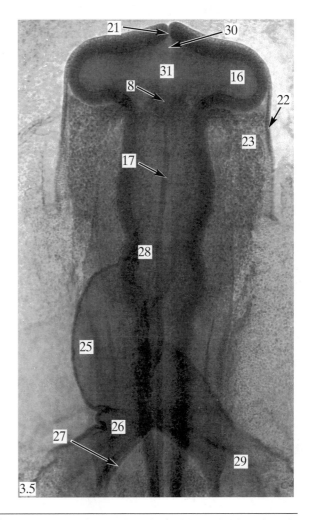

Photos 3.4, 3.5. 33-hour chick embryo whole mounts (enlargment of the head region). Photo 3.4 is a dorsal views; Photo 3.5 is a ventral view.

Photos 3.6-3.9

Chick Embryos

Legend

1. Diencephalon
2. Future lens ectoderm
3. Optic vesicle
4. Infundibulum
5. Foregut
6. Subcephalic pocket
7. Proamnion
8. Head mesenchyme

9. Dorsal aorta
10. First aortic arch
11. Ventral aorta
12. Oral membrane
13. Notochord
14. Mesencephalon
15. Coelom
16. Splanchnic mesoderm

17. Endoderm
18. Conotruncus
19. Somatic mesoderm
20. Skin ectoderm
21. Somatopleure
22. Metencephalon
23. Neural crest cells

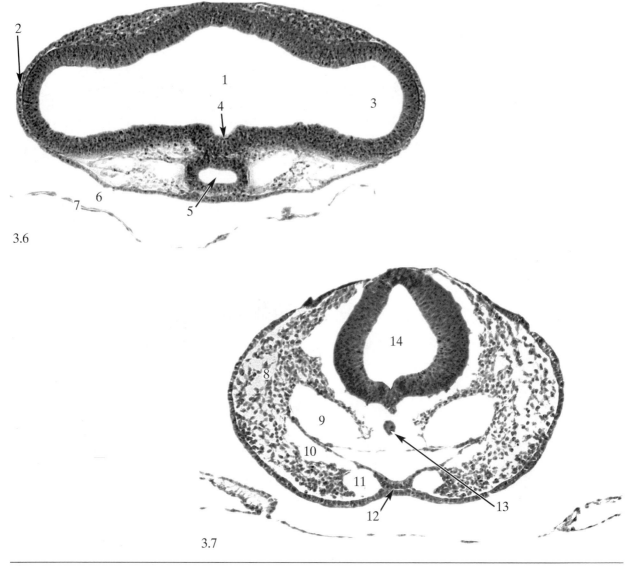

3.6

3.7

Photos 3.6, 3.7. 33-hour chick embryo serial transverse sections numbered in anterior to posterior sequence.

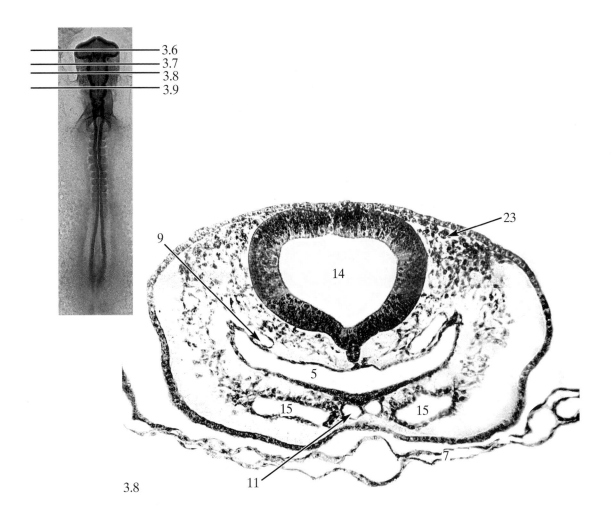

3.8

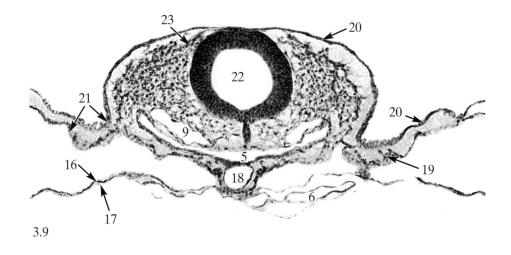

3.9

Photos 3.8, 3.9. Continuation of 33-hour chick embryo serial transverse sections numbered in anterior to posterior sequence.

Photos 3.10-3.14

Chick Embryos

Legend

1. Skin ectoderm
2. Foregut
3. Ventricle
4. Pericardial cavity region of the coelom
5. Myocardium
6. Endocardium
7. Thyroid rudiment
8. Dorsal aorta
9. Myelencephalon
10. Auditory placode
11. Junction between somatic and splanchnic mesoderm
12. Sinoatrial region of heart
13. Roof plate
14. Spinal cord
15. Floor plate
16. Pronephric cord
17. Splanchnopleure of the lateral plate
18. Notochord
19. Endoderm
20. Splanchnic mesoderm
21. Somite
22. Neural groove
23. Neural folds
24. Segmental plate
25. Primitive groove
26. Primitive ridges
27. Primitive streak
28. Somatic mesoderm
29. Somatopleure of the lateral plate
30. Nephrotome (buds off pronephric cords)
31. Coelom

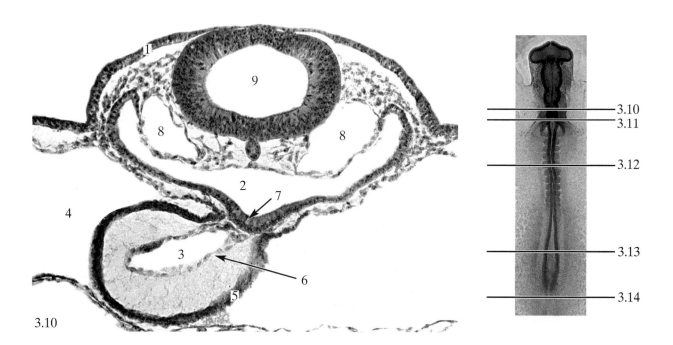

3.10

Photo 3.10. Continuation of 33-hour chick embryo serial transverse sections numbered in anterior to posterior sequence.

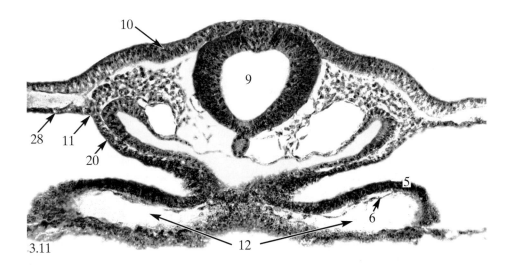

3.11

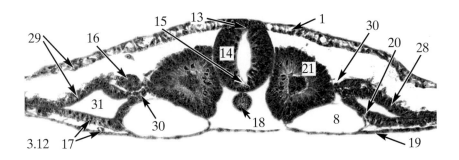

3.12

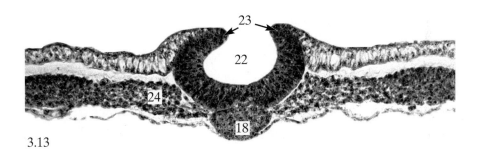

3.13

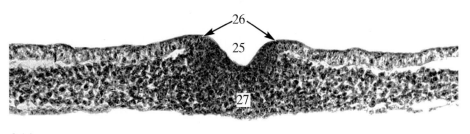

3.14

Photos 3.11-3.14.   Continuation of 33-hour chick embryo serial transverse sections numbered in anterior to posterior sequence.

Photos 3.15-3.20

Chick Embryos

Legend

1. Skin ectoderm
2. Foregut
3. Ventricle
4. Pericardial cavity region of the coelom
5. Myelencephalon
6. Sinoatrial region of heart
7. Notochord
8. Infundibulum
9. Mesencephalon
10. Metencephalon

11. Cranial intestinal portal
12. Conotruncus region of heart
13. Proamnion
14. Oral membrane
15. Subcephalic pocket
16. Prosencephalon
17. Head (cranial) end of embryo
18. Auditory placodes
19. Neural fold
20. Neural groove

21. Primitive groove (flanked by primitive ridges)
22. Remnant of the cranial neuropore
23. Optic vesicle (covered by skin ectoderm in Photo 3.18)
24. Ventral surface of head
25. Continuity between optic vesicle and diencephalon
26. Infundibulum
27. Future lens ectoderm

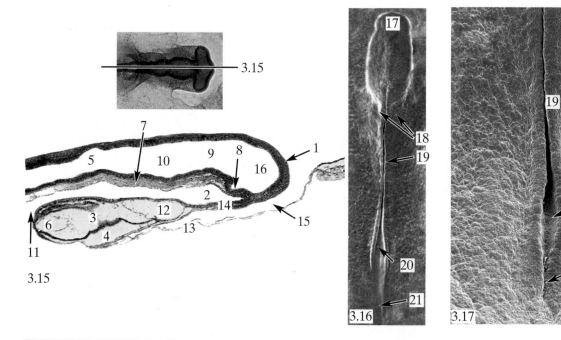

Photo 3.15. 33-hour chick embryo midsagittal section.

Photo 3.16. Scanning electron micrograph of an intact 33-hour chick embryo (dorsal view).

Photo 3.17. Scanning electron micrograph of an intact 33-hour chick embryo (dorsal view of the caudal end).

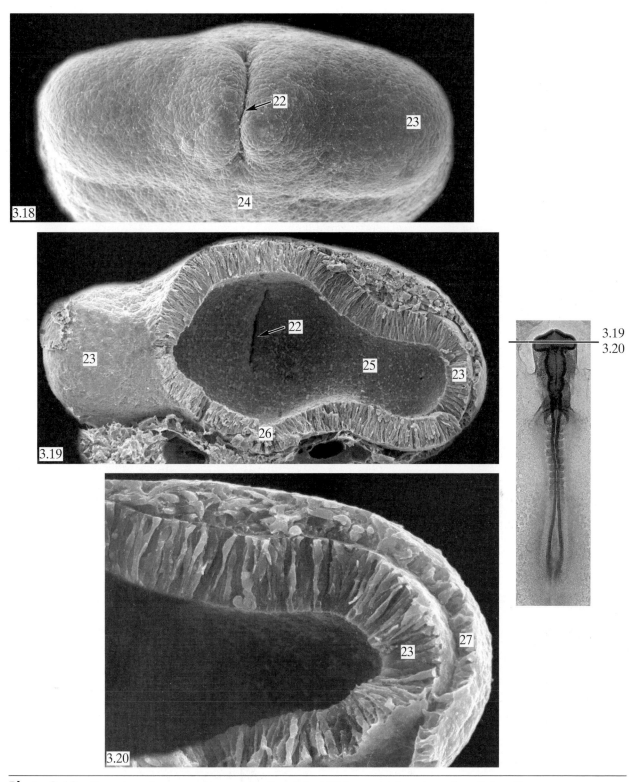

Photo 3.18. Scanning electron micrograph of an intact 33-hour chick embryo (cranial end).

Photos 3.19. Scanning electron micrograph of a transverse slice through a 33-hour chick embryo (the skin ectoderm and subjacent mesenchyme have been removed on the left side of the micrograph).

Photo 3.20. Scanning electron micrograph of an enlargement of the lateral portion of a transverse slice of a 33-hour chick embryo.

Photos 3.21-3.26

Chick Embryos

Legend

1. Mesencephalon
2. Ventral aorta
3. Oral Membrane
4. Dorsal aorta
5. Foregut
6. Metencephalon (cranial end)
7. Skin ectoderm
8. Somatic mesoderm
9. Somatopleure
10. Conotruncus region of heart
11. Dorsal mesocardium
12. Pericardial cavity
13. Ventricle of heart
14. Splanchnopleure
15. Metencephalon (caudal end)
16. Sinoatrial region of heart
17. Myelencephalon
18. Auditory pit (invaginated auditory placode)
19. Notochord
20. Thyroid rudiment
21. Endocardium of ventricle
22. Myocardium of ventricle
23. Cardiac jelly

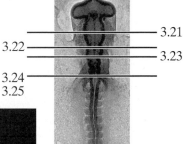

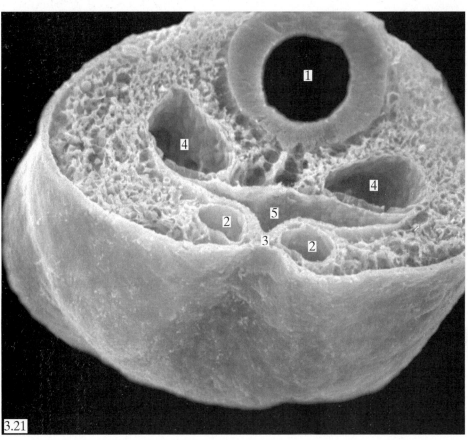

Photo 3.21. Scanning electron micrograph of the cut surface of a block (tilted to show the ectodermal side of the oral membrane) from a 33-hour chick embryo sectioned transversely.

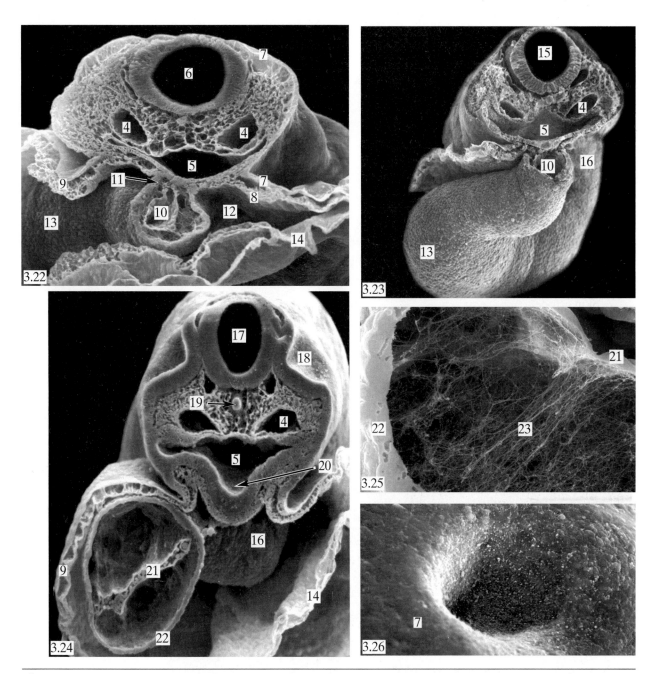

Photo 3.22. Scanning electron micrograph of the cut surface of a block from a 33-hour chick embryo sectioned transversely.

Photo 3.23. Scanning electron micrograph of a transverse slice (tilted to show the heart; splanchnopleure removed) from a 33-hour chick embryo.

Photo 3.24. Scanning electron micrograph of a transverse cryofracture from a 33-hour chick embryo.

Photo 3.25. Scanning electron micrograph of an enlargement of the cardiac jelly in a transverse cryofracture from a 33-hour chick embryo.

Photo 3.26. Scanning electron micrograph of an enlargement of the auditory vesicle from an intact 33-hour chick embryo.

Photos 3.27-3.32
Chick Embryos
Legend

1. Spinal cord
2. Somite
3. Nephrotome
4. Coelom (in lateral plate mesoderm)
5. Endoderm
6. Skin ectoderm
7. Notochord
8. Dorsal aorta
9. Neural fold

10. Neural groove
11. Segmental plate
12. Primitive streak
13. Mesoderm
14. Primitive groove
15. Primitive ridge (poorly defined)
16. Diencephalon
17. Optic vesicle
18. Infundibulum

19. Mesencephalon
20. Metencephalon
21. Myelencephalon
22. Cranial intestinal portal
23. Foregut
24. Sinoatrial region
25. Oral membrane
26. Subcephalic pocket
27. Ectoderm of proamnion

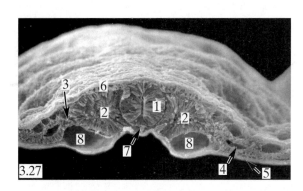

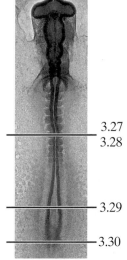

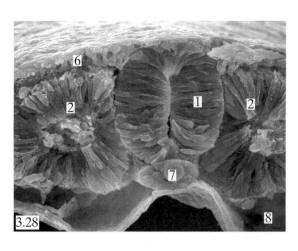

Photos 3.27, 3.28.   Scanning electron micrographs of transverse slices from a 33-hour chick embryo.

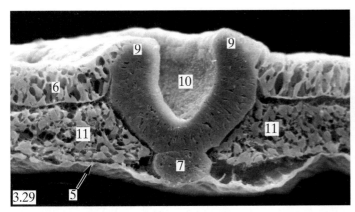

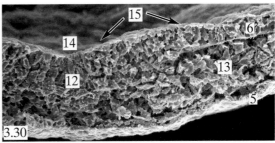

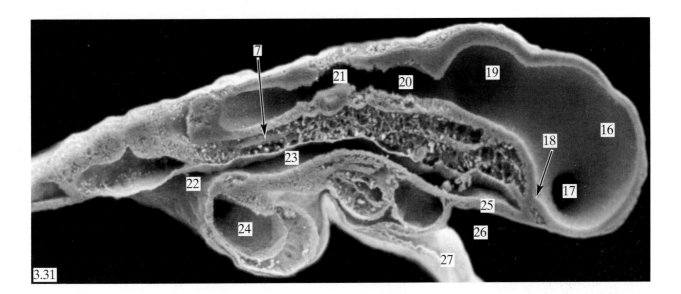

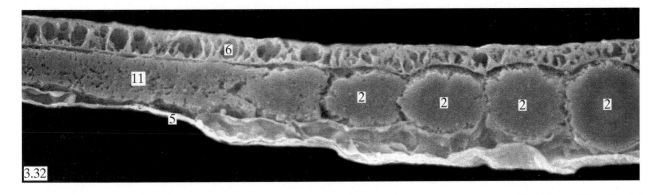

Photo 3.29. Scanning electron micrograph of a transverse cryofracture from a 33-hour chick embryo.

Photo 3.30. Scanning electron micrograph of a transverse slice from a 33-hour chick embryo.

Photo 3.31. Scanning electron micrograph of the cut surface of a block sectioned sagittally from a 33-hour chick embryo. This embryo is slightly more developed than most 33-hour embryos.

Photo 3.32. Scanning electron micrograph of a sagittal cryofracture from a 33-hour chick embryo.

E. STRUCTURE AND FUNCTION OF THE REPRODUCTIVE SYSTEM OF THE ADULT HEN

Adult hens possess a functional reproductive system (**ovary** and **oviduct**) only on the *left* side of the body. The fully developed ovary (Fig. 3.3) consists of a mass of protruding **follicles**. It usually contains 5 to 6 larger follicles and many smaller ones. The smallest follicles are called **primary follicles**. The larger ones are rapidly **growing follicles**, the largest of which is about to rupture within an area called the **stigma**. The rupture of the follicle and the release of its contained **ovum** constitute **ovulation**. Each ovulation is apparently triggered by a sudden release of **luteinizing hormone (LH)** from the **anterior pituitary gland** (**adenohypophysis**). The ovary illustrated in Fig. 3.3 also contains a **collapsed follicle** from which an ovum has just escaped (labeled as "ovum after ovulation") through the ruptured stigma.

The fully developed oviduct is a highly differentiated organ and is subdivided into five distinct regions: **infundibulum**, **magnum**, **isthmus**, **shell gland**, and **vagina**. The first subdivision of the oviduct is the **infundibulum**. It is expanded at its opened end as a delicate, funnel-shaped structure with processes called **fimbria** along its margin. The **ovum** enters the opening of the infundibulum, the **ostium**, within 15 minutes after ovulation. The **outer vitelline membrane** is added to the ovum within the infundibulum, or perhaps within the next subdivision of the oviduct, the **magnum**. The ovum remains in the infundibulum for about 15 minutes before it is carried by peristalsis to the magnum.

There is no clear line of demarcation between the infundibulum and the longest subdivision of the oviduct, the **magnum**. Principally **albumen (egg white)** is added to the ovum in the magnum. Albumen is a mixture of proteins, most of which protect the embryo from infection. It takes about three hours for the ovum to traverse the magnum.

The next region of the oviduct, the **isthmus**, secretes the **inner** and **outer shell membranes**. The ovum remains within the isthmus for about one hour and is then carried to the **shell gland**. The major function of the shell gland is formation of the calcified **shell**, a process that requires about 20 hours. The ovum is retained in the shell gland during this period by a sphincter muscle that separates the shell gland from the last region of the oviduct, the **vagina**. During laying of the bird's egg, or **oviposition**, the sphincter muscle is relaxed and the egg is rapidly forced through the vagina to the outside by contractions of the shell gland and abdominal musculature.

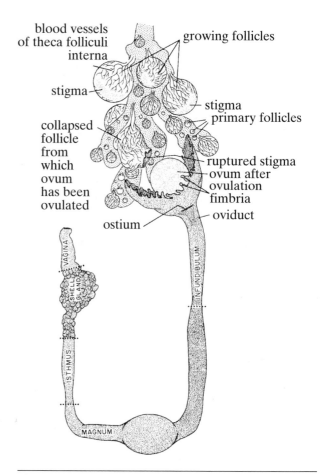

Fig. 3.3. Drawing of a hen's reproductive system. An ovum distends a portion of the magnum.

F. OOGENESIS, FERTILIZATION, AND CLEAVAGE

Development of the **ovum** (**oogenesis**) actually begins in the *embryonic* hen's left ovary prior to her hatching from the egg. Small cells, called **oogonia**, form and rapidly increase in number by mitotic division. *Before* the embryonic hen hatches from the egg, oogonia stop dividing mitotically and enlarge slightly to form **primary oocytes**, 2 mm in diameter. Each chromosome and its contained **DNA** are replicated early in the primary oocyte stage. Shortly *after* the young hen hatches from the egg, each primary oocyte becomes enclosed by a single layer of cells, called **follicle cells**, derived from the surface layer of the ovary. The composite structure, consisting of a primary oocyte enclosed by a layer of follicle cells, is called a **primary follicle**. These structures cause the surface of the ovary to bulge outward. Starting at puberty, and at intervals thereafter, a group of primary follicles begins to enlarge rapidly due to the accumulation of **yolk** by the primary oocytes. Yolk accumulation is principally stimulated by **follicle stim-**

ulating hormone (**FSH**) from the **anterior pituitary gland**. Yolk components are synthesized by the hen's liver, transported to the blood vessels in the connective tissue surrounding the follicle, the **theca folliculi interna**, and then transported to the primary oocyte via the follicle cells. Each primary follicle increases in diameter from about 5 mm to 35 mm as a result of this accumulation of yolk. As the yolk accumulates, most of the cytoplasm of the oocyte is crowded off to one side as a small circular disc, the **blastodisc**, which is approximately 3.5 mm in diameter. The blastodisc contains the enlarged nucleus of the primary oocyte, the so-called **germinal vesicle**. Meanwhile, during the rapid growth phase, the **inner vitelline membrane** forms between the **plasmalemma** of the primary oocyte and the adjacent follicle cells.

The primary oocyte undergoes the **first meiotic division** shortly before ovulation occurs. The meiotic spindle involved in this division forms within the blastodisc, and as the division is completed a tiny **first polar body** is pinched off between the inner vitelline membrane and the large **secondary oocyte**. The latter contains most of the yolk plus the blastodisc. The spindle for the **second meiotic division** forms within the blastodisc soon after the first meiotic division occurs, and the second meiotic division proceeds to the metaphase stage and then stops. The ovum (although still a secondary oocyte arrested in the metaphase stage of the second meiotic division) is sufficiently mature to be fertilized at this point and is therefore ready for ovulation. One ovum will be ovulated daily over a period of two or more days, constituting a **clutch** of ova. Ovulation will then cease for at least one day, after which it will resume in a new clutch of ova.

Following ovulation the ovum enters the infundibulum of the oviduct, the site of **fertilization**. **Sperm** must penetrate the *inner* (and often the *outer*) vitelline membrane to reach the surface of the blastodisc. To do this they pass through large gaps, or pores, in the outer vitelline membrane and then digest pathways through the inner vitelline membrane. Fertilization can no longer occur after addition of albumen in the magnum.

Recall that at ovulation the ovum is a secondary oocyte arrested in the metaphase stage of the second meiotic division. The second meiotic division is completed as a sperm contacts and penetrates the blastodisc, resulting in formation of a **second polar body** and a **mature ovum** containing the **female pronucleus**. The sperm nucleus enlarges to form the **male pronucleus** within the cytoplasm of the blastodisc, and the two pronuclei closely approach one another. The ovum remains as a secondary oocyte arrested in the metaphase stage of the second meiotic division if it is not fertilized, and the cytoplasmic cap on the yolk remains as the undeveloped blastodisc.

The blastodisc initiates **cleavage** (that is, it partially subdivides by mitosis into cells called **blastomeres**) following fertilization, and it is subsequently designated as the **blastoderm**. Cleavage occurs only within the cytoplasm (the yolk is not cleaved), and the **cleavage furrows** neither extend all the way to the periphery of the cytoplasm nor cut entirely through its thickness. Cleavage in the hen's ovum is therefore classified as **partial (meroblastic)**; it is also classified as **discoidal** because it is restricted to the circular disc of cytoplasm.

Cleavage normally begins as the fertilized ovum enters the isthmus of the oviduct, approximately 3.5 hours after ovulation. The first cleavage furrow (Fig. 3.4a) extends across the central portion of the cytoplasm, partially separating it into two blastomeres. The second cleavage furrow forms at approximately right angles to the first and about 15 minutes later (Fig. 3.4b). The result is four partially separated blastomeres. The blastoderm usually undergoes the third cleavage while the ovum is still within the isthmus (Fig. 3.4c). Two third-cleavage furrows form at approximately right angles to the second-cleavage furrow and parallel to the first, forming eight partially separated blastomeres. Later cleavages usually take place in the shell gland (Figs. 3.4d-3.4f). The size and shape of the blastomeres become variable by the 32-cell stage (Fig. 3.4d). At this time it is possible to distinguish **central blastomeres**, which appear to be completely bounded in surface view, from **marginal blastomeres**, which still are incompletely sep-

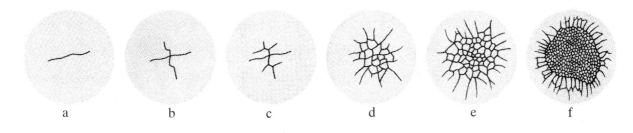

a   b   c   d   e   f

Fig. 3.4. Drawings of early cleavage stages of the chick blastoderm removed from the yolk and viewed from the upper surface. (a) 2-cell stage showing first cleavage furrow; (b) 4-cell stage showing first and second cleavage furrows; (c) 8-cell stage showing two third-cleavage furrows in addition to the first and second; (d) approximate 32-cell stage; (e) and (f) progressively more advanced cleavage stages.

arated from one another in surface view. The central blastomeres are separated from the yolk by a cavity, the **subgerminal cavity**, whereas the marginal blastomeres contact the yolk. More advanced cleavage stages (Figs. 3.4e, 3.4f) possess increased numbers of smaller blastomeres, some formed by division of marginal blastomeres. The central portion of the blastoderm will probably be more than one cell thick by the stage illustrated by Fig. 3.4f. This is because cleavage furrows also form horizontally in this region, separating outer cells from inner cells.

G. FORMATION OF THE HYPOBLAST

While the developing ovum is still present in the shell gland, and after formation of the subgerminal cavity beneath the central cells, two distinct regions of the blastoderm can be readily distinguished in surface view: the lighter, central **area pellucida**, and the darker, peripheral **area opaca**. A lower layer of large, yolky cells begins to form beneath the area pellucida at about the time that the egg is laid. This layer of cells is called the **hypoblast** (Fig. 3.5a). The surface layer of the area pellucida is called the **epiblast** after formation of the hypoblast.

The mechanism of hypoblast formation is unclear. However, observations and experiments suggest that hypoblast cells arise from two sources. First, some cells separate from the epiblast in the area pellucida and move inward (that is, they **delaminate** or **polyingress** toward the **subgerminal cavity**), contributing cells to the more central portion of the hypoblast. Second, other cells migrate cranially from **Koller's sickle**, a thickening of the epiblast (often poorly defined) at the caudal area pellucida/area opaca interface. Cells from the two sources ultimately join one another to form the sheet-like hypoblast. Thus hypoblast formation is initiated in the caudal half of the area pellucida, and the hypoblast then progressively extends more cranially. Additional details on the formation of the hypoblast are given in Chapter 6, Exercise 3.2.

The fertilized egg is usually laid during the stage of hypoblast formation. Completion of hypoblast formation and further development of the egg requires incubation.

H. GASTRULATION

Gastrulation, the process of germ layer formation, occurs after formation of the hypoblast has occurred. Our understanding of the events of gastrulation has been clarified by classical experiments using vital dyes (see Chapter 6, Exercise 3.4) as a **cell marker**. **Tritiated thymidine** has also been used for this purpose. Tritiated thymidine is incorporated specifically in **DNA**, thus ra-

dioactively labeling cells. Specific sectors of the epiblast from labeled donor embryos are used to replace corresponding sectors of unlabeled host embryos of the same stage. Movements of labeled cells can then be followed precisely during gastrulation and their fates determined.

To avoid the use of radioactivity, another technique has been employed in which sectors of *quail* epiblasts are transplanted to *chick* epiblasts to form **chimeras** (that is, mixtures of different cell types) produced from two species. Transplanted quail cells and their descendants can be distinguished from chick cells in histological sections treated with the **Feulgen procedure** to stain **DNA**. Quail cells contain a central mass of nucleolar-associated **heterochromatin**; in contrast, heterochromatin is finely dispersed in chick cells. Recently, an antibody specific for quail nuclei has been used to distinguish chick and quail cells in chimeras. Finally, another technique has been developed in which a small bolus of a fluorescent dye, one of the most popular of which is **DiI**, is injected into the epiblast. This dye labels groups of cells, which can be followed over time by viewing blastoderms with a fluorescence microscope and the appropriate wavelength of light. Information obtained by the use of these techniques is used to construct **prospective fate maps**.

1. Location and migration of the prospective endoderm

Labeling experiments have clearly demonstrated that the **prospective endoderm** is located in an oval-shaped region of the epiblast of prestreak stages, around and within the cranial half of the initial **primitive streak** (Fig. 3.5a). The primitive streak is a midline thickening of the epiblast. At the initial primitive streak stage, prospective endoderm is migrating inward through the cranial half of the primitive streak. The inward migration of epiblast cells through the primitive streak is designated as **ingression**. The ingressed **endoderm** enters the **hypoblast** layer and begins to displace the hypoblast craniolaterally (Fig. 3.5a). Ingression of prospective endoderm and displacement of the hypoblast continue rapidly in the intermediate primitive streak stage (Fig. 3.5b). By the definitive primitive streak stage, the only prospective endoderm still in the epiblast is located in the cranial end of the primitive streak (Fig. 3.5c). The primitive streak has attained its maximal length at this stage, extending about two-thirds the length of the area pellucida. Elongation of the primitive streak is responsible for the change in outline of the area pellucida from circular (Fig. 3.5a) to pear shaped (Fig. 3.5c). By the head process stage, the epiblast no longer contains any prospective endoderm because all of it has undergone ingression into the interior (Fig. 3.5d). The displaced hypoblast becomes condensed into a crescent-shaped area at the craniolateral margin of the area pellucida. This area is designated as the **germ cell crescent**

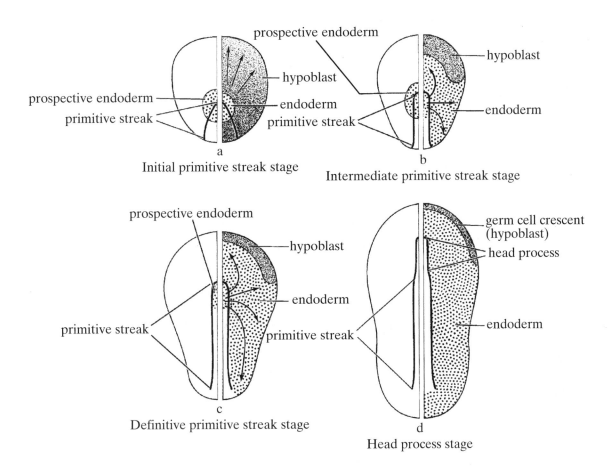

Fig. 3.5. Schematic drawings of the area pellucida showing the formation of the ingressed endoderm and the displacement of the hypoblast. The epiblast is shown on the left side of each diagram, but not on the right side. Arrows indicate the direction of displacement of the hypoblast in Fig. 3.5a and directions of migration of the ingressed endodermal cells in Figs. 3.5b and 3.5c. At all stages illustrated, *prospective mesoderm* is contained within the epiblast. By the stage shown in Fig. 3.5b, *mesoderm* is beginning to migrate into the interior of the blastoderm, between the overlying epiblast and underlying hypoblast/endoderm. For the sake of clarity in illustrating formation of the endoderm, prospective mesoderm and ingressed mesoderm have been omitted from this figure.

because it contains the **primordial germ cells**, presumably of hypoblast origin. The primordial germ cells *later* undergo extensive migration and enter the embryonic gonads where they form **oogonia** in the female (see Chapter 3, Section F) and **spermatogonia** in the male.

2. Location and migration of the prospective mesoderm

By the intermediate primitive streak stage (Fig. 3.6a) ingression of the **prospective extraembryonic mesoderm**, through the caudal part of the primitive streak, and the craniolateral internal spreading of the ingressed **extraembryonic mesoderm** have begun. This mesoderm is designated as extraembryonic because it does not contribute to the body of the embryo. Located around and within the cranial end of the intermediate primitive streak are the **prospective head mesenchyme** and **prospective notochord**. These subdivisions of **prospective embryonic mesoderm** have not yet started their

ingression at this stage. Located lateral to and within the middle part of the primitive streak is the **prospective heart mesoderm**. Ingression of **prospective heart mesoderm** and its craniolateral spreading, is also under way at the intermediate primitive streak stage.

At the definitive primitive streak stage (Fig. 3.6b), ingression of the **prospective extraembryonic mesoderm** through the caudal part of the primitive streak continues, and the internal spreading of the ingressed **extraembryonic mesoderm** is now caudolateral as well as craniolateral. All the **prospective notochord** now occupies the cranial end of the primitive streak, as does the **prospective head mesenchyme**. The latter is undergoing ingression, and the ingressed **head mesenchyme** is spreading cranially. **Prospective heart mesoderm** has completed its ingression through the middle part of the primitive streak by the definitive primitive streak stage and all **heart mesoderm** is located internally. **Prospective segmental plate mesoderm** and **prospective lateral plate mesoderm** are located in

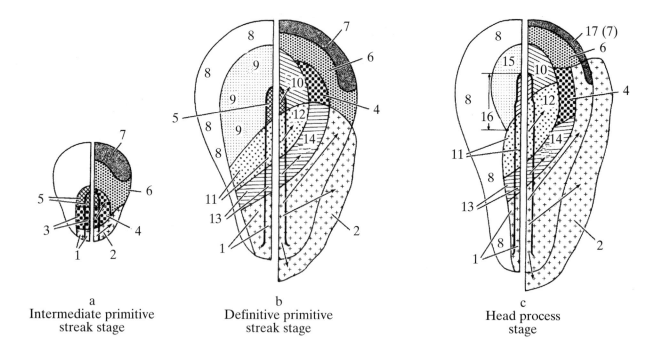

Fig. 3.6. Schematic drawings of the area pellucida showing the formation of the subdivisions of the ingressed mesoderm. The epiblast is shown on the left side of each diagram, but not on the right side. Arrows indicate the directions of migration of the ingressed mesodermal cells. The boundaries of the primitive streak and head process are indicated by a heavy black line.

1. Prospective extraembryonic mesoderm
2. Extraembryonic mesoderm
3. Prospective heart mesoderm
4. Heart mesoderm
5. Prospective head mesenchyme, prospective notochord
6. Endoderm
7. Hypoblast
8. Skin ectoderm
9. Neural plate
10. Head mesenchyme
11. Prospective segmental plate mesoderm
12. Segmental plate mesoderm
13. Prospective lateral plate mesoderm
14. Lateral plate mesoderm
15. Prosencephalon level of neural plate
16. Mesencephalon, rhombencephalon, and spinal cord levels of neural plate
17. Germ cell crescent

the epiblast at this stage, just lateral to the middle part of the primitive streak. These two subdivisions of the prospective embryonic mesoderm are already ingressing through the primitive streak and are spreading internally at the definitive primitive streak stage.

At the head process stage (Fig. 3.6c), ingression of the **prospective extraembryonic mesoderm** and the internal spreading of the ingressed **extraembryonic mesoderm** are still under way. The **prospective head mesenchyme** has now completed its ingression. The **prospective notochord** has already started ingressing through the cranial end of the primitive streak by this stage, forming the so-called **head process**. The head process is a mesodermal tongue of cells that subsequently forms the cranial part of the **notochord**.

Prospective segmental plate mesoderm and **prospective lateral plate mesoderm** are still ingressing just caudal to the prospective notochord, and the ingressed **segmental plate mesoderm** and **lateral plate mesoderm** are spreading internally.

3. Location and changes of the prospective ectoderm

The region of the epiblast that does not undergo ingression remains on the surface as the **ectoderm**. The prospective ectoderm expands, as ingression of prospective endoderm and mesoderm occurs, spreading toward the primitive streak (that is, it undergoes **epiboly**) to replace those areas of the epiblast that have ingressed. By the definitive primitive streak stage (Fig. 3.6b), the ectoderm overlying the head mesenchyme and segmental plate mesoderm has thickened, forming the **neural plate**, the major rudiment of the central nervous system; the more peripheral ectoderm is now called the **skin ectoderm**. Formation of the neural plate is induced in part by prospective mesoderm (and perhaps endoderm) within the cranial end of the primitive streak, and later by mesoderm that comes to lie directly beneath the early neural plate. The approximate locations of the **prosencephalon** level of the **neural plate**, as well as the more caudal levels (**mesencephalon**, **rhombencephalon**, and **spinal cord**), which are not yet demarcated from one another, are indicated in Fig. 3.6c. Note that mesoderm has not yet spread at this stage into the cranial portion of the area pellucida. This area lacking mesoderm is called the **proamnion**; its chief characteristic is that it consists of only ectoderm and endoderm.

4. Summary of changes in the epiblast during gastrulation

Initial primitive streak stage

Prospective endoderm is ingressing through the cranial end of the primitive streak.

Intermediate primitive streak stage

Prospective endoderm continues to ingress through the cranial end of the primitive streak.

Prospective heart mesoderm is ingressing through the middle part of the primitive streak.

Prospective extraembryonic mesoderm is ingressing through the caudal part of the primitive streak.

Definitive primitive streak stage

Prospective endoderm continues to ingress through the cranial end of the primitive streak.

Prospective head mesenchyme is ingressing through the cranial end of the primitive streak.

Ingression of *prospective heart mesoderm* is now complete.

Prospective segmental plate mesoderm and *prospective lateral plate mesoderm* are ingressing through the middle part of the primitive streak.

Prospective extraembryonic mesoderm continues to ingress through the caudal part of the primitive streak.

Ectoderm has subdivided into *skin ectoderm* and *neural plate*.

Head process stage

Ingression of *prospective endoderm* is now complete.

Ingression of *prospective head mesenchyme* is complete.

Prospective notochord is ingressing through the cranial end of the primitive streak.

Prospective segmental plate mesoderm and *prospective lateral plate mesoderm* are still ingressing through the middle part of the primitive streak.

Prospective extraembryonic mesoderm continues to ingress through the caudal part of the primitive streak.

The *neural plate* is beginning to subdivide into its various craniocaudal levels.

5. Location of epiblast areas prior to gastrulation

Cell-marking experiments have demonstrated the fates of different areas of the epiblast during gastrulation. A **prospective fate map** is a schematic representation of the location of these areas prior to the onset of gastrulation. Note that prospective extraembryonic mesoderm is positioned at the level of formation of the caudal half of the primitive streak, through which it later ingresses (Fig. 3.7). All remaining subdivisions of prospective mesoderm and prospective endoderm are positioned at the level of formation of the cranial half of the primitive streak, through which they later ingress. Prospective ectoderm is located peripheral to the regions of the epiblast that undergo ingression. The chick fate map should be carefully compared with the ones for the frog (Fig. 2.4), sea urchin (Fig. 1.1), and mouse (Fig. 4.2). For a comparison of gastrulation in vertebrates and invertebrates, see Chapter 1, Section H.

I. 18-HOUR CHICK EMBRYOS

1. Whole mounts

A diagrammatic surface and cross-sectional view of the cranial half of the blastoderm and definitive primitive streak is illustrated by Fig. 3.8. The cranial end of the primitive streak is thickened, constituting the **primitive knot** (**Hensen's node**). It partially surrounds a depression, the **primitive pit**, which is continuous caudally with the **primitive groove**. The **primitive ridges** lie lateral to the primitive groove. Note that cells of the **epiblast** are tightly apposed to one another, forming an

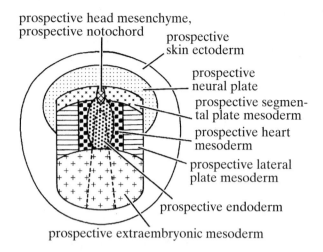

prospective head mesenchyme,
prospective notochord
prospective skin ectoderm
prospective neural plate
prospective segmental plate mesoderm
prospective heart mesoderm
prospective lateral plate mesoderm
prospective endoderm
prospective extraembryonic mesoderm

Fig. 3.7. Prospective fate map of the chick epiblast. The approximate site of primitive streak formation is indicated by dashed lines. Prospective nephrotome arises from the interface between prospective segmental plate mesoderm and prospective lateral plate mesoderm, but it is too small to illustrate; it ingresses in conjunction with these two mesodermal subdivisions.

epithelium. The outer ends of the epiblast cells become narrowed within the primitive groove, and the inner ends widen, forming the so-called **bottle (flask) cells**. **Filopodia** form at the broad inner ends of the bottle cells as the narrow outer connections to neighboring epiblast cells are lost. The ingressed cells then migrate away from the primitive streak, apparently through the action of their filopodia. The ingressed mesoderm consists of loosely packed, irregularly shaped cells (that is, mesenchyme) rather than an epithelium.

Examine whole mounts of embryos resembling Photos 3.33-3.35. Identify all parts of the **primitive streak** and the **head process** when present. Although not visible in whole mounts, an ectodermal thickening, the **neural plate**, overlies the head process and adjacent ingressed mesoderm. Identify the **area pellucida** and **area opaca**.

Examine whole mounts of embryos resembling Photos 3.36, 3.37. Note the distinct curved line just cranial to the tip of the head process. This line is the beginning of the **head fold of the body**, a ventrally directed fold of ectoderm and endoderm that establishes the cranial boundary of the head. It very quickly seems to undercut the head, forming two cavities: the ectoderm-lined **subcephalic pocket** and the endoderm-lined **foregut**. **Neurulation** is underway at this stage and a short **neural groove** or **neural tube** may have formed in the head region (Photo. 3.37). Additionally, **blood islands** may be forming (Photo. 3.37) as well as **somites** and **intersomitic furrows** (Photo. 3.38).

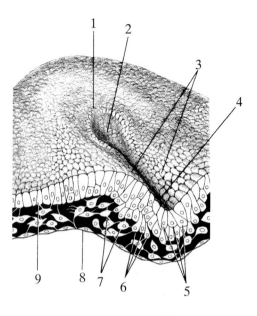

Fig. 3.8. Drawing of the cranial half of a blastoderm at the definitive primitive streak stage. The blastoderm has been cut transversely.

1. Primitive knot
2. Primitive pit
3. Primitive ridges
4. Primitive groove
5. Ingressing prospective mesodermal cells
6. Bottle cells
7. Filopodia
8. Endodermal cell
9. Epiblast cell

2. Serial transverse and sagittal sections

Examine serial transverse and sagittal sections of 18-hour embryos if available (Photo 3.39). Identify the **head fold of the body** (if present), **neural plate**, **head process**, **primitive knot**, **primitive pit**, and **primitive streak**.

3. Scanning electron microscopy

Dorsal views of intact embryos viewed with scanning electron microscopy reveal the subtle changes that occur in morphology during development from the definitive primitive streak stage to the head process stage and finally, to the early head fold/neural groove stage (Photos 3.40-3.43). At the definitive primitive streak stage (Photo 3.40), the **primitive streak** occupies the caudal two-thirds of the embryo, and the **primitive knot**, **pit**, **groove**, and **ridges** can be identified. Cellular debris fills the primitive pit and cranial half of the primitive groove. At the head process stage (Photo 3.41), the position of the **head process** beneath the **neural plate** is indicated by a shallow trough. At the early head fold stage (Photo 3.42), the boundaries of the neural plate and the beginning of the **head fold of the body** can be recognized. Finally, at the early neural groove stage (Photo 3.43), the **head fold of the body**, the **neural folds**, and the **neural groove** can be identified.

Slices through the blastoderm show the simple structure of the area of the embryo cranial to the primitive streak (Photos 3.44, 3.45). In transverse slices rostral to the head process, identify the **neural plate** and underlying **endoderm** (Photo 3.44). Within the midline, the mesoderm and endoderm collectively constitute the **prechordal plate** (Photo 3.44); this structure is involved in patterning the head of the embryo. More caudally at the level of the head process (Photo 3.45), identify the **neural plate**, more lateral **ectoderm**, midline **head process**, more lateral **mesoderm**, and the underlying **endoderm**.

The early **head fold of the body** is indicated in midsagittal slices by a shallow depression that is beginning to undercut the cranial end of the **neural plate** (Photo 3.46). The head fold consists of only two germ layers (an outer layer of **skin ectoderm** and an inner layer of ingressed **endoderm**), but just caudal to the head fold, three layers are present. These layers can best be seen in transverse slices (Photo 3.45). The outer layer of the blastoderm is thickened medially as the **neural plate** but remains thin laterally as the **skin ectoderm**. The mesoderm is organized into a midline **head process**, which is flanked by loosely organized cells of the ingressed **mesoderm**. The ingressed **endoderm** forms a single, flattened layer of cells beneath the ingressed mesoderm. In slightly older embryos, the **head fold of the body** is more advanced and the beginning of the **foregut** can be identified in midsagittal slices (Photo 3.47). Also visible is the **neural fold**, flanking the left side of the forming **neural groove**.

Transverse slices through the primitive streak at any of these early stages show the morphology of the **primitive knot** (Photo 3.48), **primitive pit** (Photo 3.49), and **primitive groove** and **ridges** (Photo 3.50). Identify these structures as well as the **epiblast**, ingressed **mesoderm**, ingressed **endoderm**, and **ingressing prospective mesodermal cells**.

J. PHOTOS 3.33-3.50: 18-HOUR CHICK EMBRYOS

Photos 3.33-3.50 depict the 18-hour chick embryos discussed in Chapter 3. These photos and their accompanying legends begin on the following page.

Photos 3.33-3.39

Chick Embryos

Legend

1. Area pellucida
2. Primitive streak
3. Head process (future cranial part of the notochord)
4. Primitive knot

5. Primitive groove
6. Primitive ridge
7. Primitive pit
8. Proamnion
9. Head fold of the body

10. Blood islands
11. Neural groove/neural tube
12. Somite
13. Intersomitic furrow
14. Neural plate

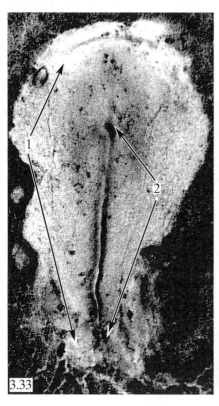

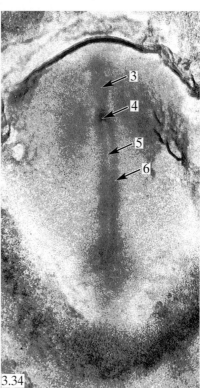

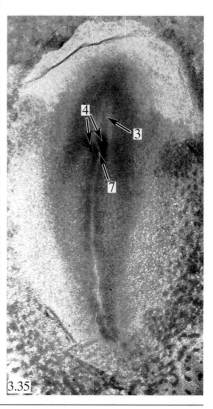

Photo 3.33. 18-hour chick embryo whole mount at the definitive primitive streak stage.

Photos 3.34, 3.35. 18-hour chick embryo whole mounts at the head process stage.

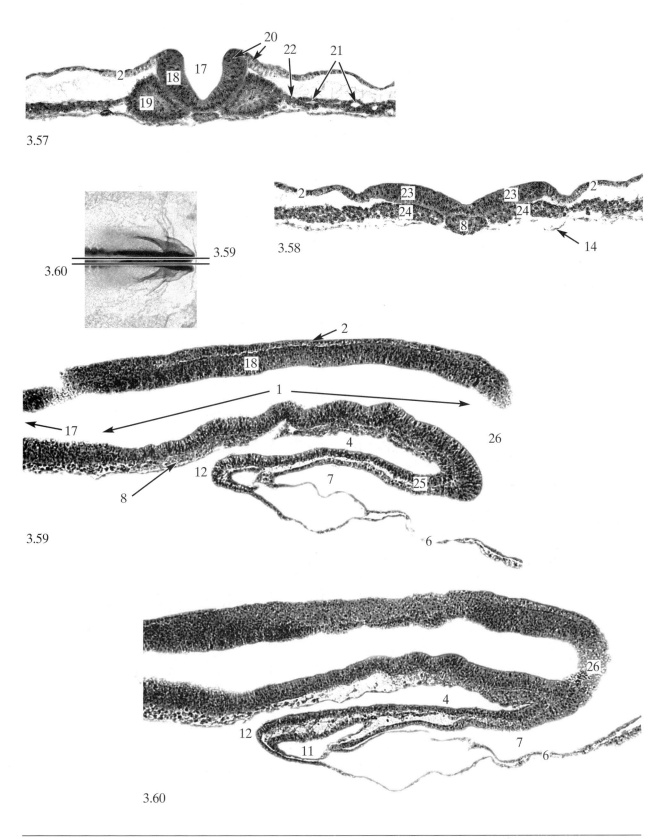

3.57

3.58

3.60 3.59

3.59

3.60

Photos 3.57, 3.58. Continuation of 24-hour chick embryo serial transverse sections numbered in anterior to posterior sequence.

Photos 3.59, 3.60. 24-hour chick embryo sagittal sections. Photo 3.59 shows a midsagittal section. Photo 3.60 shows a sagittal section just lateral to the midline.

Photos 3.61-3.64

Chick Embryos

Legend

1. Head fold of the body
2. Neural plate
3. Skin ectoderm
4. Primitive pit
5. Primitive groove
6. Primitive ridges
7. Neural groove of the future mesencephalon region

8. Neural fold
9. Primitive streak
10. Cranial neuropore
11. Neural tube of the mesencephalon region
12. Lateral extent of the head fold of the body

13. Neural groove of the future spinal cord region
14. Foregut
15. Cranial intestinal portal
16. Subcephalic pocket

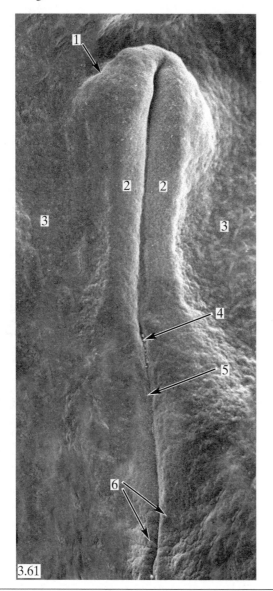

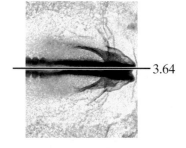

3.64

Photo 3.61. Scanning electron micrograph of an intact 21-hour chick embryo (dorsal view).

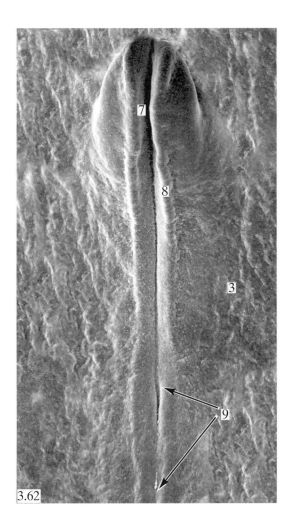

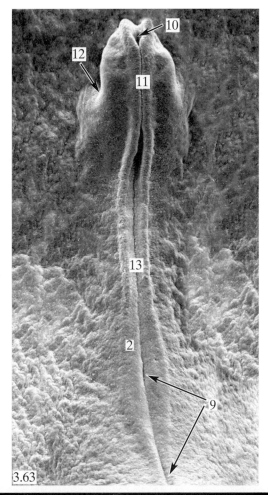

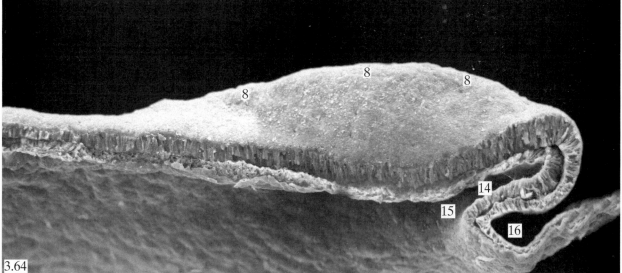

Photos 3.62, 3.63.   Scanning electron micrographs of intact 24-hour chick embryos (dorsal views).

Photo 3.64.   Scanning electron micrograph of a midsagittal slice of a 24-hour chick embryo.

Photos 3.65-3.70

Chick Embryos

Legend

1. Neural fold
2. Cranial neuropore
3. Neural groove
4. Notochord
5. Head mesenchyme
6. Foregut
7. Cranial intestinal portal
8. Skin ectoderm

9. Lateral plate mesoderm (split into somatic and splanchnic mesoderm)
10. Endoderm
11. Subcephalic pocket
12. Proamnion
13. Neural tube (just about to form at level shown in Photo 3.67)

14. Coelom
15. Splanchnopleure
16. Somatopleure
17. Somite
18. Nephrotome
19. Lateral plate mesoderm (not yet split into somatic and splanchnic mesoderm)

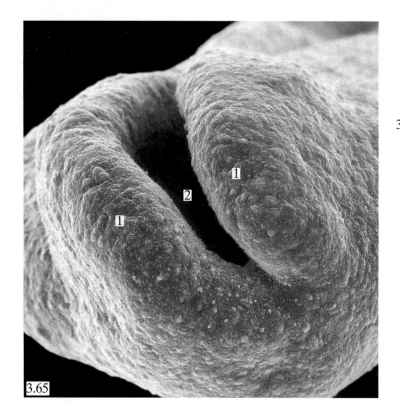

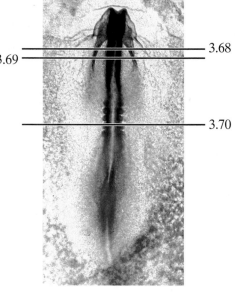

Photo 3.65.   Scanning electron micrograph of an intact 24-hour chick embryo viewed from its cranial end.

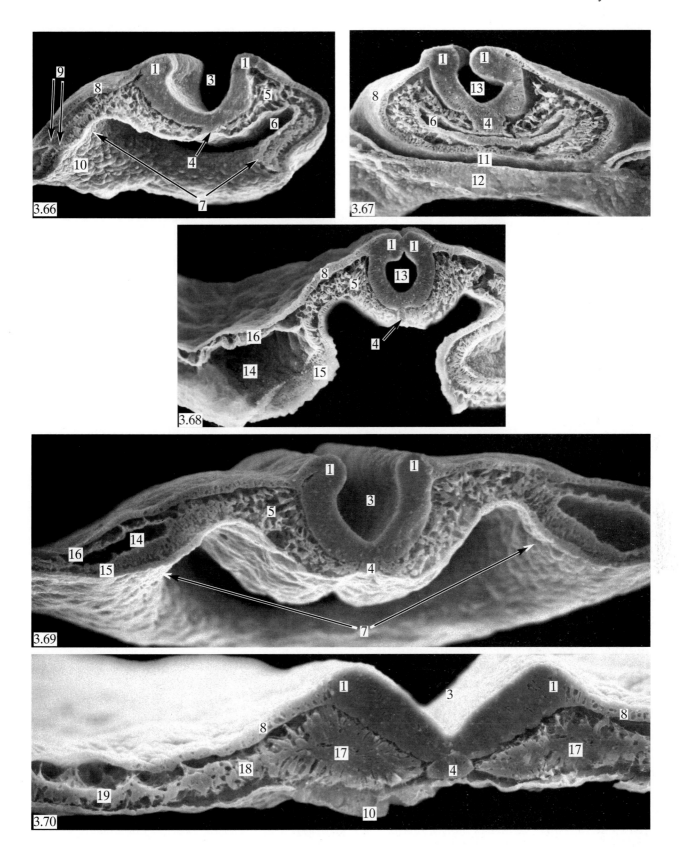

Photos 3.66-3.70. Scanning electron micrographs of transverse cryofractures of 24-hour chick embryos. Photos 3.66 and 3.67 are located at about the same craniocaudal level, but the embryo shown in Photo 3.66 is from a less-developed 24-hour embryo than is the one shown in Photo 3.67. Photos 3.66 and 3.67 show less-developed 24-hour embryos than the one shown in the orientator. Thus, the levels of these photos are not indicated on this orientator.

M. SUMMARY OF DEVELOPMENT OF THE YOUNG CHICK EMBRYO

The chick embryo develops in craniocaudal sequence from germ layers established during gastrulation. Early developmental events following gastrulation can be summarized as follows (Photos 3.71-3.74):

1. Regional specialization of the mesoderm

The ingressed mesoderm quickly becomes specialized. The mesoderm in the midline of the embryo forms a compact rod of cells, the **notochord**. Cranially, the rest of the mesoderm remains loosely packed as part of the **head mesenchyme**. More caudally, the mesoderm lateral to the notochord forms three distinct zones. The mesoderm immediately adjacent to each side of the notochord forms a thick band of cells, the **segmental plate**, which subsequently subdivides to form a row of **somites**. The mesoderm lateral to each row of somites forms a thin band of cells, the **nephrotome**. The lateromost mesoderm forms a thick plate of cells, the **lateral plate**, which splits into **somatic** and **splanchnic** layers bounding the **coelom**.

2. Formation of the neural tube

The first indication of the formation of the central nervous system is the appearance of an ectodermal thickening, the **neural plate**. The **neural folds** form at the lateral margins of the neural plate. Each fold consists of an inner layer of **neural ectoderm** and an outer layer of **skin ectoderm**. As the neural folds form and elevate, a distinct **neural groove** forms. The neural folds progressively approach each other and fuse to form the **neural tube**. After fusion of the neural folds occurs, **neural crest cells** form and begin their migration, contributing to the **head mesenchyme**.

3. Formation and consequences of the body folds

At first there is no indication as to how much of the blastoderm is destined to be *embryonic* and how much is *extraembryonic*. The first boundary of the embryo to become visible is the cranial boundary, which arises in the following way. Just cranial to the neural plate the ectoderm and endoderm of the proamnion begin to fold, forming the head fold of the body. The neural plate is rapidly expanding forward, and as it does so the **head fold of the body** seems to undercut the neural plate. To make this transformation clear, take two sheets of paper and superimpose them, considering the upper one the ectoderm and the lower one the endoderm. Hold the top edges of the papers firmly against a desk with your left hand; with your right hand grasp the bottom edges with your thumb beneath the sheets and your other fingers above and pointing toward the top of the sheets. Then lift your right hand slightly above the desk and move it toward the top of the sheets. Both sheets of paper will elevate and project forward. This movement creates an internal bay in the endoderm, the **foregut**, and an external bay in the ectoderm, the **subcephalic pocket**. These bays are turned in opposite directions, the internal one opening into the **subgerminal cavity** caudally via the **cranial intestinal portal**, and the external one opening cranially onto the surface of the blastoderm. Simultaneously the cranial boundary of the head is created.

The lateral boundaries of the embryo are formed by the action of the **lateral body folds**, which are direct continuations of the head fold of the body. The lateral body folds differ from the head fold of the body in that they consist of **somatopleure** and **splanchnopleure**. The somatopleural component of these folds is involved in lengthening the subcephalic pocket. The splanchnopleural component is involved in lengthening the foregut and in swinging together the bilateral **heart rudiments** beneath the lengthening foregut. In addition, the splanchnopleural component brings together the paired rudiments of the **pericardial cavity**.

N. PHOTOS 3.71-3.74: 18- TO 33-HOUR CHICK EMBRYOS

Photos 3.71-3.74 depict a series of 18- to 33-hour chick embryos discussed in Chapter 3. These photos and their accompanyng legends begin after the following page.

Photos begin on the following page. Use the space below for notes.

Photos 3.71-3.74

Chick Embryos

Legend

1. Notochord
2. Head mesenchyme
3. Neural plate
4. Neural fold
5. Neural ectoderm
6. Skin ectoderm

7. Neural groove
8. Cranial intestinal portal
9. Neural tube
10. Somatopleuric component of the lateral body fold
11. Splanchnopleuric component of the lateral body fold

12. Foregut
13. Rudiment of the pericardial cavity
14. Neural crest cells
15. Heart rudiments

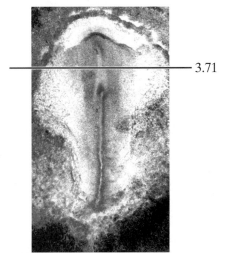

— 3.71

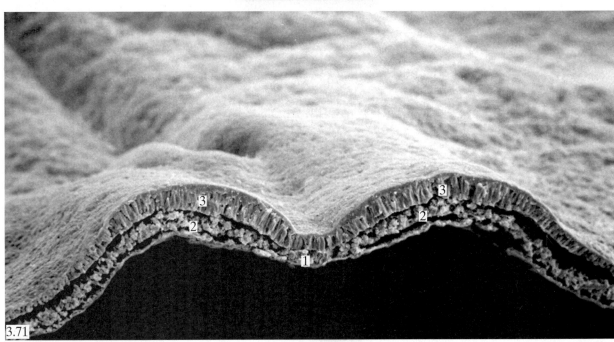

Photo 3.71. Scanning electron micrograph of a transverse slice of the head level of an 18-hour chick embryo.

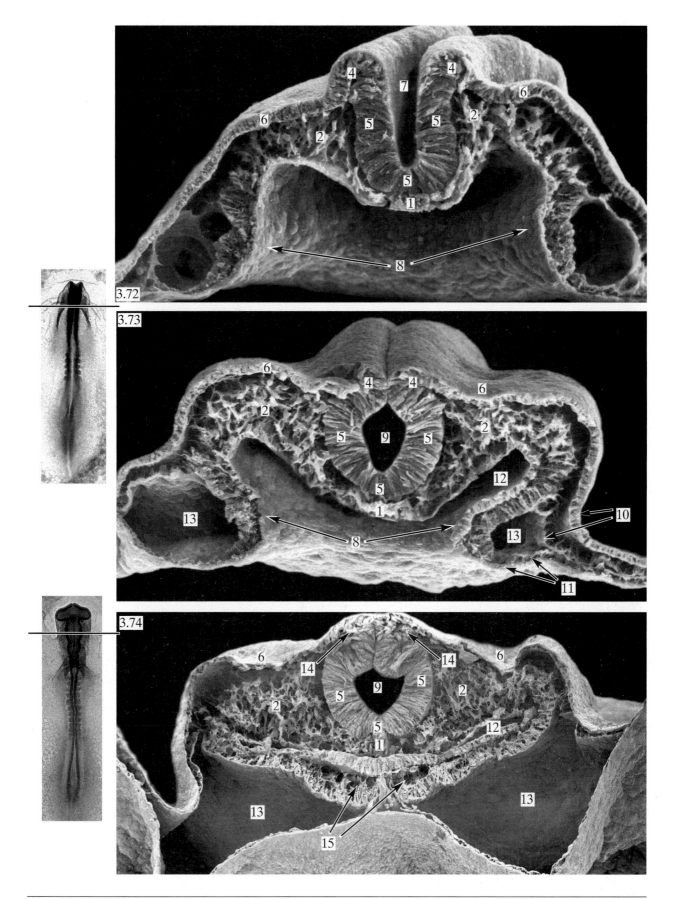

Photos 3.72-3.74. Scanning electron micrographs of transverse slices of the head level of 24- to 33-hour chick embryos. Photos 3.72 and 3.73 illustrate an early and later 24-hour embryo, respectively.

O. COMPARISON OF EARLY AMPHIBIAN AND AVIAN DEVELOPMENT

If you have completed your studies of the 4-mm frog embryo and the 33-hour chick embryo, you have undoubtedly noticed several structural similarities between them. This is because developmental events occur similarly among the various vertebrate classes. They certainly do not occur identically, as major differences exist among species. The purpose here is to emphasize two fundamental similarities between amphibian and avian development.

The first of these similarities concerns the structure of the **egg**. In both amphibians and birds, the egg consists of a large, yolky mass covered by a **plasmalemma** and enclosed by a **vitelline membrane** (or by vitelline membranes). **Cleavage** subdivides the yolky mass into smaller units called **blastomeres**. The entire mass cleaves in amphibian embryos, whereas in avian embryos a cytoplasmic cap (the **blastodisc**) forms at one pole of the egg. It is only this disc that cleaves (to form the multicellular **blastoderm**). A space forms as cleavage progresses in both amphibian and avian embryos. In amphibian embryos, this space is completely surrounded by blastomeres and is called the **blastocoel**. It allows cells to involute over the blastopore lips, where they move into the interior to form the **archenteron** and **mesoderm**. A similar space appears during avian development, but it is initially covered only dorsally by blastomeres. This similar space is the **subgerminal cavity**; it lies between the blastoderm and yolk. This space, like the blastocoel in amphibians, allows cells to move into the interior. As such cell movements occur, two layers can be identified: the dorsal, or outermost, **epiblast** and the ventral, or innermost, **hypoblast**. A space separates these two layers; it is appropriate to call this space the blastocoel, based on comparative embryology. Why is this so? The blastocoel in amphibian eggs is displaced toward the **animal pole**. It "separates" the animal cap cells, which form **ectoderm**, **mesoderm**, and some **endoderm**, from **vegetal pole** cells, which form **endoderm**. Similarly, the space between the epiblast and hypoblast "separates," respectively, cells that form **ectoderm**, **mesoderm**, and **endoderm** from cells forming **extraembryonic endoderm**. Thus, although developing amphibian and avian embryos have a very different appearance, the fundamental events occurring during development are quite similar in the two organisms.

The second major similarity occurs in the general **body plan** of the embryo. Vertebrate embryos have a **tube-within-a-tube body plan**. That is, they consist of an outer ectodermal tube forming the **skin** and an inner endodermal tube forming the **gut**. For example, compare Photo 2.14 with Photo 3.12. The tube-within-a-tube body plan is obvious in the frog embryo: there is an outer skin ectoderm and an inner gut (midgut). The mesoderm, subdivided into notochord, somites, nephrotome, and lateral plate, occupies the space between the inner and outer tubes, as does the neural tube (spinal cord), which originated from the outer ectodermal layer as an invagination. The tube-within-a-tube body plan is less obvious in avian embryos. This is because cleavage is restricted to the cytoplasmic cap at one pole of the yolky mass, rather than including all the mass as in amphibian embryos. Nevertheless, there is an outer **skin ectoderm** and an inner **gut** (in Photo 3.12, the midgut lies beneath the endoderm, which forms its roof, and is open and continuous with the subgerminal cavity; the lateral body folds have not yet formed at this level). The mesoderm and its subdivisions, as well as the neural tube (spinal cord), occupy the space between the inner and outer layers. As a learning exercise, trace Photo 3.12 on a separate sheet of paper. Label the four layers at the left edge of the section with the following letters: skin ectoderm, A; somatic mesoderm, B; splanchnic mesoderm, C; and endoderm, D. Also label the four layers at the right edge of the section with the following letters: skin ectoderm, a; somatic mesoderm, b; splanchnic mesoderm, c; and endoderm, d. Sketch two or three stages in which the four layers on the left and right sides are extended toward the *ventral midline,* where they meet one another. Compare the final sketch to a sketch of Photo 2.14; now note the considerable similarity in the body plan of the two organisms.

P. TERMS TO KNOW: 18-33 HOURS OF DEVELOPMENT

You should know the meaning of the following terms, which appeared in boldface in the preceding discussion of the embryology of the chick through the 33-hour stage.

adenohypophysis	archenteron	auditory placodes
albumen	area opaca	auditory pits or vesicles
animal pole	area pellucida	blastocoel
anterior intestinal portal	area vasculosa	blastoderm
anterior neuropore	area vitellina	blastodisc
anterior pituitary gland	atrium	blastomeres

blood islands

body

body plan

bottle cells

brain

cardiac jelly

cardiac primordia

caudal end

cell marker

central blastomeres

central nervous system

cerebrospinal fluid

chimeras

cleavage

cleavage furrows

clutch

coelom

collapsed follicle

conotruncus

cranial end

cranial intestinal portal

cranial neuropore

delaminate

diencephalon

DiI

discoidal cleavage

DNA

dorsal aortae

dorsal mesocardium

ectoderm

egg

egg white

endocardial tube

endocardium

endoderm

endothelium

epiblast

epiboly

extraembryonic endoderm

extraembryonic mesoderm

eyes

female pronucleus

fertilization

Feulgen procedure

filopodia

fimbria

first aortic arches

first meiotic division

first polar body

flask cells

floor plate

follicle cells

follicle stimulating hormone (FSH)

follicles

foregut

gastrulation

germ cell crescent

germ layer

germinal vesicle

growing follicles

gut

head end

head fold of the body

head mesenchymal cells

head mesenchyme

head process

heart

heart mesoderm

Hensen's node

heterochromatin

hypoblast

infundibulum

ingressing prospective mesodermal cells

ingression

inner shell membrane

inner ears

inner vitelline membrane

intermediate mesoderm

intersomitic furrows

isthmus

Koller's sickle

lateral body folds

lateral plate mesoderm

lateral plates

lenses

luteinizing hormone (LH)

magnum

male pronucleus

marginal blastomeres

mature ovum

meroblastic cleavage

mesencephalon

mesoderm

mesonephric duct rudiment

metencephalon

morphogenetic movements

mouth opening

myelencephalon

myocardium

nephrotomes

neural crest cells

neural ectoderm

neural folds

neural groove

neural plate

neural tube

neural tube defects

neurohypophysis

neuromeres

neurulation

notochord

oogenesis

oogonia

optic vesicles

oral membrane

ostium of oviduct

otic placodes

outer shell membranes

outer vitelline membrane

ovary

oviduct

oviposition

ovulation

ovum

partial cleavage

pericardial cavity

plasmalemma

polyingress

posterior pituitary gland

prechordal plate

primary follicles

primary oocytes

primitive blood cells

primitive folds

primitive groove

primitive knot

primitive pit

primitive ridges

primitive streak

primordial germ cells

proamnion

pronephric cord

pronephric duct rudiments

pronephric kidneys

prosencephalon

prospective embryonic
mesoderm

prospective endoderm

prospective extraembryonic
mesoderm

prospective fate maps

prospective head mesenchyme

prospective heart mesoderm

prospective lateral plate
mesoderm

prospective notochord

prospective prosencephalon

prospective segmental plate
mesoderm

rhombencephalon

roof plate

rudiments of the pericardial
cavity

second meiotic division

second polar body

secondary oocyte

segmental plate mesoderm

segmental plates

shell

shell gland

sinoatrial region

sinus terminalis

sinus venosus

skin

skin ectoderm

somatic mesoderm

somatopleure

somites

sperm

spermatogonia

spinal cord

splanchnic mesoderm

splanchnopleure

stigma

subcephalic pocket

subgerminal cavity

surface ectoderm

tail end

telencephalon

theca folliculi interna

thyroid rudiment

tritiated thymidine

tube-within-a-tube body plan

vagina

vegetal pole

ventral aortae

ventricle of heart

vitelline blood vessels

vitelline membrane

vitelline veins

yolk

Q. 48-HOUR CHICK EMBRYOS

1. The 33-hour to 48-hour transition

Before you begin studying the 48-hour chick embryo, you need to be aware of some changes that occur in the positioning of the embryo between the 33-hour and 48-hour stages. At the 33-hour stage, the embryo has a straight craniocaudal body axis and lies with its ventral side facing the yolk (Photo 3.75). Between the 33-hour and 48-hour stage, two changes occur to reorient the embryo in relation to the yolk: **torsion** and **flexion** (Photo 3.76). Torsion involves a rotation of the head such that the left side of the head now faces the yolk. Flexion involves a kinking of the head at the level of the **mesencephalon** such that the craniocaudal body axis is no longer straight. These changes in orientation are important to understand and will be further discussed in the following sections.

2. Whole mounts: injected and uninjected

First examine your uninjected whole mount slide with the naked eye. Superimpose this slide on the 33-hour whole mount and note the increased size of the embryo. The entire diameter of the **blastoderm** is not seen because it was trimmed during preparation of the slides. Position your slide on the microscope stage so that when viewed through the microscope, the embryo is oriented as in Photos 3.77-3.79; examine the slide under low magnification. As stated above, you will note a change in the axis of the body at this stage. The cranial end of the body now lies on its *left* side (that is, it has undergone twisting, or **torsion**). In addition, the **prosencephalon** is bent toward the **rhombencephalon,** forming the **cranial (cephalic) flexure**, a sharp bend in the floor of the **mesencephalon**.

Examine the body of the embryo and note that it is more opaque than at the 33-hour stage, indicating its greater thickness. *Care must be taken not to focus the objective down into the elevated coverslip over these older embryos.* First examine the caudal half of the body and identify the thickened lateral walls of the **neural tube (spinal cord** level). Follow these two parallel lines forward until they curve toward the left and fade out. Then focus down slightly and you will see that, due to torsion, your view of the neural tube shifts from a dorsal view of the caudal half to a lateral view of the right side of the cranial half. Identify the following structures: **telencephalon** (the cranial neuropore is now completely closed); **diencephalon**; **infundibulum** (often difficult to identify); **optic cups** (formed by invagination of the optic vesicles); **lens vesicles** (formed by thickening and invagination of the ectoderm overlying the optic vesicles at the 33-hour stage); **optic fissures** (ventral gap in the optic cups; Photo 3.78); **mesencephalon** (rounded area at the apex of the embryo); **metencephalon**; **isthmus** (a prominent constriction of the neural tube between mesencephalon and metencephalon); **myelencephalon** (note its thin **roof plate**); **auditory vesicles**, or **otocysts** (formed by invagination of the auditory placodes); **notochord** (usually best seen beneath the mesencephalon and rhombencephalon); **foregut**; **cranial intestinal portal**; **heart**; **cranial liver rudiment** (an evagination of the floor of the foregut that extends toward the heart); **somites** (count the number of pairs present); and **segmental plates**.

Observe that the cranial half of the embryo seems to be veiled by some coverings that the caudal half lacks. The distinct curvature between covered and uncovered parts is the **boundary of the amniotic folds**. The amniotic folds arch over the head, as continuous **cranial** (**anterior**) and **lateral amniotic folds**, forming two **extraembryonic membranes**: the outer **chorion** and the inner **amnion**. These two membranes cannot be readily distinguished from one another in whole mounts.

Note that the boundaries of the body are now quite distinct in the cranial half, whereas caudally they are probably only slightly indicated. The **tail fold of the body** may have established the caudal boundary in some embryos. Following formation of the tail fold of the body, the caudal end of the embryo is separated from the underlying blastoderm by the **subcaudal pocket**. A curved line, with its concavity facing cranially, may be visible just cranial to the caudal boundary of the body (Photo 3.77). This curvature is the beginning of the **caudal (posterior) intestinal portal**. The neural groove is no longer present in most embryos, and the neural tube merges caudally with a dark mass of cells, the **tail (end) bud**, which is derived from the cranial part of the primitive streak.

Observe three somewhat irregular white lines fanning out from the heart toward each auditory vesicle. The most cranial of these lines is the **first branchial groove**, the next one is the **second branchial groove**, and the last one is the **third branchial groove** (Photos 3.77, 3.78). Cranial to the first groove is a darkly stained mass of cells, the **first branchial arch**. This arch is partially split into two processes by another white line, the **stomodeum** (not to be confused with the first branchial groove). The dark area just cranial to the stomodeum is the **maxillary process** of the first branchial arch. The dark area just caudal to the stomodeum is the **mandibular process** of the first branchial arch. Between the first and second grooves is the **second branchial arch**, and

between the second and third grooves is the **third branchial arch**. *This series of structures causes no end of trouble in studying sections through this region, so their relationships should be mastered now.*

Examine the **heart** in uninjected embryos and in embryos whose circulatory system has been injected with india ink (Photos 3.80, 3.81). Between the 33- and 48-hour stages a major change occurs in the orientation of the heart: a process called **looping**. At the cranial end of the heart, identify the **conotruncus** (**bulbus cordis**), which is continuous, without a distinct boundary, with the next region of the heart, the **ventricle**. Notice that the ventricle is U-shaped because of looping, and that it extends craniad and to the left, underneath the conotruncus, to become continuous with a broad saccular region, the **atrium**. This is continuous, without a distinct boundary, with the **sinus venosus**. Observe that the heart consists of two layers: the inner **endocardium**, enclosing blood cells (and india ink in injected embryos), and the outer **myocardium**.

Note in the injected embryos that the conotruncus is also continuous with a very narrow region, the **aortic sac** (Photo 3.81), which was formed by fusion of the paired ventral aortae. Continuous with the aortic sac are two or three pairs of blood vessels, the **aortic arches**. The **first aortic arch**, on each side of the body, runs through the mandibular process and then curves sharply around the stomodeum. Similarly, on each side, the **second aortic arch** runs through the second branchial arch, and the **third aortic arch**, if present, runs through the third branchial arch. The aortic arches are continuous on each side with a **dorsal aorta**. The paired dorsal aortae are fused more caudally as the **descending aorta** (Photo 3.80). Numerous small blood vessels are continuous with the descending aorta dorsally; these are the **intersegmental arteries**. Paired **dorsal aortae** are again present caudal to the descending aorta.

Try to identify the **precardinal** (**cranial**, or **anterior**, **cardinal**) **vein** lying above each dorsal aorta in the region of the aortic arches (Fig. 3.9; Photo 3.81). This vessel is usually poorly injected. The precardinal vein on each side extends caudad from the **plexus of head blood vessels** and joins two other vessels (usually not readily visible): the **postcardinal** (**caudal**, or **posterior**, **cardinal**) **vein**, which extends further caudad, and the **common cardinal vein**, which joins the sinus venosus.

Identify the extensive plexus of **vitelline blood vessels** (Photo 3.80). They are contained within the **yolk sac**, an **extraembryonic membrane** that overgrows the yolk. Identify the paired **vitelline arteries** continuous with the caudal dorsal aortae at approximately right angles. The vitelline arteries and dorsal aortae are continuous at about the 22nd somite level. Also identify the paired **vitelline veins** continuous with the sinus venosus.

The circulatory system is well developed, and blood circulates between the yolk sac and body of the embryo mainly in the following sequence: vitelline veins ⇒ sinus venosus ⇒ atrium ⇒ ventricle ⇒ conotruncus ⇒ aortic sac ⇒ aortic arches ⇒ cranial dorsal aortae ⇒ descending aorta ⇒ caudal dorsal aortae ⇒ vitelline arteries ⇒ plexus of vitelline blood vessels ⇒ vitelline veins. Some of the blood in the cranial dorsal aortae supplies the plexus of head blood vessels via small branches. The precardinal veins drain this plexus, returning blood to the sinus venosus via the common cardinal veins. Some of the blood in the descending aorta enters the intersegmental arteries. This blood is returned to the heart in the following sequence: postcardinal veins ⇒ common cardinal veins ⇒ sinus venosus.

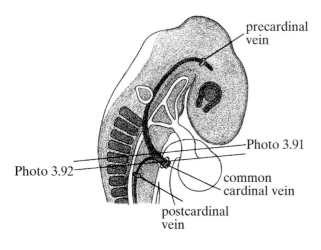

precardinal vein

Photo 3.91

Photo 3.92

common cardinal vein

postcardinal vein

Fig. 3.9. Drawing of the right side of the cranial half of a 48-hour chick embryo.

3. Serial transverse sections

Position your slide on the microscope stage so that when viewed through the microscope, each section is oriented as in Photos 3.82-3.99. Begin tracing sections in anteroposterior sequence under low magnification. The first brain sections are probably cut through the **mesencephalon**, due to the cranial flexure. These first sections will actually be frontal (that is, cut lengthwise in the right-left plane), rather than transverse. Note the oval shape of the mesencephalon sections as you continue caudally. Sections then begin to lengthen, and a constriction appears in the brain. Sections at this level cut across not only the mesencephalon, but also the **metencephalon** and **myelencephalon** (Photo 3.82). The constricted region between metencephalon and mesencephalon is the **isthmus**. The myelencephalon can be readily identified by its thin **roof plate**. The brain separates into two parts as you continue to trace sections posteriorly (Photo 3.83). The upper part is the myelencephalon (or possibly the metencephalon and myelencephalon; the boundary between these two structures is indistinct); the lower part is now the **diencephalon**. A dark accumulation of cells appears on each side of the body at about this level, the **semilu-**

nar ganglia of the **trigeminal (V) cranial nerves**. These ganglia originate from ectodermal **neural crest cells**, as well as from a thickening of the lateral skin ectoderm on each side. These ectodermal thickenings are called **epibranchial placodes**; they usually cannot be identified.

The **notochord** appears shortly after the brain separates into two parts. It is usually cut frontally due to the cranial flexure, and it quickly separates into two parts, one beneath each brain section. The portion of the notochord beneath the diencephalon disappears as you continue to follow sections posteriorly (Photo 3.84). Note an accumulation of cells on each side of the myelencephalon at about this level; these are the **acousticofacialis ganglia** of the **facial (VII)** and **auditory (VIII) cranial nerves**. The acoustic part of these ganglia originates from the ectodermal **auditory placodes** present earlier. The facialis part originates from both **neural crest cells** and from a second pair of epibranchial placodes (see Photo 3.86). These **epibranchial placodes** may not be readily visible in your embryo. Note the close relationship between the **auditory vesicles** and the acousticofacialis ganglia (Photos 3.85, 3.86).

Return to the level where the portion of the notochord beneath the diencephalon disappears and trace sections posteriorly. Observe that the diencephalon becomes more and more elongated; the narrow end of the diencephalon that projects toward the myelencephalon is the **infundibulum** (Photos 3.85, 3.86). The **optic cups** appear at about this level. They are not connected to the diencephalon in the first sections encountered. The **lens vesicles** appear in more posterior sections and are nestled within the double-layered optic cups (Photos 3.86, 3.87). The layer of each optic cup that is next to the lens vesicle is thickened as the **sensory retina**; the much thinner layer is the **pigmented retina**. The space between these two layers, the **opticoel**, is continuous with the cavity of the diencephalon. A small accumulation of cells appears above each auditory vesicle at about this level; these are the **superior ganglia** of the **glossopharyngeal (IX) cranial nerves** (Photos 3.86, 3.87). These ganglia originate from **neural crest cells**.

As you continue to trace sections posteriorly, the optic cups become continuous with the diencephalon via the **optic stalks** (Photo 3.89). The **telencephalon** appears in more posterior sections. Note that the skin ectoderm lateral to the telencephalon is thickened as the **nasal (olfactory) placodes** (Photo 3.90). The nasal placodes are induced to form by the adjacent regions of the telencephalon. They form the lining of the **nasal cavities**. The telencephalon disappears a few sections more posteriorly.

Return to the level of the auditory vesicles (Photos 3.85, 3.86) and trace sections posteriorly. The transition from myelencephalon to **spinal cord** is so gradual that the boundary between the two is indistinguishable. Find a section closely resembling Photo 3.97 and examine the spinal cord carefully. Identify the **roof** and **floor plates**. Try to identify **neural crest cells**. Continue to trace sections posteriorly. The caudal end of the neural tube often contains multiple cavities. In sections slightly posterior to this level, the notochord seems to fuse to the floor of the neural tube, and a solid mass of cells then appears in the midline. This mass is the **tail bud**. The caudal portion of the neural tube is formed by cavitation (hollowing out) of a portion of the tail bud, resulting in formation of multiple cavities that subsequently fuse, forming a single cavity.

Return to the level where the brain separates into two parts (Photo 3.83) and trace sections posteriorly, noting that the **foregut** soon appears. It rapidly becomes somewhat triangular in shape (Photo 3.85). Sections now cut through the **pharynx** and **first pharyngeal pouches**. The first pharyngeal pouches project laterad from the pharynx to contact the adjacent regions of the skin ectoderm, which they induce to invaginate, forming the **first branchial grooves**. The double-layered membranes formed by the endoderm of the pouches and the ectoderm of the grooves are the **first closing plates**.

Return to the first section through the foregut (Photo 3.84) and again trace sections posteriorly. A small vesicle soon appears between the infundibulum and foregut (Photo 3.85). This is **Rathke's pouch**, the rudiment of the **anterior pituitary gland**. Recall that the posterior pituitary gland is formed from the infundibulum. Note the close relationship between the infundibulum and Rathke's pouch. Rathke's pouch opens in more posterior sections into a transverse, slitlike space, the **stomodeum** (Photo 3.86), which is lined by ectoderm. Rathke's pouch develops as an outgrowth from the stomodeum, perhaps due to induction by the infundibulum. The stomodeum lies between two rounded masses next to the foregut, the **mandibular processes** of the first branchial arches, and two rounded or slightly flattened masses on either side of Rathke's pouch, the **maxillary processes** of the first branchial arches. Note that the stomodeum dips between the two mandibular processes and is separated from the foregut only by a thin membrane of ectoderm and endoderm, the **oral membrane** (Photo 3.86). This membrane will *later* rupture, forming the **mouth opening**. The outer part of the mouth is therefore derived from ectoderm of the stomodeum, and the inner part from foregut endoderm. Examine models, if available, to fully understand and visualize these relationships.

Continue tracing sections posteriorly. The pharynx becomes somewhat rounded and then suddenly expands laterad at the level of the **second pharyngeal pouches** (Photo 3.89). Identify the **second branchial grooves** and the **second closing plates**. The considerable depression in the floor of the pharynx at this level is the **thyroid rudiment**. Continue tracing sections posteriorly and identify the very broad **third pharyngeal pouches**, the **third closing plates**, and the **third branchial grooves**, if

formed (Photo 3.90).

Sections posterior to the third pharyngeal pouches cut through the portion of the **foregut** caudal to the pharynx. The foregut endoderm at this level is surrounded by splanchnic mesoderm (Photo 3.91). In more posterior sections, the ventrolateral portions of the foregut may be slightly evaginated. These evaginations are the beginnings of the **lung buds** (Photo 3.92). The paired portions of the coelom lateral to the developing lung buds and continuous with the **pericardial cavity** are the **pleural cavities**. The pleural and pericardial cavities are separate, caudal to this level (Photo 3.93).

Notice a small mass of cells ventral to the foregut at about this level; this is the **cranial liver rudiment**. Continue caudally, noting that the cranial liver rudiment becomes continuous with the foregut (Photo 3.94). This region of the foregut is the **duodenum**. Your embryo may also contain a **caudal liver rudiment**. If so, it will appear as a branch (or as branches) of the ventral part of the cranial liver rudiment. The paired portions of the coelom lateral to the duodenum are the **rudiments of the peritoneal cavity**. These paired rudiments continue caudally for many sections and *later* fuse, forming a single **peritoneal cavity** surrounding the gut. The **cranial intestinal portal** is encountered almost immediately as sections are traced posteriorly from the level of the duodenum. More caudally, the embryo progressively flattens dorsoventrally, and endoderm forms its lower layer (Photo 3.97). An endoderm-lined cavity, the **allantois rudiment**, may appear in more posterior sections beneath the tail bud (Photo 3.99). The exact mechanism by which this rudiment forms is unknown. It will *later* greatly enlarge and become covered externally with splanchnic mesoderm, forming the **allantois**, one of the **extraembryonic membranes**. In slightly more *anterior* sections, the floor of the allantois rudiment disappears; this is the region of the **caudal intestinal portal** (Photo 3.98). Thus the allantois rudiment opens into the **subgerminal cavity**.

Return to the level of the first somite pair (Photo 3.90) and trace sections posteriorly. The structure of the cranial **somites** is now quite different from that of the 33-hour stage. Examine them closely (Photo 3.91, 3.92). Each somite is now subdivided into three parts: **dermatome** (a plate of darkly stained cells lying just beneath the skin ectoderm—source of some of the **dermis**); **myotome** (a plate of lightly stained cells lying just medial to the dermatome—source of some of the **skeletal muscles**); **sclerotome** (a diffuse mass of mesenchymal cells lying between myotome, neural tube, notochord, and dorsal aorta or descending aorta—source of the **vertebral column** and **ribs**). The manner of formation of these somite subdivisions can be studied by comparing the most caudal somites, in which the subdivision is just beginning, with more cranial somites.

Identify the **segmental plates** caudal to the last somite

pair and note that their caudal ends are continuous with the **tail bud**. The tail bud is a mass of mesenchymal cells located at the caudal end of the embryo. It contributes cells to the neural tube, segmental plates, and mesenchyme of the tail. The notochord of the tail is *not* derived from the tail bud. Instead, the notochord extends into the developing tail from more cranial regions.

Return to the level of the cranial liver rudiment (Photo 3.93) and trace sections posteriorly. Lateral to the descending aorta or dorsal aortae, identify the paired **mesonephric duct rudiments** (Photo 3.95-3.97). They were initially solid longitudinal structures derived from the pronephric cords, but they have extended caudad and are now undergoing cavitation in most embryos. Medial to the mesonephric duct rudiments, identify the usually solid **mesonephric tubule rudiments**. The latter are induced to form from the **nephrotome** by the mesonephric duct rudiments. The mesonephric tubule rudiments undergo cavitation to form the **mesonephric tubules** of the paired **mesonephric kidneys**.

Return to the first section through the foregut (Photo 3.84). A large blood vessel, the dorsal aorta, is cut frontally on each side of the foregut. Trace sections posteriorly and note that two vessels appear in the place of each **dorsal aorta** (Photo 3.85). The one that lies above the first pharyngeal pouch is still the dorsal aorta; the one that lies beneath the first pouch is the **first aortic arch** (see Photo 3.81). Note, as sections are traced further posteriorly, that the first aortic arches become located within the mandibular processes (Photo 3.86, 3.87). Continue to trace sections posteriorly until the first aortic arches approach one another and become continuous with the **aortic sac**. Continue tracing the aortic sac caudally. Note a vessel extending downward from each dorsal aorta, just caudal to the first pharyngeal pouches. These vessels are the **second aortic arches** (Photo 3.87-3.89). Their ventral ends become continuous with the aortic sac a few sections more posteriorly. The second aortic arches are contained within the second branchial arches. Continue tracing sections posteriorly. Just caudal to the second pharyngeal pouches, another pair of downward extensions from the dorsal aortae may be seen; these are the **third aortic arches**, which are usually just developing. The ventral ends of these vessels become continuous with the aortic sac more caudally. The third aortic arches are contained within the third branchial arches (Photo 3.90).

Follow the aortic sac caudally, noting the appearance of **endocardium** clearly separated from **myocardium**, indicating the beginning of the **conotruncus** of the heart (Photo 3.90). Identify the **ventricle** and **atrium** in more posterior sections (Photo 3.91). Recall that these regions lie *beneath* the conotruncus in surface view. Notice that the endocardium of the atrium is in close contact with the myocardium, whereas these two layers are widely separated in the ventricle and conotruncus. The two dorsal aortae have fused at about this level, forming the **descending aorta**. In more posterior sections, a portion of the heart is attached to the foregut by the **dorsal mesocardium** (Photo 3.92). This portion is the **sinus venosus**. The conotruncus and ventricle become continuous at about this level. Slightly more caudally, the sinus venosus is continuous with a blood vessel on each side; these vessels are the **common cardinal veins** (Photo 3.93).

Continue caudally and note that the sinus venosus seems to divide into two vessels, one on each side of the cranial liver rudiment (Photo 3.94). These two vessels are the **vitelline veins**. Trace the vitelline veins caudally, noting that they shift laterad within the splanchnic mesoderm. Shift your attention to the descending aorta and trace it caudally. Identify small blood vessels continuous with the descending aorta dorsally; these are the **intersegmental arteries** (see Photo 3.94). Note that the descending aorta has not yet formed at more caudal levels, and paired **dorsal aortae** are again present. Continue to trace the dorsal aortae caudally. They become continuous with the **vitelline arteries**, which extend laterad within the splanchnic mesoderm (Photo 3.97).

Return to the section where the common cardinal veins are continuous with the sinus venosus (Photo 3.93). Each common cardinal vein is contained within a bridge of mesoderm connecting the lateral body wall to the sinus venosus. Trace sections *anteriorly*, observing that the lateral body walls separate from the heart (Photo 3.92). Note a vessel located within the lateral body wall on each side near the heart at this level. These vessels are located progressively more dorsally within the lateral body walls in more *anterior* sections. They are still the **common cardinal veins** near the heart (Photo 3.92), but they are the **precardinal veins** at a level approximately lateral to the foregut (Photo 3.91). Thus, there is a gradual transition between these two vessels on each side, as sections are traced anteriorly from the level of the heart. Continue to trace sections *anteriorly*. The precardinal veins lie progressively more dorsally and come to lie ventrolateral to the myelencephalon (Photo 3.87). They are cut frontally in more *anterior* sections due to the cranial flexure (Photo 3.83).

Return to a level similar to Photo 3.90 and trace sections *posteriorly, carefully* observing the precardinal veins. They become located progressively more ventrally and suddenly, at about a level lateral to the foregut, each seems to separate into two vessels. The dorsomost vessel, usually very small, is the **postcardinal vein**; the ventromost vessel is the **common cardinal vein** (Photo 3.92). See Fig. 3.9. Continue to trace sections posteriorly and note again that the common cardinal veins become continuous with the sinus venosus. The postcardinal veins become located progressively more dorsally and eventually lie dorsal to the developing kidneys (Photos 3.96, 3.97). Small blood vessels are continuous dorsally with the postcardinal veins in some embryos. These are the **intersegmental veins**.

Four **extraembryonic membranes** are in the process of formation. Two originate from splanchnopleure, the **allantois** and **yolk sac**, and two from somatopleure, the **amnion** and **chorion**.

The development of the **allantois** is barely initiated.

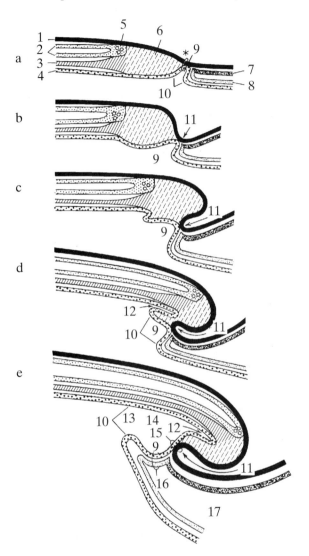

Only a small endoderm-lined cavity, the **allantois rudiment**, has formed (Photo 3.99). Splanchnic mesoderm will later surround this endodermal lining to form the wall of the allantois. Consult Fig. 3.10 to help you understand the early changes involved in formation of the allantois rudiment and wall of the allantois, as well as their relationships to later changes involved in formation of the hindgut, cloaca, and tail gut. The allantois stores nitrogenous waste products when it is fully developed, and it will eventually fuse with the chorion to form the **chorioallantoic membrane**.

The **yolk sac** is formed from the splanchnopleure that overgrows the yolk. At cranial levels where **torsion** is complete, it lies between the amnion (covering the *left* side of the embryo) and the yolk (Fig. 3.11a; Photos 3.83, 3.89). At these levels the yolk sac has been separated from the embryo by the action of the splanchnopleur-

1. Ectoderm
2. Neural tube
3. Notochord
4. Endoderm
5. Area cavitating to form neural tube
6. Tail bud
7. Somatic mesoderm
8. Splanchnic mesoderm
9. Allantois rudiment
10. Caudal intestinal portal
11. Tail fold of the body
12. Tail gut
13. Hindgut
14. Cloaca
15. Cloacal membrane
16. Wall of the allantois
17. Extraembryonic coelom

Fig. 3.10. Schematic drawings of midsagittal sections of the caudal ends of chick embryos between 48 (Fig. 3.10a) and 72 (Fig. 3.10e) hours of incubation. The tail fold of the body changes the positions and orientations of the endodermal allantois rudiment and the caudal intestinal portal so that at 48 hours (Figs. 3.10a, 3.10b) the caudal intestinal portal opens cranioventrally, whereas at intermediate stages (Figs. 3.10c, 3.10d) and at 72 hours (Fig. 3.10e) it opens cranially. The allantois rudiment receives an outer investment of splanchnic mesoderm as it changes orientation, whereupon it is renamed the allantois (Figs. 3.10d, 3.10e), and a distinct tail gut forms as an evagination of the endoderm (Figs. 3.10d, 3.10e). At 48 hours the caudal intestinal portal is the opening of the allantois rudiment into the subgerminal cavity (Fig. 3.10a). However, at 72 hours the caudal intestinal portal is the opening of the endodermal hindgut (formed by the tail fold of the body) into the subgerminal cavity (Fig. 3.10e); the allantois then opens dorsally into a portion of the hindgut called the cloaca. Note that the endodermal floor of the cloaca contacts the skin ectoderm caudal to the allantois (Fig. 3.10e); this area of contact is the cloacal membrane. Although this membrane is generally not named prior to formation of the cloaca at the 72-hour stage, note that it actually forms at the 48-hour stage (Fig. 3.10a, asterisk), where the endodermal allantois rudiment contacts the overlying skin ectoderm.

al component of the **lateral body folds**. More caudally the yolk sac has not yet been separated from the developing gut (Photo 3.95). Further caudally, **torsion** has not yet occurred, and the ventral surface of the embryo as well as the yolk sac lie above the yolk (Fig. 3.11c; Photo 3.96). Notice the numerous **vitelline blood vessels** within the splanchnic mesoderm of the yolk sac. The endodermal cells of the yolk sac digest the yolk, the products of this digestion enter the plexus of vitelline blood vessels, and these products are transported to the

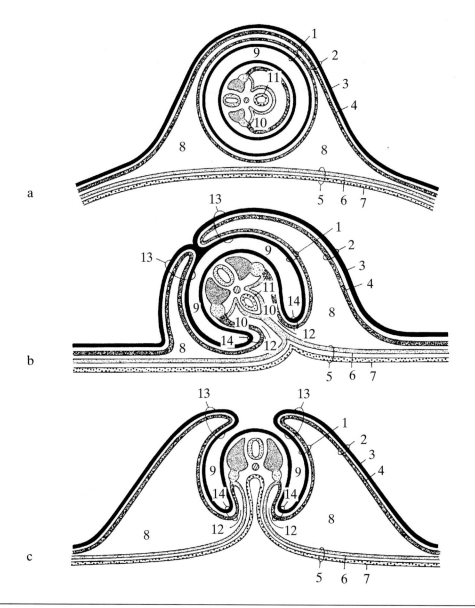

Fig. 3.11. Schematic drawings of transverse sections of 48-hour chick embryos showing the formation of the amnion, chorion, and yolk sac. Fig. 3.11a is more cranial than Fig. 3.11b, which in turn is more cranial than Fig. 3.11c. Torsion of the embryo occurs as the lateral amniotic folds elevate and fuse.

1. Amnion
2. Chorion
3. Ectoderm
4. Somatic mesoderm
5. Yolk sac
6. Splanchnic mesoderm
7. Endoderm
8. Extraembryonic coelom
9. Amniotic cavity
10. Intraembryonic coelom
11. Gut endoderm
12. Continuity between intra- and extraembryonic coeloms
13. Lateral amniotic fold
14. Lateral body fold

developing embryo via the vitelline veins.

The **amnion** and **chorion** are formed simultaneously by elevation and fusion of amniotic folds. Select a section resembling Fig. 3.11c and Photo 3.96, showing the **lateral amniotic folds**. Note that the amniotic folds consist of somatopleure and are continuous with the lateral body walls. The outer wall of each fold consists of *ectoderm on the outside,* with an adjacent layer of *somatic mesoderm on the inside* and forms the **chorion**. The inner wall of each fold consists of *somatic mesoderm on the outside* and *ectoderm on the inside* and forms the **amnion**. The somatic mesoderm-lined cavity within each fold is the **extraembryonic coelom**; it is continuous with the **intraembryonic coelom** (that is, the pericardial, peritoneal, and pleural cavities). The extra- and intraembryonic coeloms become separated due to the action of the somatopleural component of the **lateral body folds**. The cavity developing between the amnion and embryo is the ectoderm-lined *amniotic cavity.* Trace sections *anteriorly* and observe the region of fusion of the amniotic folds (Fig. 3.11b). In more anterior sections the outer chorion is separate from the inner amnion. The amniotic cavity is later filled with **amniotic fluid**. The embryo then floats freely (Fig. 3.11a) within this protective fluid. The extraembryonic coelom later contains the enlarging allantois. Fusion occurs wherever the splanchnic mesoderm of the allantois contacts the somatic mesoderm of the chorion, forming the **chorioallantoic membrane**. The latter lies immediately beneath the inner shell membrane and functions in respiration and absorption of calcium from the shell.

4. Serial sagittal sections

Only about the cranial one-third of the embryo is actually cut sagittally; the remaining caudal regions are cut frontally. Try to find a section that is approximately a midsagittal section through the **brain** and **notochord** (Photo 3.100). Identify all regions of the brain, including the **infundibulum**. Also identify the **pharynx**, **oral membrane**, and **preoral gut** (**Seessel's pouch**). The latter is that part of the foregut that lies cranial to the oral membrane; it later degenerates. Identify the ectodermal **stomodeum** and its evagination, **Rathke's pouch**, which extends toward the infundibulum.

Now focus your attention on parasagittal sections in which you can recognize the following structures (Photo 3.101): **optic cup** (including the **sensory retina** and **pigmented retina**), **lens vesicle**, and **optic fissure**. Identify the **optic stalk** connecting the optic cup to the **diencephalon**. The optic fissure is formed by ventral invagination of the optic cups and stalks. At this stage the optic fissure is probably formed only in the optic cup, or at most in the optic cup and in the part of the optic stalk immediately adjacent to the optic cup. The optic fissure is important in development because **optic nerve**

fibers from the retina grow back to their proper connections within the diencephalon through the wall of the fissure. Moreover, the cavity of the fissure provides a pathway for blood vessels to enter the optic cup. This fissure will later close.

5. Summary of the contributions of the germ layers to structures present in the 48-hour chick embryo but not present in the 33-hour chick embryo

Ectoderm

auditory vesicles

branchial grooves

cranial ganglia:

> V. semilunar

> VII-VIII. acousticofacialis

> IX. superior

epibranchial placodes

lens vesicles

nasal placodes

optic cups

optic stalks

Rathke's pouch

stomodeum

Mesoderm

dermatomes

mesonephric duct rudiments

mesonephric tubule rudiments

myotomes

sclerotomes

Endoderm

allantois rudiment

cranial liver rudiment

caudal liver rudiment

duodenum

lung buds

pharynx

pharyngeal pouches

preoral gut

Ectoderm and somatic mesoderm

amnion

chorion

Endoderm and splanchnic mesoderm

yolk sac

Ectoderm and endoderm

closing plates

Ectoderm, mesoderm, and endoderm

branchial arches

6. Scanning electron microscopy

A scanning electron micrograph of the developing brain from a block sectioned transversely is shown in Photo 3.102. Identify the **myelencephalon, diencephalon, infundibulum, Rathke's pouch, optic cups** (with **sensory** and **pigmented retinas**), and **lens vesicles**. Note the continuity between each lens vesicle and the adjacent **skin ectoderm**. The **acousticofacialis ganglia** lie ventrolateral to the floor of the myelencephalon, just dorsolateral to the **precardinal veins**. Also identify the **pharynx, first pharyngeal pouch, first closing plate, first branchial groove, dorsal aortae, first aortic arches, first and second branchial arches** (first arches are not split at this level into maxillary and mandibular processes), and **notochord**. Three extraembryonic membranes can be identified: the **chorion, amnion,** and **yolk sac**. Note the locations of the **amniotic cavity** and **extraembryonic coelom**.

Three representative transverse views of the spinal cord examined by scanning electron microscopy are shown in Photos 3.103-3.105. Near the level of the cranial portion of the spinal cord (Photo 3.103), identify the **skin ectoderm, spinal cord** (note that its lumen is partially occluded), **notochord, dorsal aortae, coelom, endoderm,** and **lateral body folds**. Note the **amniotic folds** (continuous **cranial** and **lateral amniotic folds**). These folds are seen in cross section at the top right of Photo 3.103. Also note the **yolk sac, amnion,** and **amniotic cavity**. The slice shown in Photo 3.103 passes through the level of a pair of **intersomitic furrows**. Therefore, the subdivisions of the somites are not recognizable. Instead, endothelial cells forming **intersegmental arteries**, and possibly **veins**, can be identified (note on the right side the continuity of an intersegmental artery and a dorsal aorta). Also identify **neural crest cells** leaving the roof of the spinal cord at this level.

Identify near the mid-portion of the spinal cord (Photo 3.102) the incipient **lateral body folds** (somatopleural component) and **amniotic folds**. The thickening of somatopleure on each side of the embryo at this level represents the rudiments of the **wing buds**. Note that the embryo is relatively flat at this level (the splanchno-pleuric component of the lateral body folds has not yet formed here) and contains an open **midgut**. Also note the various subdivisions of the **mesoderm: somites** (subdividing into **dermatome, myotome,** and **sclerotome**), **nephrotome,** and **lateral plate mesoderm** (subdivided into **somatic mesoderm**, associated with the **ectoderm**, and **splanchnic mesoderm**, associated with the **endoderm**; the somatic and splanchnic mesoderm are separated from one another by the **coelom**).

Identify near the caudal end of the spinal cord (Photo 3.105) the **skin ectoderm, spinal cord, notochord, dorsal aortae, coelom, somites** (not yet subdivided at this level into dermatome, myotome, and sclerotome), **endoderm, allantois rudiment,** and **somatopleure** and **splanchnopleure** of the **lateral plate**.

Photos 3.106-3.109 examine the development of the caudal end of the embryo using scanning electron microscopy. Photo 3.106 shows a whole mount of a young 48-hour embryo; it was shown already in Photo 3.76 and is used here for orientation. Note that the **tail fold of the body** is just forming and the **tail bud** is becoming delineated near the caudal end of the closed neural tube (**spinal cord**). Photo 3.107 shows a parasagittal slice through the caudal end of the embryo. Note the **tail fold of the body, allantois rudiment, tail bud, skin ectoderm, endoderm, spinal cord** and **notochord**. Photo 3.108 shows a transverse cryofracture through the **tail bud**. Note the epithelial **skin ectoderm** and **endoderm** covering its mesenchymal core. Photo 3.109 is just cranial to Photo 3.108 and shows a transverse cryofracture through the caudal end of the **spinal cord**. The spinal cord at this level develops by cavitation of the tail bud. Note the incipient **notochord, skin ectoderm, endoderm, segmental plate mesoderm, nephrotome, lateral plate mesoderm,** and **dorsal aorta.**

A scanning electron micrograph of the cut surface of a block sectioned parasagittally is shown in Photo 3.110 (compare Photo 3.110 with Photo 3.100; the section shown in Photo 3.100 was cut midsagittally). Identify the **telencephalon, diencephalon, optic stalk, infundibulum, mesencephalon, metencephalon, myelencephalon, dorsal aorta, pharynx, first aortic arch, mandibular process,** and **ventricle**.

R. PHOTOS 3.75-3.110: 48-HOUR CHICK EMBRYOS

Photos 3.75-3.110 depict the 48-hour chick embryos discussed in Chapter 3. These photos and their accompanying legends begin after the following page.

Photos begin on the following page. Use the space below for notes.

Photos 3.75-3.76

Chick Embryos

Legend

1. Cranial end
2. Bulge indicating optic vesicle

3. Auditory vesicle
4. Brain level
5. Spinal cord level

6. Caudal end
7. Cranial flexure

Photos 3.75, 3.76. Facing page. Scanning electron micrographs of 33-hour to early 48-hour chick embryo whole mounts (dorsal views).

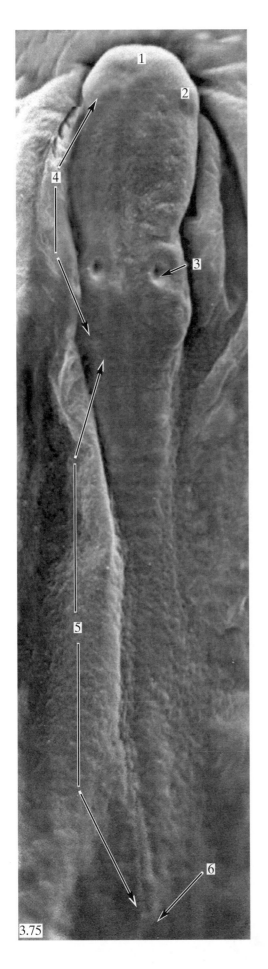

3.75

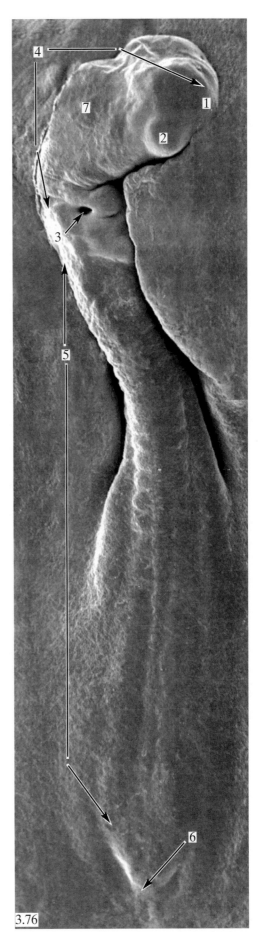

3.76

Photos 3.77-3.80
Chick Embryos
Legend

1. Stomodeum
2. Optic cup
3. Maxillary process
4. Notochord
5. Isthmus
6. Mandibular process
7. Auditory vesicle
8. Second branchial groove
9. Third branchial groove
10. Caudal intestinal portal
11. Caudal boundary of the body

12. Cranial intestinal portal
13. Cranial liver rudiment
14. Ventricle
15. Conotruncus
16. Optic fissure
17. Metencephalon
18. Roof plate of myelencephalon
19. First branchial groove
20. Second branchial arch
21. Third branchial arch

22. Boundary of amniotic folds
23. Spinal cord
24. Telencephalon
25. Foregut
26. Diencephalon
27. Mesencephalon
28. Myelencephalon
29. Somites
30. Vitelline artery
31. Tail bud

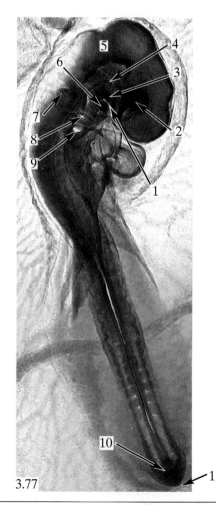

Photo 3.77. 48-hour chick embryo whole mount.

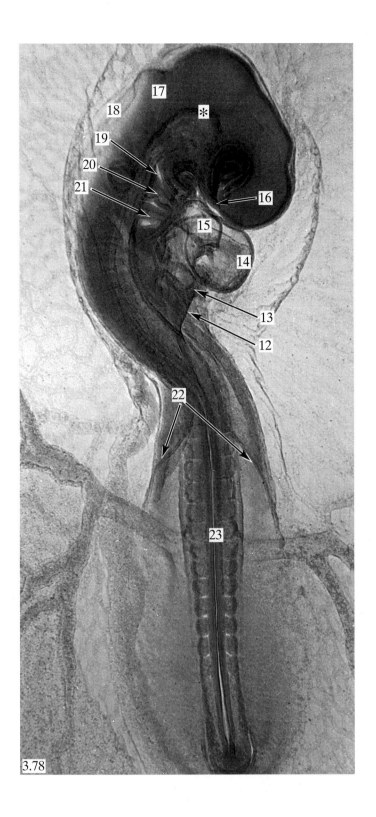

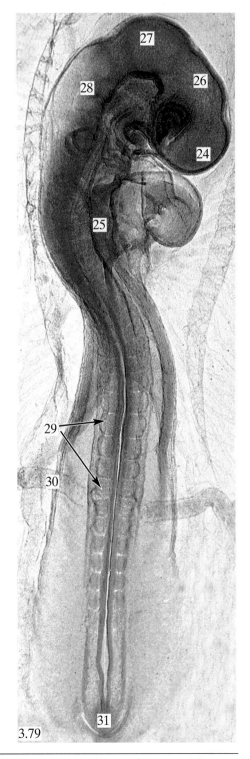

Photos 3.78, 3.79. 48-hour chick embryo whole mounts. Asterisk in Photo 3.78 indicates the cranial flexure.

Photos 3.80-3.81

Chick Embryos

Legend

1. Right vitelline vein
2. Plexus of head blood vessels
3. Left vitelline vein
4. Plexus of vitelline blood vessels
5. Descending aorta
6. Right dorsal aorta
7. Right vitelline artery

8. Right vitelline vein entering sinus venosus
9. Sinus venosus
10. Common cardinal and left vitelline veins entering sinus venosus
11. Ventricle
12. Conotruncus
13. Atrium

14. Aortic sac
15. First aortic arch
16. Continuity between dorsal aorta and first aortic arch (indicated by line)
17. Precardinal vein
18. Second aortic arch
19. Third aortic arch
20. Intersegmental arteries

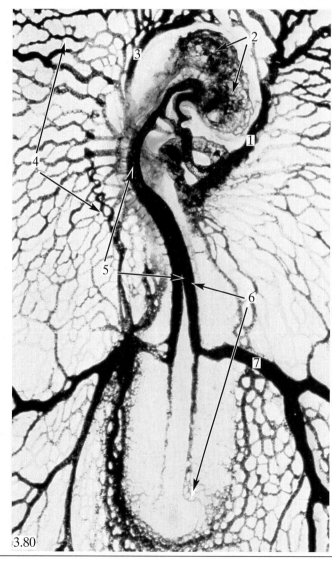

Photo 3.80. 48-hour chick embryo whole mount injected with India ink. The cranial half of the embryo in Photo 3.80 is enlarged in Photo 3.81.

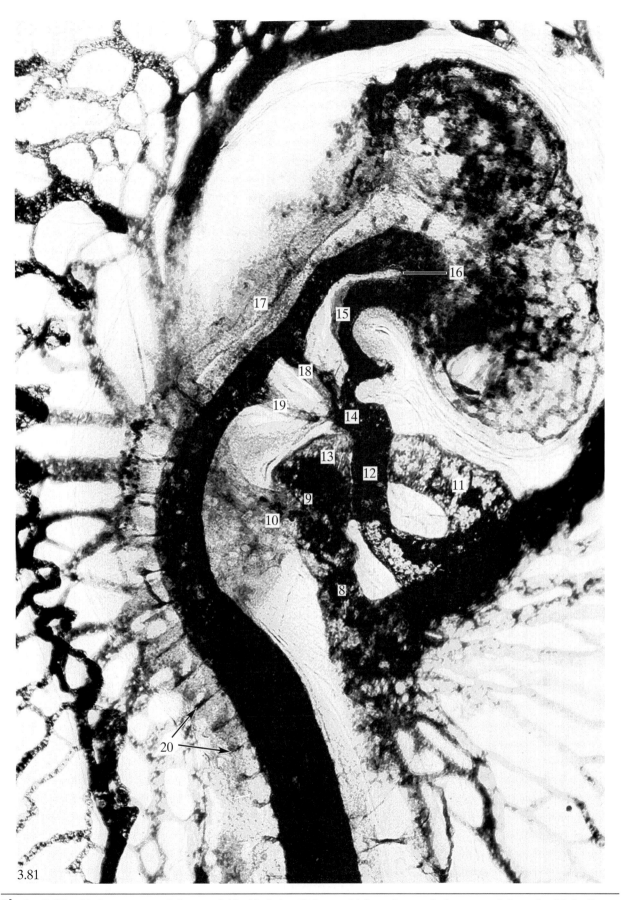

3.81

Photo 3.81. Enlargement of the cranial half of the 48-hour chick embryo whole mount injected with India ink and shown in Photo 3.80.

Photos 3.82-3.85

Chick Embryos

Legend

1. Myelencephalon
2. Metencephalon
3. Isthmus
4. Head mesenchyme
5. Mesencephalon
6. Diencephalon
7. Roof plate of myelencephalon
8. Semilunar ganglion

9. Notochord
10. Amnion
11. Yolk sac
12. Precardinal vein
13. Acousticofacialis ganglion
14. Dorsal aorta
15. Foregut endoderm
16. Auditory vesicle

17. First pharyngeal pouch
18. First branchial groove
19. Pharynx
20. First aortic arch
21. Chorion
22. Rathke's pouch
23. Infundibulum
24. First closing plate

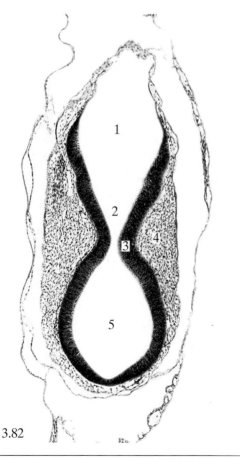

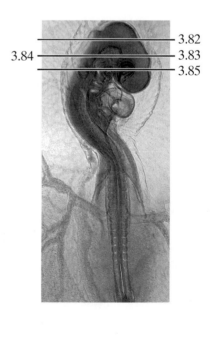

Photo 3.82.   48-hour chick embryo serial transverse section.

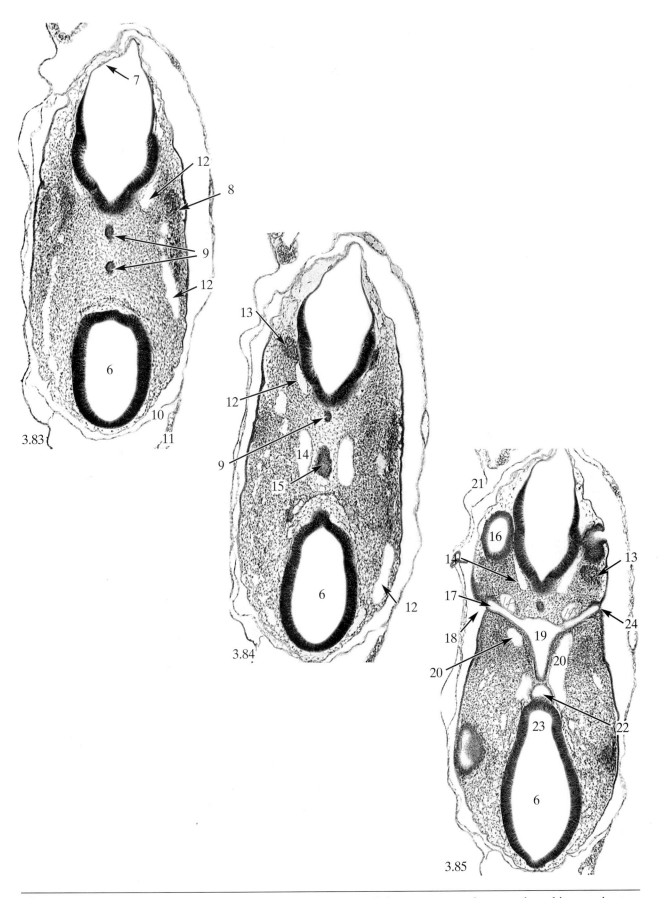

Photos 3.83-3.85. Continuation of 48-hour chick embryo serial transverse sections numbered in anterior to posterior sequence.

Photos 3.86-3.89

Chick Embryos

Legend

1. Myelencephalon
2. Superior ganglion
3. Auditory vesicle
4. Acousticofacialis ganglion
5. First pharyngeal pouch
6. Pharynx
7. First branchial groove
8. First aortic arch
9. Mandibular process
10. Maxillary process
11. Oral membrane
12. Stomodeum
13. Rathke's pouch (continuous with stomodeum)
14. Infundibulum
15. Optic cup
16. Epibranchial placode (contributes cells to facialis part of the acousticofacialis ganglion)
17. Opticoel
18. Precardinal vein
19. Notochord
20. Dorsal aorta
21. Second aortic arch
22. Pigmented retina
23. Continuity between skin ectoderm and lens vesicle
24. Sensory retina
25. Skin ectoderm
26. Second pharyngeal pouch
27. Aortic sac
28. Vitelline blood vessel
29. Second closing plate
30. Chorion
31. Amnion
32. Yolk sac
33. Optic stalk
34. Thyroid rudiment
35. Diencephalon

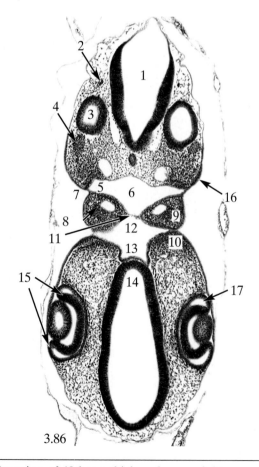

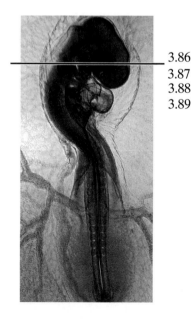

3.86
3.87
3.88
3.89

Photo 3.86. Continuation of 48-hour chick embryo serial transverse sections.

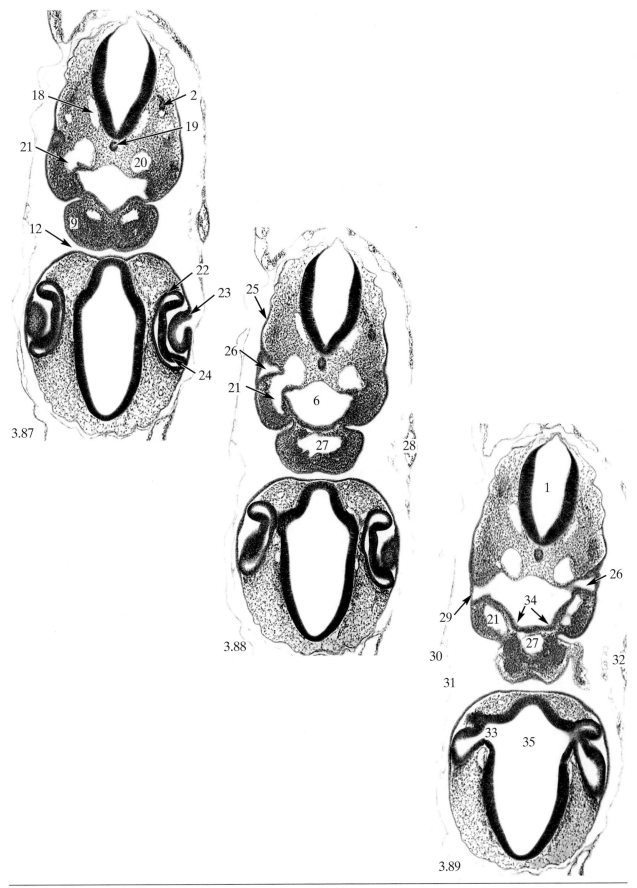

3.87

3.88

3.89

Photos 3.87-3.89. Continuation of 48-hour chick embryo serial transverse sections numbered in anterior to posterior sequence.

Photos 3.90-3.93

Chick Embryos

Legend

1. Somite #1
2. Chorion
3. Precardinal vein
4. Dorsal aortae
5. Amniotic cavity
6. Third pharyngeal pouch
7. Third closing plate
8. Pharynx
9. Third aortic arch
10. Endocardium of conotruncus
11. Myocardium of conotruncus
12. Aortic sac

13. Nasal placode
14. Telencephalon
15. Dermatome
16. Myotome
17. Sclerotome
18. Atrium
19. Descending aorta
20. Foregut
21. Amnion
22. Splanchnic mesoderm
23. Pericardial cavity
24. Conotruncus

25. Ventricle
26. Spinal cord
27. Pleural cavity
28. Postcardinal vein
29. Common cardinal vein
30. Beginning of lung bud
31. Dorsal mesocardium
32. Sinus venosus
33. Vitelline blood vessel
34. Cranial liver rudiment

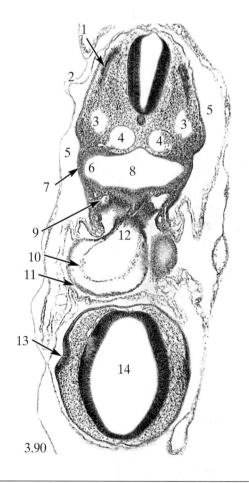

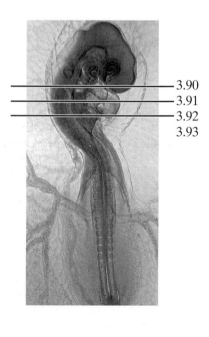

Photo 3.90. Continuation of 48-hour chick embryo serial transverse sections.

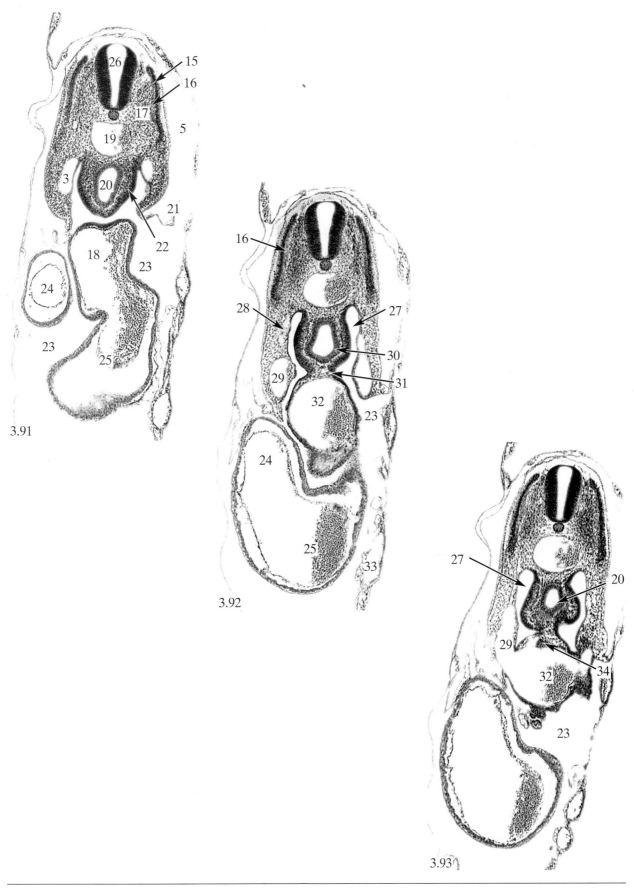

Photos 3.91-3.93. Continuation of 48-hour chick embryo serial transverse sections numbered in anterior to posterior sequence.

Photos 3.94-3.97
Chick Embryos
Legend

1. Spinal cord
2. Notochord
3. Intersegmental artery
4. Descending aorta
5. Postcardinal vein
6. Duodenum
7. Cranial liver rudiment
8. Vitelline vein
9. Caudal liver rudiment
10. Rudiment of the peritoneal cavity
11. Mesonephric duct rudiment (cavitated)
12. Cranial intestinal portal
13. Yolk sac

14. Mesonephric tubule rudiment
15. Ectoderm of lateral amniotic fold
16. Somatic mesoderm of lateral amniotic fold
17. Chorion
18. Amnion
19. Amniotic cavity
20. Dorsal aortae (near level of fusion to form the descending aorta)
21. Extraembryonic coelom
22. Mesonephric duct rudiment (noncavitated)
23. Sclerotome

24. Myotome
25. Dermatome
26. Vitelline blood vessel
27. Floor plate
28. Roof plate
29. Dorsal aorta
30. Vitelline artery
31. Ectoderm of lateral body fold
32. Somatic mesoderm of lateral body fold
33. Endoderm

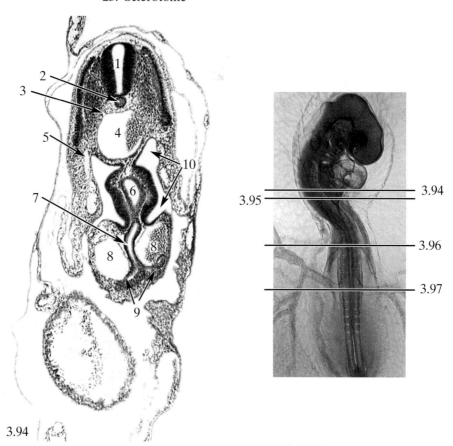

3.94

Photo 3.94. Continuation of 48-hour chick embryo serial transverse sections.

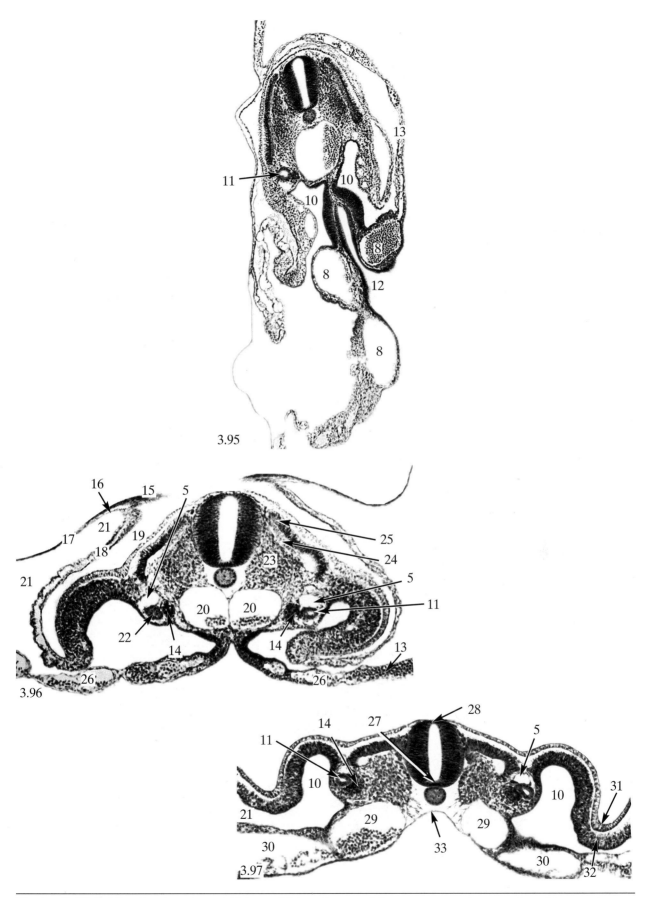

Photos 3.95-3.97. Continuation of 48-hour chick embryo serial transverse sections numbered in anterior to posterior sequence.

<h2>𝒫𝒽𝑜𝑡𝑜𝑠 3.98-3.101</h2>

Chick Embryos

𝓛𝑒𝑔𝑒𝓃𝒹

1. Tail bud
2. Somatopleure
3. Vitelline blood vessel
4. Caudal intestinal portal
5. Skin ectoderm
6. Extraembryonic coelom
7. Splanchnopleure (of yolk sac)
8. Allantois rudiment
9. Roof plate of myelencephalon
10. Isthmus
11. Myelencephalon
12. Metencephalon
13. Mesencephalon
14. Notochord
15. Infundibulum
16. Rathke's pouch
17. Stomodeum
18. Diencephalon
19. Preoral gut
20. Oral membrane
21. Pharynx
22. Mandibular process
23. First aortic arch
24. Telencephalon
25. Conotruncus
26. Ventricle
27. Aortic sac
28. Sensory retina
29. Pigmented retina
30. Lens vesicle
31. Opticoel (continuous with cavity of diencephalon)
32. Optic fissure
33. Head mesenchyme

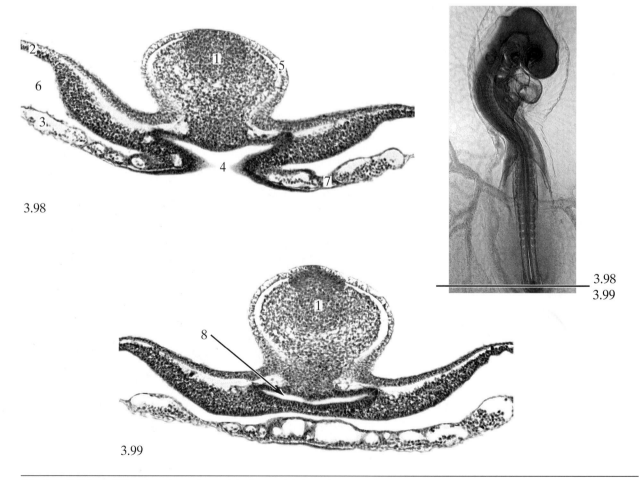

3.98

3.98
3.99

3.99

Photos 3.98, 3.99. Continuation of 48-hour chick embryo serial transverse sections numbered in anterior to posterior sequence.

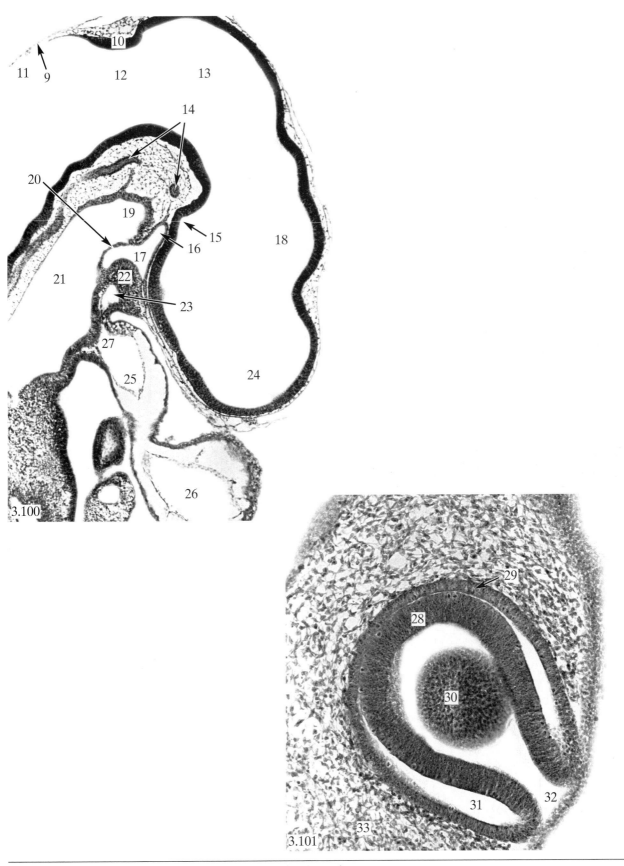

Photo 3.100. 48-hour chick embryo midsagittal section.

Photo 3.101. 48-hour chick embryo parasagittal section at the level of the optic cup.

Photo 3.102

Chick Embryos

Legend

1. Myelencephalon
2. Acousticofacialis ganglion
3. Yolk sac
4. Amnion
5. Notochord
6. Precardinal vein
7. Dorsal aorta
8. Chorion
9. Extraembryonic coelom

10. Amniotic cavity
11. Pharynx
12. First pharyngeal pouch
13. First branchial groove
14. Second branchial arch
15. First branchial arch
16. First aortic arch
17. Rathke's pouch

18. Infundibulum
19. Diencephalon
20. Optic cup
21. Sensory retina
22. Pigmented retina
23. Lens vesicle
24. Continuity between skin ectoderm and lens vesicle
25. Skin ectoderm

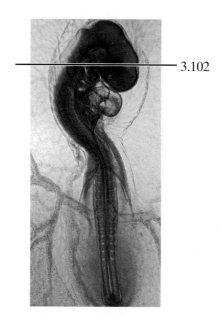

— 3.102

Photo 3.102. Facing page. Scanning electron micrograph of the cut surface of a block sectioned transversely from a 48-hour chick embryo.

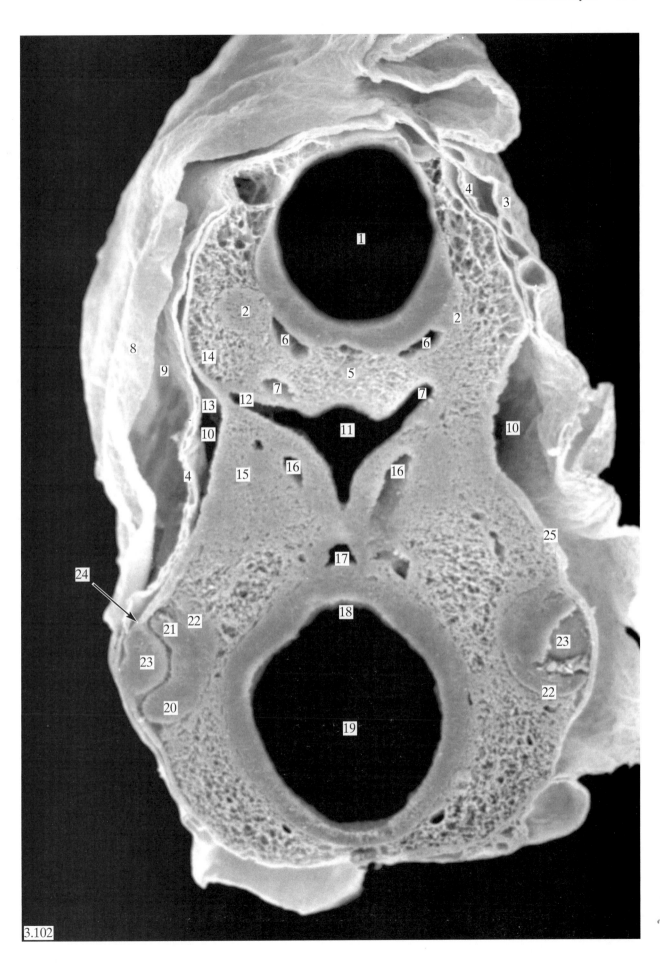

3.102

Photos 3.103-3.105
Chick Embryos

Legend

1. Skin ectoderm
2. Spinal cord
3. Notochord
4. Dorsal aorta
5. Coelom (rudiment of the peritoneal cavity)
6. Endoderm
7. Lateral body fold
8. Caudal extent of the continuous cranial and lateral amniotic folds
9. Yolk sac (splanchnopleure)

10. Amnion (somatopleure)
11. Amniotic cavity
12. Endothelial cells forming an intersegmental artery in an intersomitic furrow
13. Neural crest cells
14. Postcardinal vein
15. Lateral amniotic fold
16. Lateral body fold (somatopleuric component)
17. Wing bud rudiment
18. Somite

19. Mesonephros
20. Lateral plate mesoderm
21. Somatic mesoderm
22. Splanchnic mesoderm
23. Open midgut
24. Allantois rudiment (opening cranially as the caudal intestinal portal)
25. Somatopleure of lateral plate
26. Splanchnopleure of lateral plate

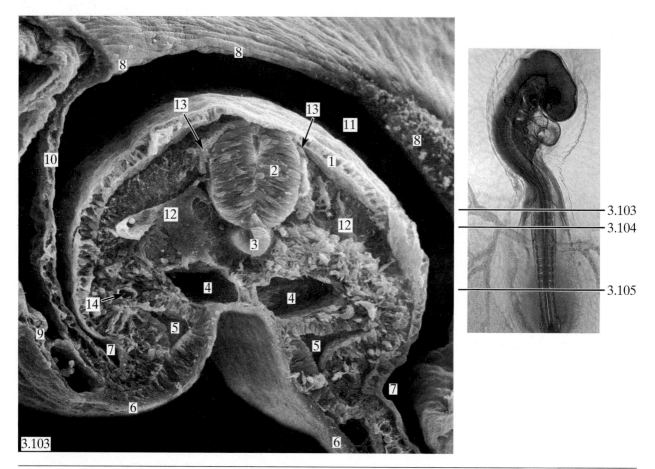

Photo 3.103. Scanning electron micrograph of a transverse slice (cranial portion of spinal cord, just cranial to the level shown in Photo 3.96; the background shows more *cranial* levels).

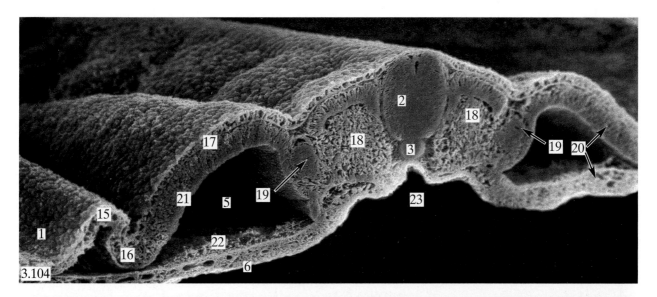

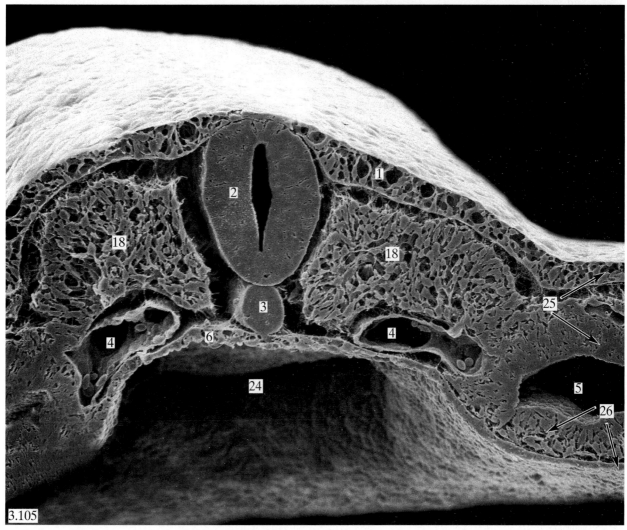

Photo 3.104. Scanning electron micrograph of a transverse cryofracture (mid-portion of spinal cord; the background shows more caudal levels) from a 48-hour chick embryo.

Photo 3.105. Scanning electron micrograph of a transverse cryofracture (caudal portion of spinal cord; the background shows more *caudal* levels) from a 48-hour chick embryo.

Photos 3.106-3.109
Chick Embryos

Legend

1. Tail fold of the body
2. Tail bud
3. Caudal end of spinal cord
4. Allantois rudiment
5. Skin ectoderm

6. Endoderm
7. Spinal cord
8. Notochord
9. Cavitating caudal spinal cord
10. Incipient notochord

11. Segmental plate mesoderm
12. Lateral plate mesoderm
13. Dorsal aorta

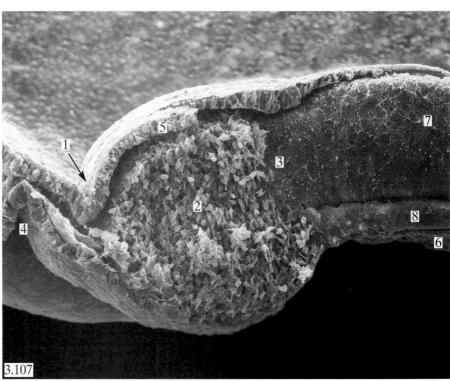

Photo 3.106. Scanning electron micrograph of a whole mount of a young 48-hour chick embryo. The vertical line indicates the level of the parasagittal slice shown in Photo 3.107.

Photo 3.107. Scanning electron micrograph of a parasagittal slice of the caudal end of a 48-hour chick embryo at the level shown in Photo 3.106 by the vertical line.

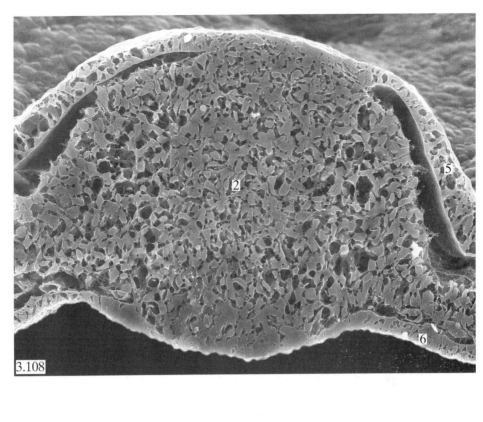

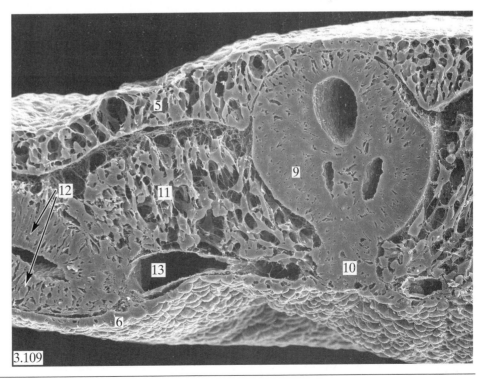

Photos 3.108, 109. Scanning electron micrographs of transverse cryofractures through the caudal end of 48-hour chick embryos.

Photo 3.110

Chick Embryos

Legend

1. Telencephalon
2. Diencephalon
3. Optic stalk
4. Infundibulum
5. Mesencephalon

6. Metencephalon
7. Myelencephalon
8. Dorsal aorta
9. First aortic arch

10. Precardinal vein
11. Pharynx
12. Mandibular process
13. Atrium

Photo 3.110. Facing page. Scanning electron micrograph of the cut surface of a block sectioned sagittally from a 48-hour chick embryo.

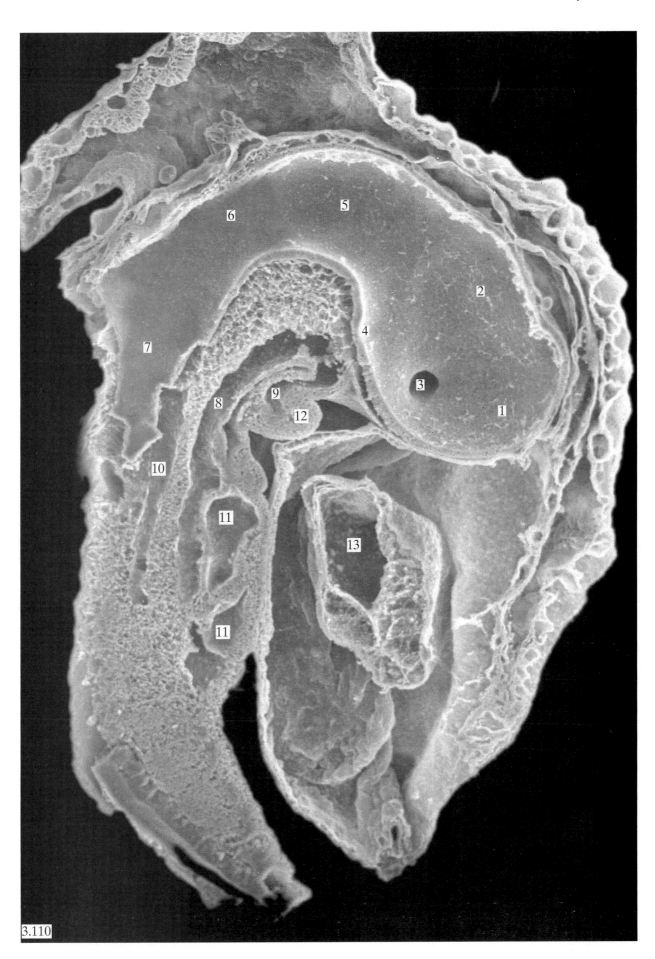

S. 72-HOUR CHICK EMBRYOS

1. Whole mounts: injected and uninjected

First examine your uninjected whole mount slide with the naked eye. Superimpose this slide on the 48-hour uninjected whole mount slide and note the changes that have occurred in the shape of the embryo. In addition to the **cranial flexure**, two other flexures are probably well developed. These are the **cervical flexure**, at the level of the first several somites (Photo 3.111), and the **tail flexure**, at the caudal end of the embryo (Photo 3.112). Note that the embryo is sharply bounded and that it lies on its *left* side throughout a considerable part of its length. This means that it has undergone (or is undergoing) **torsion** along most of its length. A saclike structure, the **allantois** (Photos 3.112, 3.113), is somewhat encircled by the **tail**. Paired **wing** and **leg buds** can be identified, but they are not sharply bounded (Photos 3.111, 3.112). Determine the approximate somitic levels of the limb buds. The entire embryo is usually covered by **amnion** and **chorion**, although the oval or circular boundary of the continuous **cranial, lateral,** and **caudal (posterior) amniotic folds** still may be visible (Photo 3.111).

Examine your slide under low power and identify the following structures: **telencephalon** (with its lateral oval-shaped expansions, the **cerebral hemispheres**); **nasal pits** (formed by invagination of the nasal placodes); **diencephalon; pineal gland** or **epiphysis** (a small dorsal evagination of the diencephalon); **optic cups; optic fissures; lens vesicle; infundibulum; Rathke's pouch** (examine it under high power and note its close relationship to the infundibulum; Photo 3.114); **mesencephalon; metencephalon; isthmus; myelencephalon** (note its thin **roof plate**); **auditory vesicles; endolymphatic ducts** (formed by evagination of the dorsal part of the auditory vesicles); **spinal cord; cranial intestinal portal;** and **caudal intestinal portal,** if visible.

Identify the **acousticofacialis ganglia** just cranial to the auditory vesicles (Photo 3.112). They lie above the **second branchial arches**. The facialis part of these ganglia is the source of the **sensory nerve fibers** of the **facial nerves,** which innervate the *second* branchial arches. Identify the **semilunar ganglia** lying above the **first branchial arches** (Photo 3.112). These ganglia are the source of **sensory nerve fibers** of the **trigeminal nerves,** which innervate the *first* branchial arches.

Observe that the first branchial arches are partially split by the **stomodeum** into **maxillary processes** (which will form the lateral portions of the **upper jaw** and most of the **cheeks**) and **mandibular processes** (which will form the **lower jaw**). The maxillary processes may partially overlap the mandibular processes in some embryos. Identify the **first branchial grooves,** just caudal to the mandibular processes; the **second branchial grooves,** be-

tween the **second** and **third branchial arches**; and the **third branchial grooves**.

In favorably injected embryos the **precardinal, postcardinal,** and **common cardinal veins,** as well as the **intersegmental veins** (continuous with the postcardinal veins dorsally), can be identified. The common cardinal veins return blood to the **sinus venosus**. Identify the following structures (see Photos 3.111-3.113, 3.115, 3.116): **atrium; ventricle; conotruncus; aortic sac; first, second, third,** and possibly **fourth aortic arches** (the first aortic arches are degenerating and may be reduced in diameter); **dorsal aortae; internal carotid arteries** (cranial extensions of the dorsal aortae); **descending aorta;** and **intersegmental arteries**. Also identify the **vitelline arteries,** which are continuous with the dorsal aortae at approximately the 22nd somite level, and **vitelline veins**. The latter are fused near the level of the heart to form a single vessel, the **ductus venosus,** which is continuous with the sinus venosus. Try to identify the ductus venosus.

2. Serial transverse sections

Position your slide on the microscope stage so that when viewed through the microscope, each section is oriented as in Photos 3.117-3.136. Trace sections in anteroposterior sequence unless otherwise specified. The first brain sections are probably cut through the **myelencephalon** (and possibly through the **metencephalon** as well) due to the **cervical flexure**. A variable amount of tissue may appear to be lying freely within the cavity of the myelencephalon. This tissue is the thin **roof plate** of the myelencephalon and the adjacent **skin ectoderm**. Continue tracing sections posteriorly. **Endolymphatic ducts** soon appear alongside the walls of the myelencephalon. These become continuous with the **auditory vesicles** a few sections more posteriorly. At about this level, sections begin to cut across a small, rounded region lying beneath the metencephalon. This region is the **mesencephalon;** it is continuous with the metencephalon in more posterior sections (Photo 3.117). Note the segmental enlargements of the walls of the rhombencephalon; these are the **neuromeres**.

As sections are traced posteriorly from about the level at which the auditory vesicles are first identified, two groups of nerve fibers seem to emerge from neuromeres cranial to each auditory vesicle. They quickly become continuous with two prominent ganglia on each side (Photo 3.117). The ganglia lying against the cranial wall of the auditory vesicles are the **acousticofacialis ganglia** (future **acoustic** and **geniculate ganglia**). The other very large ganglia are the **semilunar ganglia**. At about this level, note on the caudal side of each auditory vesicle a poorly circumscribed, very small, rounded accumulation of cells, the **superior ganglia**. The neural tube separates into two parts in slightly more posterior sections. The upper part is either the myelencephalon or the **spinal cord;** the lower part is principally the mesencephalon.

As you continue tracing sections posteriorly, the superior ganglia are continuous, without a distinct boundary, with **sensory nerve fibers** of the **glossopharyngeal (IX) nerves** (Photo 3.118). Similarly, the acousticofacialis ganglia are directly continuous with the **sensory nerve fibers** of the **facial (VII) nerves**. The facial nerves can be traced into the **second branchial arches**, where they merge on each side with a distinct **epibranchial placode**. These placodes contribute cells to the acousticofacialis ganglia and are the source of the geniculate ganglia. Trace the glossopharyngeal nerves into the **third branchial arches**, where each similarly merges, on each side, with an **epibranchial placode** (Photo 3.120). These placodes are the source of the **petrosal ganglia** of the glossopharyngeal nerves. *If you trace the facial and glossopharyngeal nerves into their respective branchial arches, correct identification of these arches becomes relatively simple.* At about the level where each glossopharyngeal nerve merges with an epibranchial placode, try to identify an accumulation of **neural crest cells** on each side, the **jugular ganglion** of the **vagus (X) cranial nerve** (Photo 3.120). The jugular ganglia fade out within the **fourth branchial arches**.

Return to sections through the semilunar ganglia (Photo 3.117) and note that each one seems to subdivide caudally into three branches, the branches of the **trigeminal (V) nerve** (Photo 3.118). The medial branch is the **maxillary branch**; the lateral branch closest to the mes-encephalon is the **ophthalmic branch**; the lateral branch above the ophthalmic is the **mandibular branch**. These branches are contained within the region of the **first branchial arches**; they fade out as they are traced caudally.

Shift your attention to the mesencephalon and follow it posteriorly in your sections (cranially within the embryo). Nerve fibers (axons) seem to emerge from the floor of the mesencephalon to either side of the midline. These will probably *not* be cut symmetrically. The accumulations of **neural ectodermal cells** from which these nerve fibers arise lie within the floor of the mesencephalon. Each accumulation is designated as a **motor nucleus**. **Motor nerve fibers** grow out from the motor nuclei toward areas they will later innervate; collectively these nerve fibers constitute the **oculomotor (III) cranial nerves**. These nerves will innervate four pairs of extrinsic eye muscles, which develop *later*.

The following events occur in the establishment of *sensory* portions of cranial nerves. (1) Cranial ganglia, consisting of **cell bodies** of **young sensory neurons**, are formed adjacent to the brain from accumulations of ectodermal neural crest cells and/or cells derived from ectodermal placodes. (2) Sensory nerve fibers grow out from these ganglia and into the brain wall. These fibers are **axons**; they will eventually transmit nerve impulses *away from* their cell bodies. (3) Other sensory nerve fibers grow out from these ganglia toward the areas to be innervated. We will refer to these fibers as **dendrites**

The following chart summarizes the development of cranial nerves through the 72-hour stage.

Cranial Nerves	Cranial Ganglia Present	Origin of Cranial Ganglia	Type of Nerve Fibers Present	Regions Innervated
oculomotor	————————	————————	motor	four pairs of extrinsic eye muscles
trigeminal	semilunar	neural crest cells and epibranchial placodes	sensory (*later* also motor)	first branchial arches
facial	acousticofacialis (*later* geniculate)	neural crest cells and epibranchial placodes	sensory (*later* also motor)	second branchial arches
auditory (nerves not yet formed)	acousticofacialis (*later* acoustic)	auditory placodes	(*later* sensory)	inner ears
glossopharyngeal	superior	neural crest cells	sensory (*later* also motor)	third branchial arches
vagus (nerves not yet formed)	jugular	neural crest cells	(*later* sensory and motor)	fourth branchial arches

because they will eventually transmit nerve impulses *toward* their cell bodies.

The following events occur in the establishment of *motor* portions of cranial nerves. (1) Motor nuclei, consisting of **cell bodies** of **young motor neurons**, are formed within the ventral part of the wall of the brain from accumulations of neural ectodermal cells. (2) Motor nerve fibers, **axons**, grow out from these nuclei, leaving the wall of the brain and extending toward the areas to be innervated. (3) Other motor nerve fibers, **dendrites**, grow out from these nuclei but remain within the brain wall.

The **infundibulum** appears in sections posterior to the oculomotor nerves, marking the beginning of the **diencephalon**. Identify the **optic cups** and the **optic fissures** (Photo 3.125). The **lens vesicle** is now completely closed. It is separated from the overlying **skin ectoderm**, which *later* forms the **corneal epithelium** (Photo 3.124). The lens vesicle is beginning to differentiate into two regions: the outer, thin **lens epithelium** and the inner, thick **lens fibers**.

Continue to trace sections posteriorly until sections cut through both the diencephalon and **telencephalon** (Photos 3.128, 3.129). The telencephalon is expanded laterally as the **cerebral hemispheres**. Identify the **nasal pits** and **pineal gland** at about this level (Photo 3.129). Continue tracing sections posteriorly, noting the disappearance of the diencephalon and telencephalon.

Quickly follow the **spinal cord** throughout its length. **Neural crest cells** are now aggregated alongside its most cranial levels, forming the paired **spinal ganglia**. Segmentation of these ganglia corresponds to segmentation of the **somites**, which are the much more prominent and elongated structures lateral to each ganglion. Spinal ganglia are located beneath the **myotome** in the middle of the craniocaudal extent of each somite. **Spinal nerves** will form *later*. Examine the most caudal tip of the spinal cord, observing that the **notochord** is still distinguishable at this level. The spinal cord and notochord are usually cut frontally at this level due to the **tail flexure**.

Return to the most anterior section in which the notochord can be identified (Photo 3.118). It is cut frontally at this level due to the **cervical flexure**. Trace sections posteriorly. Sections soon cut through the **first**, and possibly the **second**, **pharyngeal pouches**, which at first appear as isolated structures (Photo 3.119). Both pairs of pouches become continuous with the **pharynx** as you continue to trace sections posteriorly (Photo 3.120). (Remember to use the facial nerves as a landmark for identifying of the second branchial arches. This will enable you to identify the first and second pharyngeal pouches with certainty.) Both the first and second pharyngeal pouches may open to the outside via the **first** and **second branchial clefts**, which arise by rupture of the corresponding **closing plates**. Identify the

shallow **first** and **second branchial grooves**. The **third pharyngeal pouches** appear in more posterior sections and almost immediately become continuous with the pharynx (Photo 3.121). Identify the **third closing plates** and the shallow **third branchial grooves**. Note that the first pharyngeal pouches fade out at about this level. In more posterior sections, identify the **fourth pharyngeal pouches**, **fourth closing plates**, **fourth branchial grooves** (very shallow), and **thyroid rudiment** (Photos 3.122, 3.123).

Return to a section similar to Photo 3.121 and identify the **preoral gut**, which is located just cranial to the pharynx. Also identify the **first branchial arches**, just lateral to the preoral gut. The **stomodeum** appears in slightly more posterior sections and separates the first branchial arches into **maxillary** and **mandibular processes**. At about this level, the cavity of the gut usually opens into the stomodeum via the **mouth opening** because the **oral membrane** has probably ruptured. Try to identify the mouth opening in your embryo. It will be seen in sections located between those illustrated by Photos 3.121 and 3.122. Identify **Rathke's pouch** in slightly more posterior sections (Photo 3.122); it quickly becomes continuous with the stomodeum (Photo 3.123). Note the close relationship between Rathke's pouch and the **infundibulum**.

Continue tracing sections posteriorly and identify the **laryngotracheal groove**, which is continuous, without a distinct boundary, with the pharynx (Photo 3.124). (The laryngotracheal groove is formed by elongation of the ventral portion of the foregut, at the level where lung bud formation was initiated earlier. See Photo 3.92.) More posterior sections cut through the **lung buds**, which at this stage are bilateral expansions of the laryngotracheal groove (Photo 3.126). Moreover, the laryngotracheal groove is beginning to constrict between the pharynx and lung buds. This constriction is complete in slightly more posterior sections, and paired lung buds lie beneath the **esophagus** (Photo 3.127). Both the esophagus and lung buds are contained within a thick mesentery composed of splanchnic mesoderm. The portion of the mesentery dorsal to the esophagus is the **mesoesophagus**; the portion dorsal to the sinus venosus is the **dorsal mesocardium**. The coelomic cavities lateral to the lung buds are the **pleural cavities**.

Continue tracing sections posteriorly and note the disappearance of the lung buds. Two narrow coelomic cavities are located ventrolateral to the esophagus at about this level. These cavities are extensions of the pleural cavities (Photo 3.128). They become continuous with the more lateral portions of the pleural cavities in slightly more posterior sections (Photo 3.129). The diameter of the gut is usually slightly larger at about this level; this is the region of the **stomach**. The mesentery dorsal to the stomach is the **dorsal mesogaster**; the one ventral to it is the **hepatogastric (gastrohepatic) ligament** (**ventral mesogaster**). Although the stomach lies medial to the pleural cavities at this stage, it will *later* descend and be enclosed by the **peritoneal cavity**.

Identify the **cranial liver rudiment** in sections at about the level of the stomach (Photo 3.129); it lies just above a large blood vessel, the **ductus venosus**. The cranial liver rudiment and gut become continuous in more posterior sections (Photo 3.130). This level of the gut is the **duodenum**. The mesentery lying dorsal to the duodenum is the **mesoduodenum**; the one ventral to it is the **hepatoduodenal (duodenohepatic) ligament**. Identify the **caudal liver rudiment** lying beneath the ductus venosus at about this level. The close relationship between the ductus venosus and liver tissue is very important. *Later*, as the liver tissue grows, it invades the ductus venosus and subdivides this vessel into smaller channels, the **hepatic sinusoids**.

Continue tracing sections posteriorly and notice that the cranial liver rudiment lengthens dorsoventrally and becomes continuous with the caudal liver rudiment (Photo 3.131). The cranial liver rudiment lies between two blood vessels at this level, the **vitelline veins**. The boundaries between the duodenum, cranial liver rudiment, and caudal liver rudiment are indistinct. Note a solid mass of cells continuous with the dorsal wall of the duodenum at this level; this is the **dorsal pancreatic rudiment**. Identify the **cranial intestinal portal** in more posterior sections (Photo 3.132).

Trace sections posteriorly and identify the paired **wing buds** (Photo 3.133), and further posteriorly, the paired **leg buds** (Photos 3.134, 3.135). Both wing and leg buds consist of a core of somatic mesoderm covered by skin ectoderm, which is thickened laterally as the **apical ectodermal ridge**. Experiments have demonstrated that this ridge is necessary for normal outgrowth and development of limb buds.

In sections through about the level of the leg buds, identify an endoderm-lined cavity, the **hindgut** (Photo 3.134), which is formed by the combined action of the **tail fold of the body** and **lateral body folds**. A portion of the hindgut, the **cloaca**, is continuous ventrally with the **allantois** in more posterior sections (Photo 3.135). The allantois is formed by expansion of the endodermal allantois rudiment, which has become surrounded by splanchnic mesoderm (Fig. 3.10). Trace the cloaca and allantois caudally, noting the disappearance of the latter and the separation of the **tail** from the **amnion** by the ectoderm-lined **subcaudal pocket** (Photo 3.136). The ventral endoderm of the cloaca usually contacts the ventral skin ectoderm at about this level, forming the double-layered **cloacal membrane**. Caudal to this membrane, sections usually cut through the much smaller **tail gut**, which will soon degenerate.

Return to a section similar to Photo 3.134 and trace sections *anteriorly*, noting the disappearance of the floor of the **hindgut**. This is the level of the **caudal intestinal portal**, the opening of the hindgut into the **subgerminal cavity**. (Recall that, at 48 hours of incubation, the caudal intestinal portal was the opening of the allantois rudiment into the subgerminal cavity. Carefully study Fig. 3.10 so that you understand why this spatial relationship has changed by 72 hours of incubation.)

Return cranially to a level showing well-formed **somites** and identify **dermatome**, **myotome**, and **sclerotome** (Photo 3.133). Trace sections *posteriorly*, noting that tail somites are considerably less advanced in their differentiation. Identify the short **segmental plates** caudal to the somites; their caudal ends are continuous with the small **tail bud**.

Return to the level where the cranial liver rudiment connects to the duodenum (Photo 3.130). A tiny duct can be seen on each side at about this level, dorsolateral to the mesoduodenum. These ducts are the **mesonephric ducts**, which formed by cavitation of the mesonephric duct rudiments. The cranial end of each mesonephric duct is now undergoing degeneration. Trace sections posteriorly and observe that **mesonephric tubules** are forming from **mesonephric tubule rudiments**, medial to each mesonephric duct (Photo 3.135). Some mesonephric tubules may have established connections with the mesonephric ducts. The mesonephric tubules collectively form the paired **mesonephric kidneys**. Continue tracing sections posteriorly. Observe that the mesonephric ducts have extended caudad beyond the developing mesonephric tubules (Photos 3.134, 3.135), and that they either join the cloaca laterally or terminate near it. The mesonephric tubules become functional slightly later than the 72-hour stage and extract nitrogenous wastes from the blood. These wastes pass through the mesonephric tubules and ducts and then to the cloaca; they then enter the allantois, where they are stored.

Return to the section in which the notochord is cut frontally (Photo 3.118) and trace sections posteriorly. Identify the paired **dorsal aortae** (Photo 3.119). They are at first cut frontally due to the **cervical flexure**. Each dorsal aorta seems to be constricted into two parts at the level of the first pharyngeal pouches. The lower part is the **internal carotid artery**, the cranial extension of the dorsal aorta. Quickly trace sections posteriorly and observe that the internal carotid arteries fade out at about the level of the diencephalon.

Return to a level similar to Photo 3.119 and trace sections posteriorly. Observe that the **second aortic arches** extend from the dorsal aortae into the second branchial arches, and subsequently the **third aortic arches** extend into the third branchial arches (Photo 3.120). If the **first aortic arches** are still present in your embryo, note that they extend from the internal carotid arteries into the first branchial arches. The small **fourth aortic arches** extend from the dorsal aortae into the fourth branchial arches in more posterior sections

(Photo 3.122). Note that the dorsal aortae have fused at about this level, forming the **descending aorta**.

Continue tracing sections posteriorly, observing the aortic arches. They approach one another in the region of the thyroid gland (Photo 3.123) and soon unite with the **aortic sac** (Photo 3.124). In more posterior sections, identify the **conotruncus**; **atrium**, which lies to the l*eft* (apparent right) of the conotruncus; and **sinus venosus**, which lies in the midline (Photo 3.126). The sinus venosus is continuous with the **common cardinal veins** in more posterior sections. Usually the *right* (apparent left) common cardinal vein is continuous with the sinus venosus more cranially than is the *left* common cardinal vein (Photo 3.128). At about this level, sections begin to cut across a small portion of the **ventricle**, which is continuous with the atrium.

Continue to trace sections posteriorly and observe the gradual disappearance of the atrium and the progressive enlargement of the ventricle in its place (Photo 3.129). Also note the gradual transition between sinus venosus and **ductus venosus**. More caudally, the conotruncus and ventricle become continuous, without a distinct boundary, and the ductus venosus seems to separate into two vessels, the **vitelline veins**, which extend laterad into the **yolk sac** (Photos 3.130-3.132). Trace sections posteriorly, noting the gradual disappearance of the ventricle.

Return to the level where the **descending aorta** first appears (Photo 3.122) and trace sections posteriorly. Identify small blood vessels, the **intersegmental arteries**, continuous with the descending aorta dorsally (Photo 3.122). Quickly trace sections posteriorly until paired **dorsal aortae** again appear (usually between the levels of the wing and leg buds). The **vitelline arteries** are continuous with the dorsal aortae just caudal to this level. The vitelline arteries extend laterad into the **yolk sac**; at this level they lie beneath (ventral to) the **vitelline veins**. Continue tracing the dorsal aortae caudally and note that they become markedly reduced in diameter (Photo 3.134). They lie just dorsal to the gut within the tail region (that is, caudal to the leg buds) and are designated as **caudal arteries**.

Return to the level where the *right* (apparent left) **common cardinal vein** is continuous with the **sinus venosus** (Photo 3.128). The right common cardinal vein, as at the 48-hour stage, is contained within a bridge of mesoderm connecting the lateral body wall to the sinus venosus. Trace sections *anteriorly*, observing that the lateral body wall separates from the heart. Two vessels are usually located within the right lateral body wall at this level (Photo 3.127). The lower vessel, nearest the sinus venosus, is still the common cardinal vein; the upper vessel, lateral to the descending aorta, is the **postcardinal vein**. These vessels approach one another and become continuous in more *anterior* sections (Photo 3.125, 3.124). At this level these vessels are cut *frontally*. (In some embryos the common cardinal and postcardinal veins may be interconnected by smaller vessels at some levels and completely separated in more anterior sections. If this is the case in your embryo, trace sections anteriorly until the common cardinal and postcardinal veins are broadly continuous.) Continue to trace sections *anteriorly*, noting the disappearance of the upper portion of this frontally cut vessel. The lower, persisting portion is now the **precardinal vein**. In more *anterior* sections, the precardinal vein first lies ventrolateral to the descending aorta (Photo 3.123), and then lateral to the dorsal aorta (Photo 3.120, 3.119); finally it is cut frontally (Photo 3.117). It disappears in more *anterior* sections.

Return to the level where the right postcardinal and common cardinal veins are broadly continuous (Photo 3.124) and trace sections *posteriorly*. The postcardinal vein quickly separates from the common cardinal vein and lies lateral to the descending aorta (Photos 3.126 , 3.127, 3.130, 3.133). Try to identify small blood vessels continuous with the postcardinal vein dorsally; these vessels are the **intersegmental veins**. Note that the postcardinal vein lies above (that is, dorsal to) the developing mesonephric kidney (Photo 3.133). Try to identify a small vessel lying beneath (that is, ventral to) each mesonephric kidney; these vessels are the **subcardinal veins**. Trace the postcardinal vein caudally until it disappears.

Try to identify and trace the precardinal, common cardinal, and postcardinal veins on the *left* side of the embryo. These vessels are usually more difficult to identify with certainty than those on the right side.

3. Serial sagittal sections

Approximately the cranial half of the embryo is actually cut sagittally; the caudal half is cut frontally. As with the 48-hour stage, find a section that is approximately a midsagittal section through the **brain** and **notochord** (Photo 3.137). Identify all regions of the brain, including the **infundibulum** and **pineal gland**. Also identify the **pharynx**, **preoral gut**, **stomodeum**, **Rathke's pouch**, and **mouth opening**. The mouth opening has formed by rupture of the oral membrane. After this rupture occurs, the cavities of the foregut and stomodeum are continuous. Try to identify as many other structures as time permits.

4. Summary of the contributions of the germ layers to structures present in the 72-hour chick embryo but not present in the 48-hour embryo

Ectoderm

cerebral hemispheres
cranial ganglia:
    X. jugular
cranial nerves:
    III. oculomotor
    V. mandibular, maxillary, and ophthalmic branches of trigeminal
    VII. facial
    IX. glossopharyngeal
endolymphatic ducts
lens epithelium
lens fibers
nasal pits
pineal gland
spinal ganglia

Mesoderm

mesenteries:
    dorsal mesogaster
    hepatoduodenal ligament
    hepatogastric ligament
    mesoduodenum
    mesoesophagus
mesonephric ducts
mesonephric tubules

Endoderm

cloaca
dorsal pancreatic rudiment
esophagus
hindgut
laryngotracheal groove

stomach
tail gut

Ectoderm and somatic mesoderm

leg buds
wing buds

Endoderm and splanchnic mesoderm

allantois

Ectoderm and endoderm

cloacal membrane

5. Scanning electron microscopy

Photos 3.138 and 3.139 show scanning electron micrographs of dorsal views of the caudal half of 72-hour chick embryos. Identify the caudal border of the **cranial** and **lateral amniotic folds**, **somites** and **spinal cord** (beneath the skin ectoderm), **tail bud**, **tail fold of the body**, **leg buds**, and **tail**.

Photos 3.140 and 3.141 show scanning electron micrographs of transverse slices. Identify in Photo 3.140 the **spinal cord**, **somites**, **notochord**, **descending aorta**, **postcardinal vein**, **intersegmental vein**, **dorsal mesocardium**, **heart** with its **atria** and **ventricles** and **pericardial cavity**. Identify in Photo 3.141 the **spinal cord**, **notochord**, **dorsal aortae**, **peritoneal cavity**, somatopleuric component of the **lateral body folds**, **amniotic cavity**, **yolk sac**, **amnion**, **chorion**, and **lateral amniotic folds**.

A scanning electron micrograph of the cut surface of a block that has been sectioned midsagittally is shown in Photo 3.142. Identify the **cerebral hemisphere** of the **telencephalon**; **diencephalon**; **optic stalk**; **infundibulum**; **Rathke's pouch**; **mesencephalon**; **isthmus**; **metencephalon**; **myelencephalon**; **notochord**; **dorsal aorta**; **pharynx**; **mouth opening**; **stomodeum**; **amniotic cavity**; **amnion**; **mandibular process**; **thyroid rudiment**; **first, second, third**, and **fourth pharyngeal pouches**; **aortic sac**; **conotruncus**; **ventricle**; and **atrium**.

T. TERMS TO KNOW: 48-72 HOURS OF DEVELOPMENT

You should know the meaning of the following terms, which appeared in boldface in the preceding discussion of the embryology of the chick at the 48- and 72-hour stages.

acoustic ganglia	anterior cardinal vein	axons
acousticofacialis ganglia	anterior pituitary gland	blastoderm
allantois	aortic arches	boundary of the amniotic folds
allantois rudiment	aortic sac	brain
amnion	apical ectodermal ridge	caudal amniotic folds
amniotic cavity	atrium	caudal arteries
amniotic fluid	auditory (VIII) cranial nerves	caudal cardinal vein
amniotic folds	auditory placodes	caudal intestinal portal
anterior amniotic folds	auditory vesicles	caudal liver rudiment

cell bodies

cephalic flexure

cerebral hemispheres

cervical flexure

cheeks

chorioallantoic membrane

chorion

cloaca

cloacal membrane

closing plates

coelom

common cardinal veins

conotruncus

corneal epithelium

cranial amniotic folds

cranial cardinal vein

cranial flexure

cranial intestinal portal

cranial liver rudiment

cranial nerves

dendrites

dermatome

dermis

descending aorta

diencephalon

dorsal aortae

dorsal mesocardium

dorsal mesogaster

dorsal pancreatic rudiment

ductus venosus

duodenohepatic ligament

duodenum

end bud

endocardium

endoderm

endolymphatic ducts

epibranchial placodes

epiphysis

esophagus

extraembryonic coelom

extraembryonic membranes

facial (VII) cranial nerves

first aortic arches

first branchial arches

first branchial clefts

first branchial grooves

first closing plates

first pharyngeal pouches

floor plate

foregut

fourth aortic arches

fourth branchial arches

fourth branchial grooves

fourth closing plates

fourth pharyngeal pouches

gastrohepatic ligament

geniculate ganglia

glossopharyngeal (IX) cranial nerves

heart

hepatic sinusoids

hepatoduodenal ligament

hepatogastric ligament

hindgut

infundibulum

internal carotid arteries

intersegmental arteries

intersegmental veins

intersomitic furrows

intraembryonic coelom

isthmus

jugular ganglion

laryngotracheal groove

lateral amniotic folds

lateral body folds

lateral plate

leg buds

lens epithelium

lens fibers

lens vesicles

looping of the heart

lower jaw

lung buds

mandibular branch of the trigeminal (V) cranial nerve

mandibular processes

maxillary branch of the trigeminal (V) cranial nerve

maxillary processes

mesencephalon

mesoderm

mesoduodenum

mesoesophagus

mesonephric duct rudiments

mesonephric ducts

mesonephric kidneys

mesonephric tubule rudiments

mesonephric tubules

mesonephros

metencephalon

motor nerve fibers

motor nucleus

mouth opening

myelencephalon

myocardium

myotome

nasal cavities

nasal pits

nasal placodes

nephrotome

neural crest cells

neural ectodermal cells

neural tube

neuromeres

notochord

oculomotor (III) cranial nerve

olfactory placodes

ophthalmic branch of the trigeminal (V) cranial nerve

optic cups

optic fissures

optic nerve fibers

optic stalks

opticoel

oral membrane

otocysts

pericardial cavity

peritoneal cavity

petrosal ganglia

pharyngeal pouches

pharynx

pigmented retina

pineal gland

pleural cavities

plexus of head blood vessels

postcardinal vein

precardinal veins

preoral gut

prosencephalon

Rathke's pouch

rhombencephalon

ribs

roof

roof plate

rudiments of the peritoneal cavity

sclerotome

second aortic arches

second branchial arches

second branchial clefts

second branchial grooves

second closing plates

second pharyngeal pouches

Seessel's pouch

segmental plates

semilunar ganglia

sensory nerve fibers

sensory retina

sinus venosus

skeletal muscles

skin ectoderm

somatopleure

somites

spinal cord

spinal ganglia

spinal nerves

splanchnopleure

stomach

stomodeum

subcardinal veins

subcaudal pocket

subgerminal cavity

superior ganglia

tail

tail bud

tail flexure

tail fold of the body

tail gut

telencephalon

third aortic arches

third branchial arches

third branchial grooves

third closing plates

third pharyngeal pouches

thyroid rudiment

torsion

trigeminal (V) cranial nerves

upper jaw

vagus (X) cranial nerves

ventral mesogaster

ventricle of heart

vertebral column

vitelline arteries

vitelline blood vessels

vitelline veins

wing

wing buds

yolk sac

young motor neurons

young sensory neurons

U. PHOTOS 3.111-3.142: 72-HOUR CHICK EMBRYOS

Photos 3.111-3.142 depict the 72-hour chick embryos discussed in Chapter 3. These photos and their accompanying legends begin on the following page.

Photos 3.111-3.113

Chick Embryos

Legend

1. Ventricle
2. Cerebral hemisphere of telencephalon
3. Pineal gland
4. Diencephalon
5. Mesencephalon
6. Isthmus
7. Metencephalon
8. Myelencephalon
9. Endolymphatic duct
10. Auditory vesicle
11. Wing bud
12. Cervical flexure

13. Vitelline artery
14. Caudal intestinal portal
15. Boundary of the amniotic folds
16. Nasal pit
17. Semilunar ganglion
18. Acousticofacialis ganglion
19. Cranial intestinal portal
20. Tail flexure
21. First branchial groove
22. Roof plate of the myelencephalon
23. Third branchial arch

24. Atrium
25. Leg bud
26. Allantois
27. Tail
28. Maxillary process
29. Optic cup
30. Mandibular process
31. Lens vesicle
32. Second branchial arch
33. Second branchial groove
34. Third branchial groove
35. Conotruncus
36. Sinus venosus

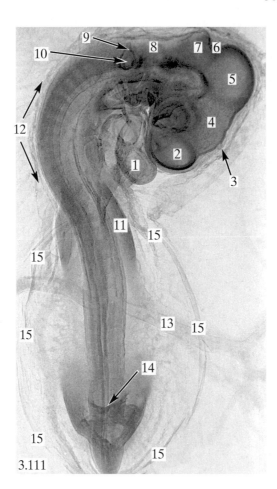

Photo 3.111. 72-hour chick embryo whole mount.

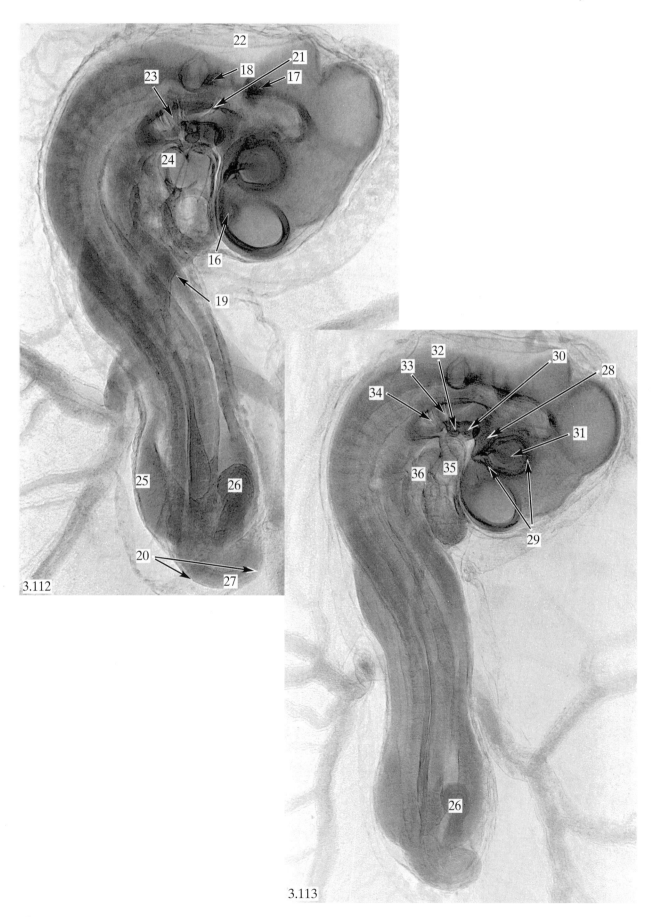

Photos 3.112, 3.113.   72-hour chick embryo whole mounts.

Photo 3.114
Chick Embryos

Legend

1. Second branchial arch
2. Mandibular process
3. Pharynx
4. Semilunar ganglion

5. Rathke's pouch
6. Infundibulum
7. Stomodeum

8. Lens vesicle
9. Diencephalon
10. Cerebral hemisphere of the telencephalon

Photo 3.114.   Facing page.   Enlargement of the head region of a 72-hour chick embryo whole mount.

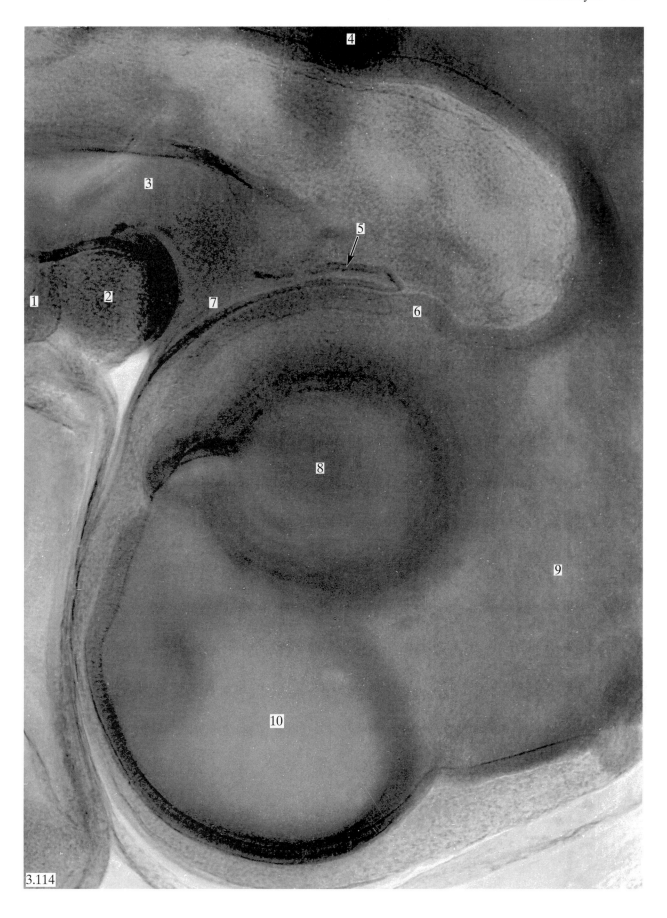

Photos 3.115-3.117

Chick Embryos

Legend

1. First aortic arch
2. Second aortic arch
3. Third aortic arch
4. Internal carotid artery
5. Dorsal aorta
6. Descending aorta
7. Vitelline artery
8. Fourth aortic arch

9. Aortic sac
10. Conotruncus
11. Ventricle
12. Yolk sac
13. Somite #1
14. Superior ganglion
15. Myelencephalon
16. Auditory vesicle

17. Acousticofacialis ganglion
18. Precardinal vein
19. Neuromeres
20. Amniotic cavity
21. Semilunar ganglion
22. Metencephalon
23. Mesencephalon

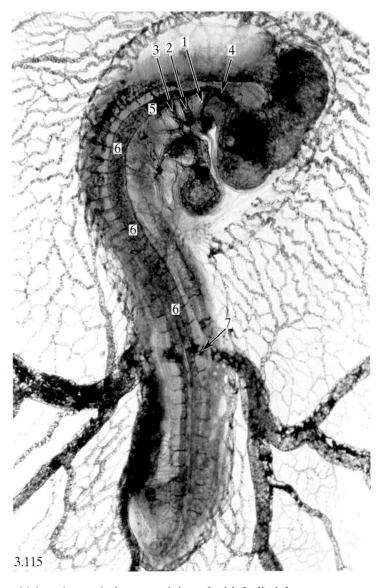

3.115

Photo 3.115. 72-hour chick embryo whole mount injected with India ink.

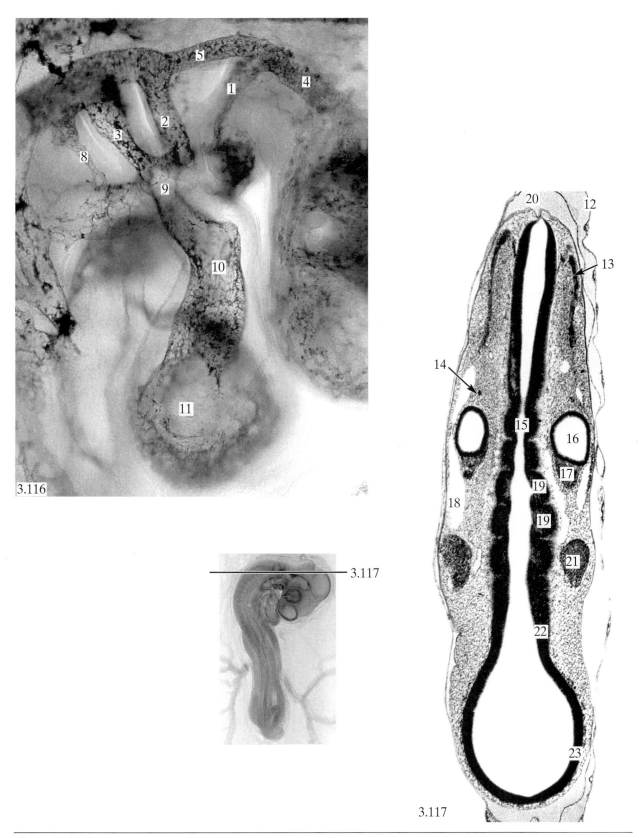

3.116

3.117

3.117

Photo 3.116. Enlargement of the heart and aortic arch region of the 72-hour chick embryo whole mount injected with India ink shown in Photo 3.115.

Photo 3.117. 72-hour chick embryo serial transverse section. This section is the most anterior one illustrated.

Photos 3.118-3.120

Chick Embryos

Legend

1. Amnion
2. Yolk sac
3. Amniotic cavity
4. Somite
5. Notochord
6. Precardinal vein
7. Glossopharyngeal nerve
8. Epibranchial placode (contributes cells to facialis part of the acousticofacialis ganglion)
9. Facial nerve
10. Trigeminal nerve—maxillary branch

11. Trigeminal nerve— mandibular branch
12. Trigeminal nerve— ophthalmic branch
13. Mesencephalon
14. Extraembryonic coelom
15. Skin ectoderm
16. Dorsal aorta
17. Second pharyngeal pouch
18. First pharyngeal pouch
19. First branchial cleft
20. Internal carotid artery

21. Third branchial arch
22. Jugular ganglion
23. Third aortic arch
24. Epibranchial placode (contributes cells to petrosal ganglion)
25. Second closing plate
26. Second aortic arch
27. Second branchial arch
28. Pharynx
29. First branchial groove
30. First closing plate
31. First aortic arch

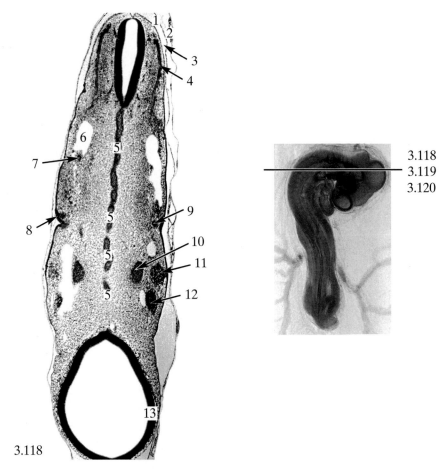

Photo 3.118. Continuation of 72-hour chick embryo serial transverse sections numbered in anterior to posterior sequence.

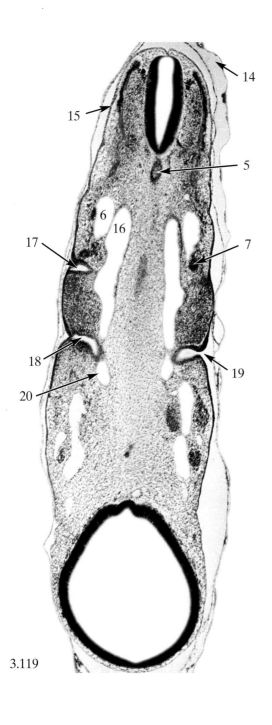

3.119

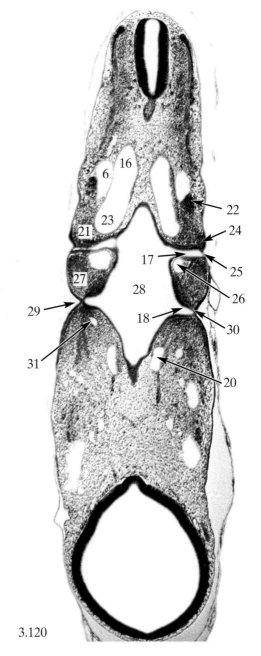

3.120

Photos 3.119, 3.120. Continuation of 72-hour chick embryo serial transverse sections numbered in anterior to posterior sequence.

Photos 3.121-3.123

Chick Embryos

Legend

1. Descending aorta
2. Dorsal aorta
3. Third pharyngeal pouch
4. Third aortic arch
5. Pharynx
6. Second pharyngeal pouch
7. Second branchial groove
8. Second aortic arch
9. First pharyngeal pouch
10. First aortic arch
11. Preoral gut
12. First branchial arch

13. Precardinal vein
14. Mesencephalon
15. Intersegmental artery
16. Fourth aortic arch
17. Third closing plate
18. Third branchial arch
19. Thyroid rudiment
20. Second branchial arch
21. Fourth branchial arch
22. Mandibular process
23. Maxillary process
24. Stomodeum

25. Rathke's pouch
26. Infundibulum
27. Spinal ganglion
28. Spinal cord
29. Notochord
30. Fourth closing plate
31. Fourth pharyngeal pouch
32. Vitelline blood vessel
33. Chorion
34. Amnion
35. Diencephalon

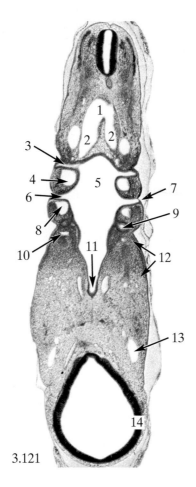

3.121

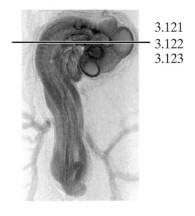

3.121
3.122
3.123

Photo 3.121. Continuation of 72-hour chick embryo serial transverse sections numbered in anterior to posterior sequence.

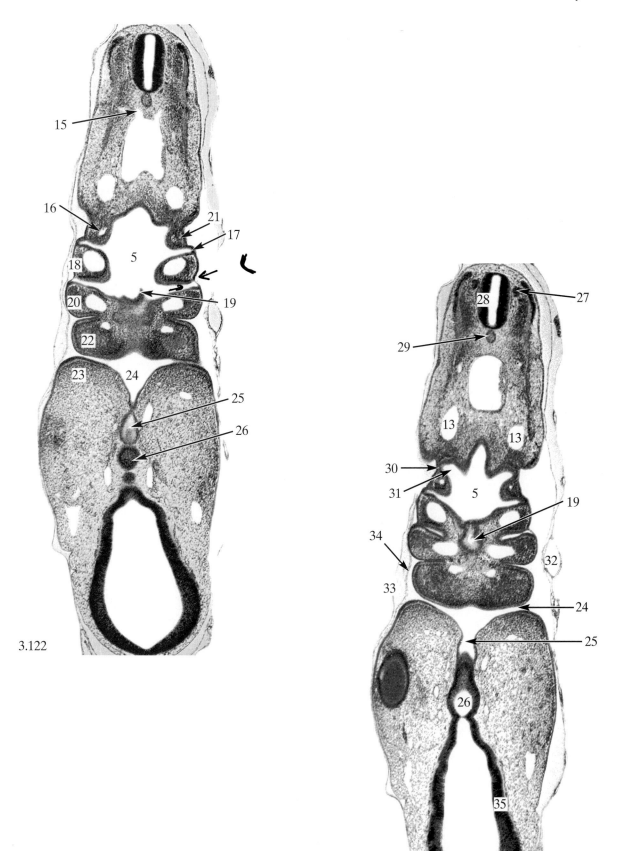

3.122

3.123

Photos 3.122, 3.123. Continuation of 72-hour chick embryo serial transverse sections numbered in anterior to posterior sequence.

Photos 3.124-3.126

Chick Embryos

Legend

1. Spinal cord
2. Postcardinal vein
3. Descending aorta
4. Common cardinal vein
5. Pharynx
6. Laryngotracheal groove
7. Aortic sac
8. Future corneal epithelium
9. Lens epithelium

10. Lens fibers
11. Ventral lip of the optic cup
12. Dorsal lip of the optic cup
13. Diencephalon
14. Spinal ganglion
15. Extraembryonic coelom
16. Atrium
17. Optic fissure (note the lack of a ventral lip of the optic cup)

18. Pigmented retina
19. Sensory retina
20. Head mesenchyme
21. Lung bud
22. Sinus venosus
23. Pericardial cavity
24. Myocardium of conotruncus
25. Endocardium of conotruncus
26. Optic stalk

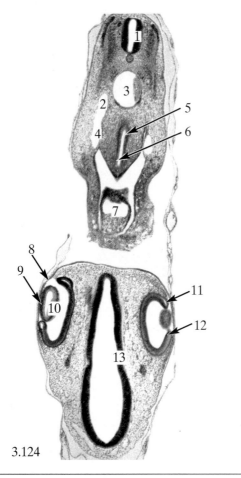

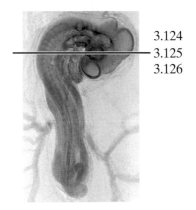

3.124

3.124
3.125
3.126

Photo 3.124. Continuation of 72-hour chick embryo serial transverse sections numbered in anterior to posterior sequence.

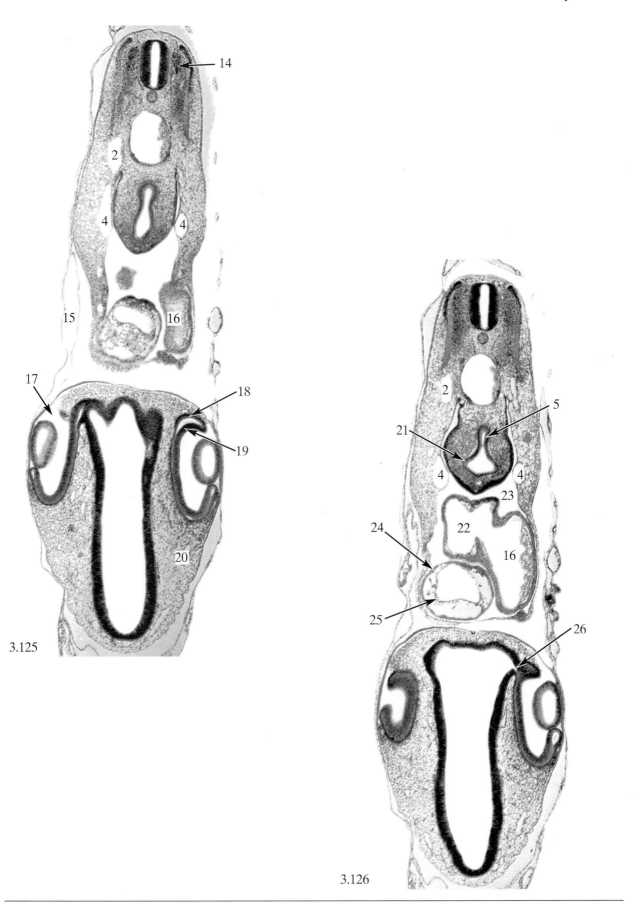

3.125

3.126

Photos 3.125, 3.126. Continuation of 72-hour chick embryo serial transverse sections numbered in anterior to posterior sequence.

Photos 3.127-3.129

Chick Embryos

Legend

1. Descending aorta
2. Postcardinal vein
3. Esophagus
4. Lung bud
5. Common cardinal vein
6. Sinus venosus
7. Atrium (continuous with the ventricle below)
8. Conotruncus
9. Diencephalon

10. Mesoesophagus
11. Pleural cavity
12. Dorsal mesocardium
13. Somite
14. Notochord
15. Pericardial cavity
16. Cerebral hemisphere of the telencephalon
17. Mesonephric duct (degenerating portion)

18. Dorsal mesogaster
19. Stomach
20. Hepatogastric ligament
21. Cranial liver rudiment
22. Ductus venosus
23. Endocardium of the ventricle
24. Myocardium of the ventricle
25. Nasal pit
26. Pineal gland

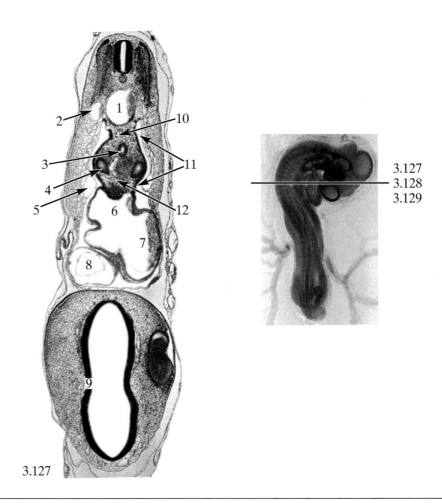

3.127

3.127
3.128
3.129

Photo 3.127. Continuation of 72-hour chick embryo serial transverse sections numbered in anterior to posterior sequence.

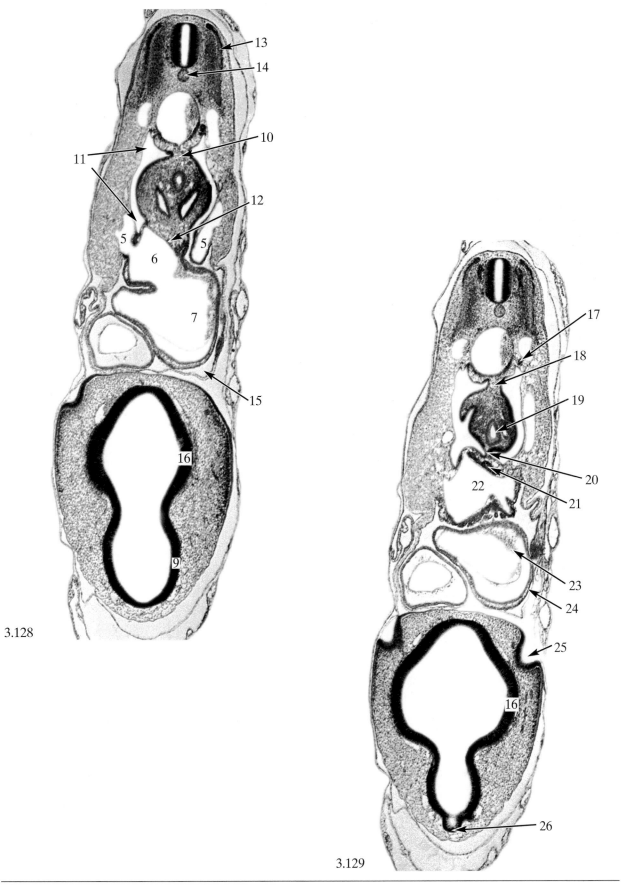

3.128

3.129

Photos 3.128, 3.129. Continuation of 72-hour chick embryo serial transverse sections numbered in anterior to posterior sequence.

Photos 3.130-3.132

Chick Embryos

Legend

1. Spinal ganglion
2. Postcardinal vein
3. Mesonephric duct (degenerating portion)
4. Mesoduodenum
5. Duodenum
6. Cranial liver rudiment
7. Peritoneal cavity
8. Ductus venosus

9. Caudal liver rudiment
10. Conotruncus
11. Ventricle
12. Cerebral hemisphere of the telencephalon
13. Hepatoduodenal ligament
14. Spinal cord
15. Yolk sac
16. Amnion

17. Descending aorta
18. Dorsal pancreatic rudiment
19. Vitelline vein
20. Chorion
21. Dermatome
22. Myotome
23. Sclerotome
24. Intersegmental artery
25. Cranial intestinal portal

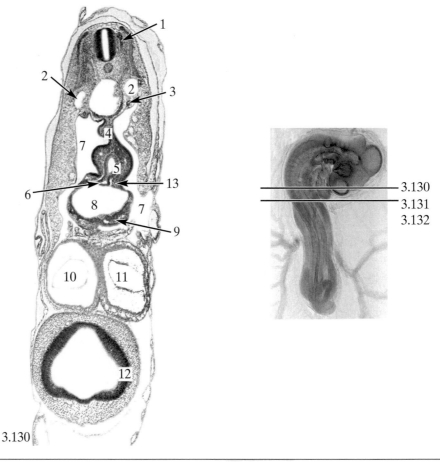

Photo 3.130. Continuation of 72-hour chick embryo serial transverse sections numbered in anterior to posterior sequence.

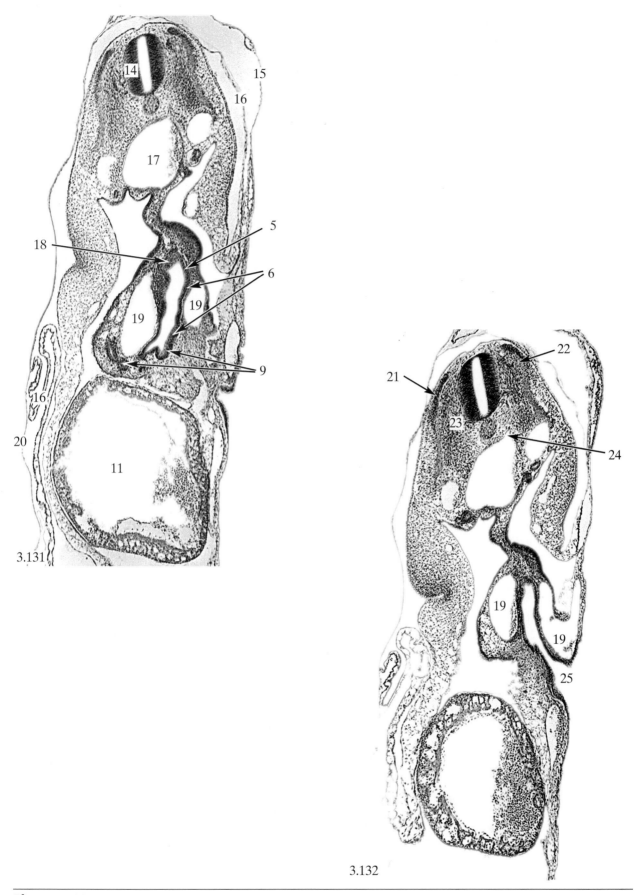

3.131

3.132

Photos 3.131, 3.132. Continuation of 72-hour chick embryo serial transverse sections numbered in anterior to posterior sequence.

Photos 3.133-3.136

Chick Embryos

Legend

1. Dermatome
2. Myotome
3. Sclerotome
4. Somatic mesoderm of the wing bud
5. Mesonephric duct
6. Mesonephric tubule
7. Mesonephric tubule rudiment
8. Descending aorta
9. Apical ectodermal ridge of the wing bud

10. Vitelline blood vessel
11. Postcardinal vein
12. Dorsal aortae
13. Somatic mesoderm of the leg bud
14. Chorion
15. Amnion
16. Amniotic cavity
17. Extraembryonic coelom
18. Yolk sac
19. Lateral amniotic fold

20. Peritoneal cavity
21. Hindgut
22. Cloaca
23. Apical ectodermal ridge of the leg bud
24. Splanchnic mesoderm of allantois
25. Endoderm of allantois
26. Cloacal membrane
27. Subcaudal pocket (amniotic cavity)

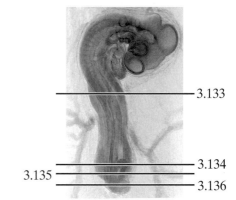

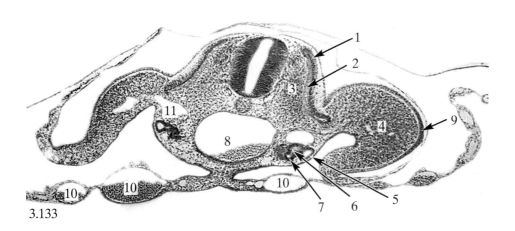

3.133

Photo 3.133. Continuation of 72-hour chick embryo serial transverse sections numbered in anterior to posterior sequence.

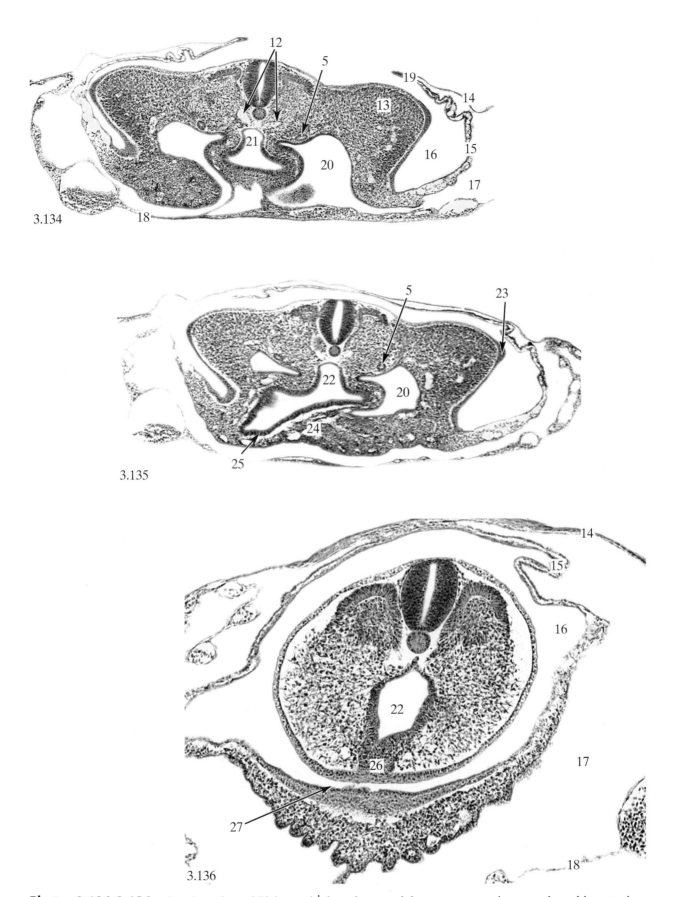

Photos 3.134-3.136. Continuation of 72-hour chick embryo serial transverse sections numbered in anterior to posterior sequence.

Photo 3.137

Chick Embryos

Legend

1. Isthmus
2. Myelencephalon
3. Notochord
4. Metencephalon
5. Mesencephalon

6. Preoral gut
7. Amniotic cavity
8. Rathke's pouch
9. Pharynx
10. Mouth opening

11. Stomodeum
12. Infundibulum
13. Diencephalon
14. Pineal gland
15. Telencephalon

Photo 3.137. Facing page. Enlargement of the head region of a midsagittal section of a 72-hour chick embryo.

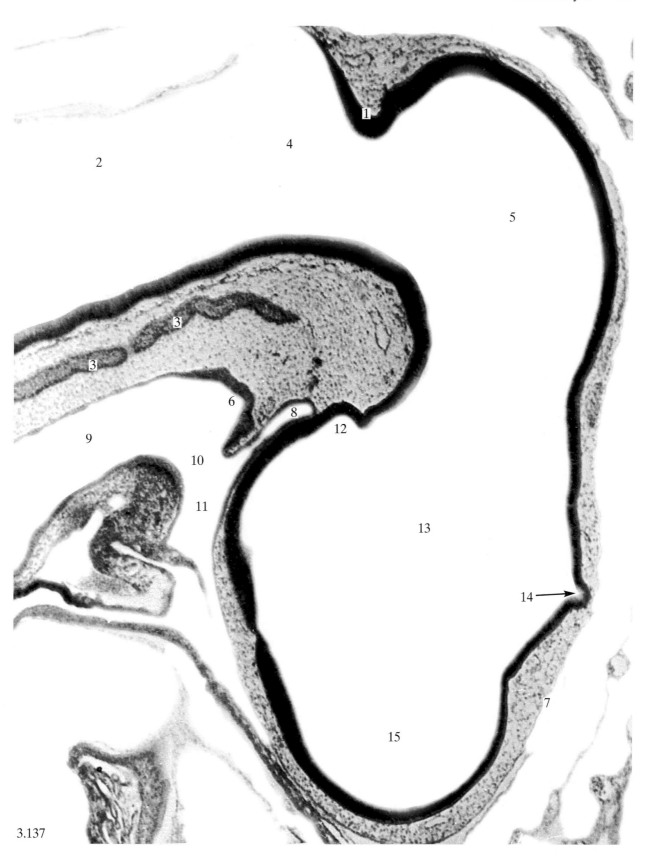

3.137

Photos 3.138-3.141
Chick Embryos
Legend

1. Amniotic folds (caudal boundary of cranial and lateral)
2. Somite
3. Spinal cord
4. Tail bud
5. Tail fold of the body
6. Leg bud
7. Tail

8. Notochord
9. Descending aorta
10. Postcardinal vein
11. Intersegmental vein
12. Dorsal mesocardium
13. Heart
14. Pericardial cavity
15. Head

16. Dorsal aorta
17. Peritoneal cavity
18. Lateral body fold (somatopleuric component)
19. Amniotic cavity
20. Yolk sac
21. Amnion
22. Chorion
23. Lateral amniotic fold

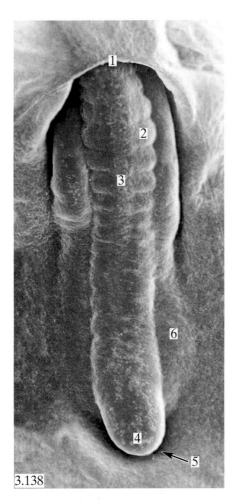

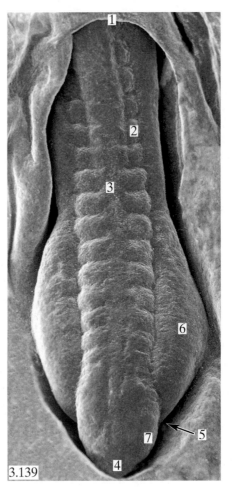

Photos 3.138, 3.139. Scanning electron micrographs of dorsal views of the caudal halves of 72-hour chick embryos.

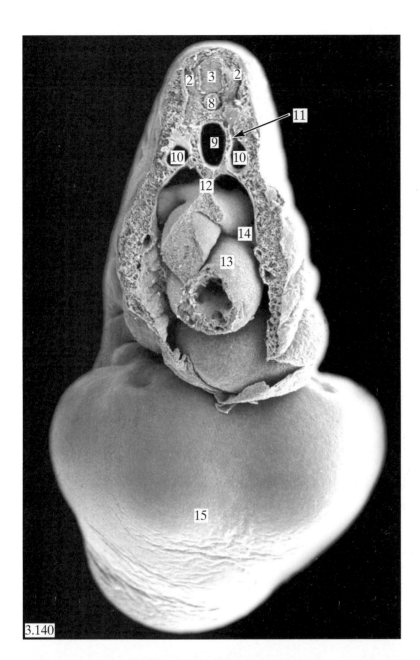

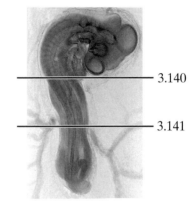

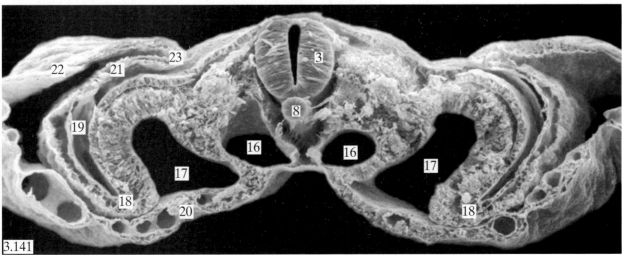

Photos 3.140, 3.141. Scanning electron micrographs of transverse slices of 72-hour chick embryos.

Photo 3.142

Chick Embryos

Legend

1. Cerebral hemisphere of the telencephalon
2. Diencephalon
3. Optic stalk
4. Infundibulum
5. Rathke's pouch
6. Mesencephalon
7. Isthmus

8. Wall of the metencephalon
9. Myelencephalon
10. Notochord
11. Dorsal aorta
12. Pharynx
13. Mouth opening
14. Stomodeum
15. Amniotic cavity

16. Amnion
17. Mandibular process
18. Thyroid rudiment
19. Pharyngeal pouches
20. Aortic sac
21. Conotruncus
22. Ventricle
23. Atrium

Photo 3.142.   Facing page.   Scanning electron micrograph of the cut surface of a block sectioned sagittally from a 72-hour chick embryo.

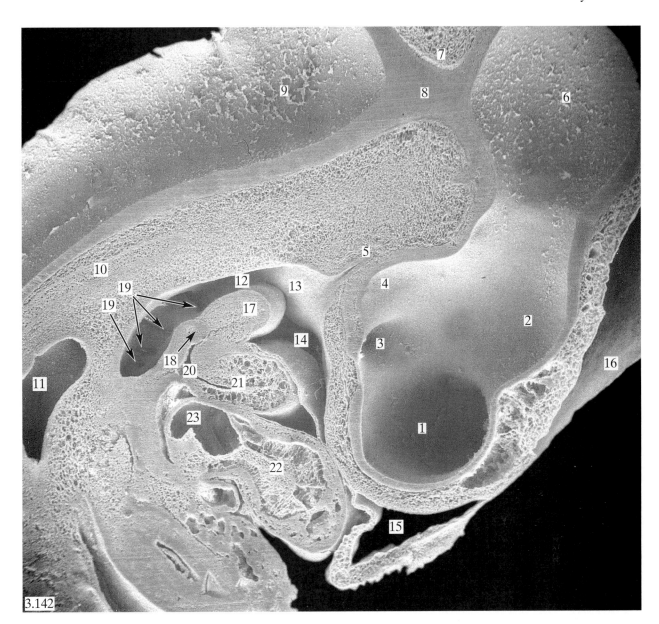

3.142

Chapter 4

Mouse Embryos

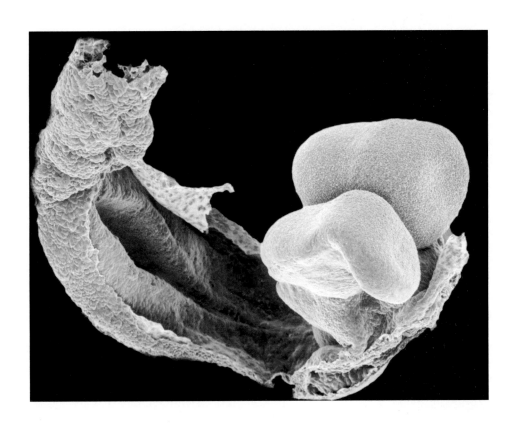

Chapter 4

Mouse Embryos

A. INTRODUCTION

The mouse embryo has become the favored mammalian model for research in developmental biology. A wealth of genetic information is available for the mouse, and several types of spontaneous mutants have been preserved for study. Recent advances in molecular genetics and molecular biology now make it possible to mutate essentially any mouse gene at will. By making **transgenic mice**, genes can be expressed ectopically (that is, they can be expressed at sites that they are not normally expressed at) to determine whether they are *sufficient* to cause a particular process to occur at the ectopic site. Also, through the process of **homologous recombination** (also known as **gene targeting**), genes can be inactivated or **knocked-out**, examining whether their function is *required* for a particular developmental process. Within the next few years, we expect that the entire mouse genome will be mapped and sequenced, further increasing the value of the mouse model.

The mouse has been avoided historically as a model for the teaching of developmental biology/embryology, because its early development is often said to occur inside-out, confusing students. What is meant by this, is that we typically think of the vertebrate embryo as having three **germ layers**, the **ectoderm**, the **mesoderm**, and the **endoderm**, with the ectoderm on the *outside*, the mesoderm in the *middle*, and the endoderm on the *inside*. In the early mouse embryo (that is at 6.5-8.5 days of development; total gestation period of 19 days), the blastoderm is cup-shaped rather than flat as in the chick blastoderm at a comparable stage (that is, during gastrulation and neurulation—typically during 18- to 33-hours of development of the chick). The **epiblast**/ectoderm forms the *inner* lining of the cup (and is, therefore, said to be on the *inside*), whereas the endoderm forms the *outer* lining of the cup (and is, therefore, said to be on the *outside*). As in the chick blastoderm, the ingressed mesoderm lies in the *middle*, between the epiblast/ectoderm and endoderm.

We think this difference between the chick and mouse blastoderm will not lead to extensive confusion if it is understood at the outset, and that the increasing importance of the mouse model to understanding human development far outweighs some initial confusion. To overcome this confusion, we suggest two things. First, study the early chick blastoderm (or review it if you have already studied it) prior to beginning your studies of the mouse blastoderm. Second, try the following exercise. Using three colors of modeling clay (preferably, red, blue, and yellow, with red representing mesoderm, blue representing epiblast/ectoderm, and yellow representing endoderm), make three circular pancakes, each about 4 inches in diameter, with each pancake consisting of one of the three colors of clay. Then, repeat this process making three more pancakes of red, blue, and yellow clay. To make a chick blastoderm, stack the colors in the following order: yellow on the bottom, red in the middle, and blue on the top. Set the stack aside. To make a mouse blastoderm, again stack the colors in the same order: yellow on the bottom, red in the middle, and blue on the top. Then shape your second stack into a cup, with the blue clay forming the *inner* lining of the cup, and the yellow clay forming its *outer* lining. We suggest that you keep the two models of the chick and mouse blastoderms nearby as you conduct your laboratory studies. Finally, you might find it useful during the course of your studies to cut the cup-shaped blastoderm into two half-cups. This will help you to understand the mouse blastoderm as you study sagittally sectioned gastrulae.

Our purpose in this chapter is not to provide a detailed description of mouse embryogenesis. The mouse embryo develops similarly to that of the chick, and later stages of **organogenesis** in the mouse are almost identical to that of the pig (Chapter 5), which are covered in detail. Rather, our purpose in this chapter is to focus on those stages in which the mouse embryo is said to be *inverted*. Our goal is to help you to understand these stages that are characteristic of mouse development, so that your understanding of experiments on this important stage in mouse development will be enhanced. As you study the material below, examine living embryos when available (Hands-On Studies, E. Exercises 4.1, 4.2) and the cited figures and photos.

B. OVERVIEW OF EARLY DEVELOPMENT

As in other vertebrates, development of the mouse begins with the process of **gametogenesis**. **Egg** and **sperm** are produced, respectively, within the **ovaries** of the female mouse and the **testes** of the male mouse. Female and male mice begin producing functional gametes at **sexual maturity** (reached at about 6 weeks of age). Within the ovary, as well as after **ovulation**, eggs (that is, **primary** and **secondary oocytes**) are surrounded by an acellular membrane, called the **zona pellucida**, which encloses not only the egg but also the **polar bodies** formed during the **meiotic divisions** accompanying **oogenesis**. The zona pellucida is secreted around the egg by the adjacent cells of the ovary. Sperm are produced during **spermatogenesis**. Each mature sperm has a prominent hook-like **head** and a long straight **tail** (Photo. 4.1).

Sexually mature female mice have an **estrus cycle.** Each cycle lasts 4-5 days and culminates in the **ovulation** of several eggs (that is, **secondary oocytes**) from the paired ovaries. Ovulated eggs, enclosed in their zona pellucida, enter the **oviduct** where **fertilization** occurs. As a result of fertilization, a **zygote** forms (Photo 4.2).

Cleavage is initiated following fertilization, leading to the formation of the **two-cell stage** (Photo 4.3), **four-cell stage** (Photo 4.4), **eight-cell stage** (Photo 4.5), and then the **morula stage** (Photo 4.6). Cells (that is, **blastomeres**) become tightly adherent to one another at the morula stage, a process called **compaction**. Compaction is followed by **cavitation,** a process that results in the formation of a **blastocyst**, consisting of an **inner cell mass** and an outer **trophoblast** (Photo 4.7). The cavity in the blastocyst formed by cavitation is called the **blastocoel**. These events from the zygote stage to the blastocyst stage in the mouse are schematized in Fig. 4.1a-e.

The blastocyst stage forms at 4 days of development, that is, the fourth day after fertilization. Between 4 and 6.5 days of development, the blastocyst undergoes a rapid elongation to form the **egg cylinder** (Fig. 4.1f-h; Photo 4.8). As this process is occurring, the blastocyst enters the **uterine cavity** and undergoes **implantation**, developing an intimate association with the **uterine wall** that is essential for the embryo's survival.

The **egg cylinder** contains both an **extraembryonic** and **embryonic part** (Photo 4.8; note that this embryo, and those shown in subsequent photos, have been dissected from the uterus; also part of the **yolk sac**, called the *parietal* yolk sac, has been removed). The extraembryonic part of the egg cylinder contains the rudiments of two **extraembryonic membranes**: the **chorion** and **allantois**. In addition, the **ectoplacental cone**, another extraembryonic structure that contributes to the **placenta**, extends from the extraembryonic part of the egg cylinder. The embryonic part of the egg cylinder contains the

blastoderm, inverted into a cup-like structure containing **epiblast/ectoderm** on the inside, **endoderm** on the outside, and **mesoderm** in between (the latter two layers are formed during gastrulation; see next section). A partition, called the **amnion**, separates the extraembryonic and embryonic parts of the egg cylinder from one another. The space below the amnion bounded by the inside of the cup-like blastoderm is called the **amniotic cavity**. At the stage illustrated in Photo 4.8, the space contained within the extraembryonic portion of the egg cylinder is called the **chorionic cavity**.

The **axes** of the blastoderm (primitive body) can be identified by 7 days of development (Photo 4.8). The *proximal* level of the blastoderm is located adjacent to the extraembryonic part of the egg cylinder. The *distal* level is located at the apex of the cup. The *rostral* end of the blastoderm is marked by a distinct depression, and the *caudal* end of the blastoderm lies at the opposite side. The **node** or **organizer** lies near the distal level of the blastoderm, displaced somewhat rostrally.

C. GASTRULATION

The embryonic part of the egg cylinder is initially composed of two layers: an upper **epiblast** and lower **hypoblast** (also called the *visceral* yolk sac). As a result of **gastrulation**, the initially two-layered blastoderm is converted into a three-layered blastoderm consisting of **ectoderm, mesoderm**, and **endoderm**. Gastrulation begins at about 6 days of development with the formation of the **primitive streak**. Epiblast cells move toward and through the primitive streak to **ingress** into the interior of the blastoderm. Ingressing cells form two populations: **endoderm**, which displaces the hypoblast away from the developing embryo toward extraembryonic regions, and **mesoderm**. The cells that remain within the epiblast instead of ingressing eventually form the **ectoderm**. The primitive streak forms mainly within the caudal half of the blastoderm. The ectoderm that spans the midline within the rostral half of the blastoderm thickens to form the **neural plate**.

Examine Fig. 4.2, which shows **prospective fate maps** of the mouse blastoderm. At the **intermediate primitive streak stage** (Fig. 4.2a), the **prospective notochord** is contained within the **node (organizer)**. **Prospective extraembryonic mesoderm** has ingressed into the interior of the blastoderm by this stage, and the **extraembryonic mesoderm** is expanding into the **extraembryonic part of the egg cylinder**. **Prospective heart** and **prospective head mesoderm** occupy the primitive streak and they are ingressing into the interior to form the **heart mesoderm** and **head mesoderm**. **Prospective somitic mesoderm** and **prospective lateral plate mesoderm** occupy the epiblast and primitive streak, and they have not yet started their ingression into the interior. At the definitive primitive streak stage (Fig. 4.2b), the **prospective**

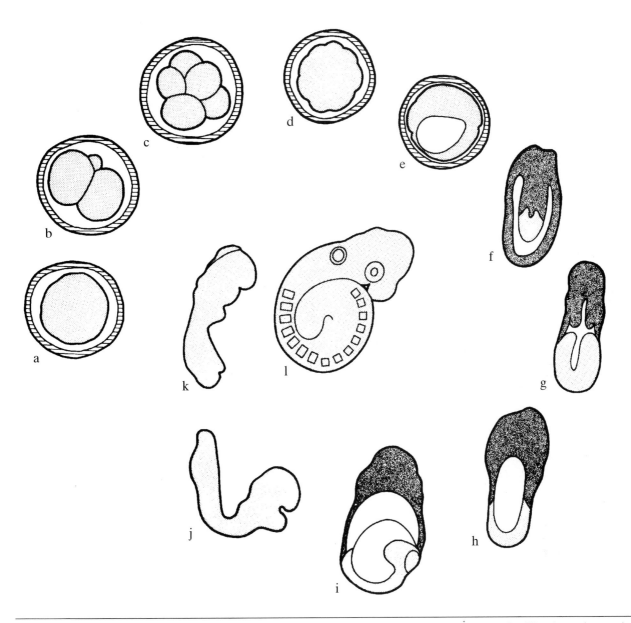

Fig. 4.1. Overview of the early stages of mouse development covered in this chapter. a, fertilized egg (zygote). The zygote through the blastocyst stage (a-e) are enclosed within the zona pellucida (horizontal lines). The blastocyst hatches from the zona pellucida within the uterus, undergoes implantation, and elongatesas the egg cylinder (f-h). b, c, cleavage stages (two-cell stage showing polar body and eight-cell stage). d, morula stage. e, blastocyst stage. f, g, elongating egg cylinder stages during implantation. h, 6.5- to 7-day embryo. i, j, 8.5-day embryo. k, 9-day embryo. l, 9.5- to 10-day embryo. j-l show embryos removed from their extraembryonic membranes, illustrating the process of turning.

notochord is still contained within the **node** (**organizer**). The **extraembryonic mesoderm** is now completely contained within the **extraembryonic part of the egg cylinder**, and the **heart mesoderm** and **head mesoderm** are moving toward the **rostral** and **proximal level of the embryonic part of the egg cylinder**. Prospective somitic **mesoderm** and **prospective lateral plate mesoderm** are ingressing into the interior of the blastoderm to form the **somites** and **lateral plate mesoderm**. Finally, at the head process stage (Fig. 4.2c), the **prospective notochord** is ingressing into the interior of the blastoderm to form the **notochord**, and the **prospective somitic mesoderm** and **lateral plate mesoderm** are continuing to ingress into the interior of the blastoderm to form, respectively, the **somites** and **lateral plate mesoderm**. The mouse fate map (Fig. 4.2) should be carefully compared with the ones for the chick (Fig. 3.7) and sea urchin (Fig. 1.1). Also, for a comparison of gastrulation in vertebrates and invertebrates, see Chapter 1, Section H.

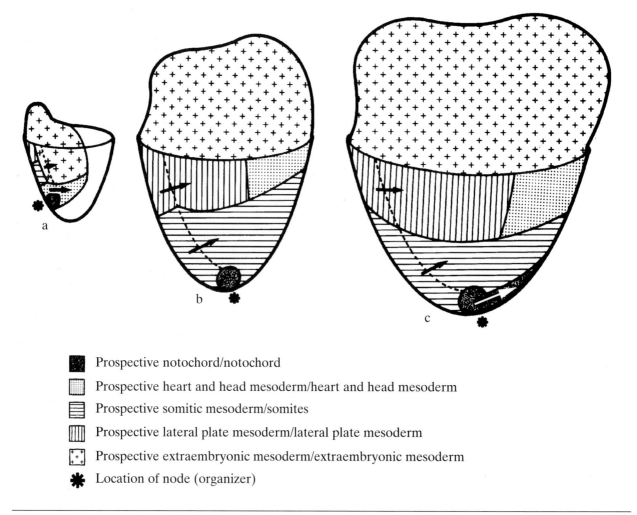

☒ Prospective notochord/notochord

▦ Prospective heart and head mesoderm/heart and head mesoderm

▤ Prospective somitic mesoderm/somites

▥ Prospective lateral plate mesoderm/lateral plate mesoderm

▦ Prospective extraembryonic mesoderm/extraembryonic mesoderm

✳ Location of node (organizer)

Fig. 4.2. Prospective fate map of the mouse gastrula at the intermediate primitive streak stage (a; 6.5 days of development), definitive primitive streak stage (b; 7.5 days of development), and head process stage (c; 8.5 days of development). The egg cylinder is viewed from its right-lateral side and the endodermal wall of the right side of the embryonic part of the egg cylinder is shown as transparent, revealing the mesoderm that has ingressed from the epiblast, through the primitive streak, and into the interior of the blastoderm. With the exception of the prospective notochord, prospective mesodermal cells ingress bilaterally as in the chick gastrula, forming right and left ingressed mesodermal wings. Only the right ingressed wing is shown at each stage. The wall of the extraembryonic part of the egg cylinder is not shown. The internal edge of the primitive streak is indicated by the dashed line at each stage. Mesodermal cells within the primitive streak are referred to as "prospective," with the same symbols marking identical mesodermal subdivisions both prior to and after ingression. Based chiefly on the data of Patrick Tam and co-workers.

Examine transverse sections of mouse embryos at 7 days of gestation, if available, as well as Photos 4.9-4.13. Examine sections in proximal to distal sequence and identify the wall of the **egg cylinder**, **chorion**, **allantois**, and **chorionic cavity** within the **extraembryonic part of the egg cylinder** (Photo 4.9), and the **neural plate, mesoderm, endoderm, epiblast, primitive streak,** and **amniotic cavity** within the **embryonic part of the egg cylinder** (Photos 4.10-4.13).

Also examine the scanning electron micrographs shown in Photos 4.14-4.19. In the whole embryo shown in Photo 4.14, identify the **extraembryonic** and **embryon-**ic parts of the egg cylinder, caudal and **rostral end of the embryo, ectoplacental cone, endoderm,** and **node (organizer)**. In the parasagittally sliced embryo shown in Photo 4.15, identify the **endoderm, chorion, chorionic cavity, amnion, amniotic cavity, neural plate, primitive streak, proximal** and **distal levels of the embryonic part of the egg cylinder,** and **intraembryonic** and **extraembryonic mesoderm**. In the dissected embryo shown in Photo 4.16, identify the **endoderm, neural plate, intraembryonic** and **extraembryonic mesoderm,** and **epiblast**. In the transversely sliced embryo shown in Photos 4.14, 4.18, identify the **endoderm, amniotic cavity, neural** plate, **primitive streak**, **intraembryonic mesoderm**, and **epiblast**. Finally, in the midsagittally sliced embryo shown in Photo 4.19, identify the **chorionic cavity**, **amnion**, **allantois**, **endoderm**, **amniotic cavity**, **primitive streak**, **neural plate**, and **node (organizer)**.

D. NEURULATION, CARDIOGEN-SIS, SOMITOGENESIS, AND BODY FOLDING

Between 7 and 7.5-days of gestation, four important events occur: **neurulation**, the formation of the **neural tube**; **cardiogenesis**, the formation of the **heart**; **somitogenesis**, the formation of the **somites**; and **body folding**, the process that sculpts the essentially two-dimensional blastoderm into a three-dimensional body. Although these four events occur simultaneously, we will discuss them separately for the sake of simplicity.

Neurulation involves the formation of an ectodermal thickening called the **neural plate**. The neural plate begins to fold along its midline, establishing the **neural groove**, and laterally at the interface between the neural plate and adjacent **skin ectoderm**, a **neural fold** forms on each side of the neural groove. The neural folds elevate toward the dorsal midline where they fuse to establish the **neural tube**, covered dorsally by skin ectoderm. As the neural folds are undergoing elevation, cells within each neural fold undergo an **epithelial-to-mesenchymal transformation** and migrate away from the folds to form the **neural crest cells**.

Cardiogenesis involves the formation of two tubes formed from **endothelial cells**. These tubes constitute the paired rudiments of the **heart**; they fuse beneath the forming **foregut** to form a single heart tube.

Somitogenesis involves the segmentation of the ingressed **presomitic mesoderm** into discrete units called **somites**. Segmentation is coordinated on the right and left sides of the axis, so that somite pairs form at each level in unison.

Body folding involves the formation of the **body folds**: the single **head fold of the body**, the paired **lateral body folds**, and the single **tail fold of the body**. The head fold of the body establishes the free **head** and creates two spaces: an ectodermally lined space called the **subcephalic pocket**, which is continuous with the **amniotic cavity**; and an endodermally lined space called the **foregut**, which is open caudally as the **caudal intestinal portal**. The lateral body folds bring together in the ventral midline the paired rudiments of the **heart**, establishing a single heart tube. Additionally, the lateral body folds establish the endodermally lined **midgut**, eventually separating the gut from the **yolk sac**. The tail fold of the body establishes the free **tail** and creates two spaces: an ectodermally lined space called the **subcau-**

dal pocket, which is continuous with the **amniotic cavity**; and an endodermally lined space called the **hindgut**, which is open cranially as the **caudal intestinal portal**.

Examine the scanning electron micrographs shown in Photos 4.19-4.24 to begin your study of the most of the events just discussed. In regards to **neurulation**, identify the **neural plate** (Photos 4.19-4.24), **neural folds** (Photos 4.20-4.22. 4.24), and **neural groove** (Photos 4.23, 4.24). In regards to **cardiogenesis**, identify the incipient heart (Photo 4.22). In regards to **body folding**, identify the **head fold of the body** and **foregut**, with its opening, the **cranial intestinal portal** (Photos 4.20-4.22). Also identify the **amnion**, **allantois**, **endoderm**, **amniotic cavity**, **primitive streak**, and **node (organizer)** (Photo 4.20); **chorionic cavity**, **amnion**, **allantois**, **endoderm**, **amniotic cavity**, **primitive streak**, and **node (organizer)** (Photo 4.21); **amnion**, **endoderm**, **amniotic cavity**, and **primitive streak** (Photo 4.22); **endoderm**, **primitive streak**, **node (organizer)**, **epiblast**, and **mesoderm** (Photo 4.23); and **amnion**, **primitive streak**, and **node (organizer)** (Photo 4.24).

Photos 4.25 and 4.26 reveal the early **heart** in the 8.5 day mouse embryo. Also identify in these photos the **ectoplacental cone**, **chorion**, **chorionic cavity**, **amnion**, **amniotic cavity**, **head**, **trunk**, **primitive streak**, and **allantois** (Photo 4.25); and **neural plate** of the **forebrain**, and **cranial intestinal portal** (Photo 4.26).

Examine transverse sections of mouse embryos at 8.5 days of gestation, if available, as well as Photos 4.27-4.37. Examine sections in proximal to distal sequence and identify the **amnion**, **amniotic cavity**, **primitive streak**, **neural plate** of the **forebrain** and **midbrain**, **hindgut**, **head mesenchyme**, **yolk sac**, **extraembryonic coelom**, **dorsal aortae**, **neural crest cells**, **neural groove** of the **forebrain**, **midbrain**, and **caudal spinal cord**, **neural folds** of the **forebrain**, **midbrain**, and **caudal spinal cord**, **optic sulcus (cup)**, **foregut**, and **first aortic arches** (Photos 4.27-4.29); **yolk sac**, **extraembryonic coelom**, **amnion**, **amniotic cavity**, **primitive streak**, **lateral body folds**, **skin ectoderm**, **somatic and splanchnic mesoderm**, **endoderm**, **neural groove** of the **spinal cord**, **midbrain**, **forebrain**, and **hindbrain**, **neural crest cells**, **foregut**, **notochord**, **ventral aortae**, **dorsal aortae**, **conotruncus (bulbus cordis)**, **ventricle**, and **sinoatrial region** of the **heart**, **pericardial cavity**, and **somites** (Photos 4.30-4.33); and **notochord**, **endoderm**, **somites**, **splanchnic and somatic mesoderm**, **intraembryonic** and **extraembryonic coeloms**, **amniotic cavity**, **amnion**, **neural groove** of the **spinal cord** level, **skin ectoderm**, **yolk sac**, open **midgut**, and **neural tube** of the **spinal cord**.

Examine the scanning electron micrographs shown in Photos 4.38-4.4.41 to continue your study of the 8.5-day mouse embryo. Identify the **allantois**, **endoderm** of the **yolk sac**, **skin ectoderm**, **neural groove**, **neural folds** of the **spinal cord**, **forebrain**, and **midbrain** levels, **neural plate**, and **optic sulcus (cup)** (Photo 4.38); **skin ectoderm**, neural folds of the forebrain and midbrain levels, optic sulcus (cup), and heart (Photo 4.39); skin ectoderm, neural tube of the spinal cord level, notochord, dorsal aortae, and endoderm (Photo 4.40); and skin ectoderm, neural groove, neural folds of the spinal cord level, notochord, dorsal aortae, endoderm, presomitic mesoderm, nephrotome, and lateral plate mesoderm (Photo 4.41).

To complete your study of the 8.5-day mouse embryo, examine Photos 4.42 and 4.43, which show changes that occur in the heart from its initial formation (Photos 4.25, 4.26) through 9 days of gestation. Note that the heart tube has become s-shaped through the process of looping. Also identify the yolk sac, amnion, neural folds of the forebrain level, and cranial intestinal portal (Photo 4.42); and neural folds of the forebrain, midbrain, and hindbrain levels, cranial intestinal portal, optic sulcus (cup), and neural groove (Photo 4.43).

E. THE TUBE-WITHIN-A-TUBE BODY PLAN

As described above, the early mouse embryo develops from an inverted, U-shaped blastoderm. Between 8.5 and 9.5 days of development, the inverted blastoderm undergoes a process called turning in which the developing body of the embryo transforms from U-shaped to C-shaped (see Fig. 4.1i-l). The process of turning is illustrated in whole mounts of living embryos in Photos 4.44-4.46, and in scanning electron micrographs in Photos 4.47-4.51.

The four main events discussed above in the previous section—neurulation, cardiogenesis, somitogenesis, and body folding—as well as turning, act collectively to established what is know as the tube-within-a-tube body plan, a body organization that is typical of vertebrate embryos. The tube-within-a-tube body plan consists of an outer tube, composed of ectoderm, part of which folds into the interior of the embryo to form the neural tube during neurulation, and part of which remains on the surface of the embryo as the skin ectoderm. Additionally, part of the skin ectoderm also folds into the interior, for example, as the auditory vesicles and lens vesicles. The outer ectodermal tube of the tube-within-a-tube body plan is separated from the ectoderm contributing to the extraembryonic membranes, the chorion and amnion, by the action of the body folds. The inner tube of the tube-within-a-tube body plan is composed of endoderm. Due to the action of the body folds, part of this endoderm is internalized within the embryo as the gut (foregut, midgut, and hindgut), and the remainder contributes to two other extraembryonic membranes: the yolk sac and allantois. Thus due to the action of the lateral body folds the embryo is separated from its extraembryonic membranes, a cylindrical body shape is established, and inner and outer tubes

are formed. The space between the inner and outer tubes is initially filled with mesoderm formed during gastrulation. This mesoderm undergoes regional subdivision within the trunk to form the notochord, somites, nephrotome, and lateral plate mesoderm. Within the head, the mesoderm become supplemented with ectodermal cells derived from the neural crest, with the two populations of cells forming the head mesenchyme. The head mesenchyme undergoes segmentation, forming the cores of the branchial arches, a series of paired structures that flank the pharynx portion of the foregut and are covered internally with endoderm and externally with ectoderm. Between adjacent pairs of branchial arches, the ectoderm invaginates to form the branchial grooves and the endoderm of the pharynx evaginates to form the pharyngeal pouches.

Return to Photos 4.44-4.51. In addition to examining the change in shape that occurs in the body of the embryo during turning, identify the following structures: neural folds of the forebrain, midbrain, and hindbrain levels, heart, cranial intestinal portal, head, trunk, allantois, somites, eyes, forebrain, midbrain, hindbrain, auditory vesicles, spinal cord, and tail (Photos 4.44-4.46); neural folds of the forebrain, midbrain, hindbrain levels, skin ectoderm, spinal cord, neural groove of the spinal cord and brain levels, allantois, head, trunk, tail, hindleg and foreleg buds, branchial arches and grooves, somites, caudal neuropore, spinal cord, neural crest cells, lateral plate mesoderm, stomodeum, neural tube of the forebrain, midbrain, hindbrain, trunk, and tail levels, and eyes (Photos 4.47-4.51).

Continue your study of the C-shaped mouse embryo by examining the whole mount shown in Photo 4.52. Identify the prosencephalon, eyes, mesencephalon, isthmus, metencephalon, myelencephalon, auditory vesicle, spinal cord, branchial grooves, somites, tail, mandibular processes of the first branchial arches, stomodeum, and infundibulum.

Next, study transverse sections of the C-shaped mouse embryo (Photos 4.53-4.64). Examine sections in anterior to posterior sequence. The first sections cut through the mesencephalon, isthmus, and metencephalon (Photos 4.53, 4.54). Also identify the head mesenchyme and skin ectoderm in these sections. In more caudal sections, the prosencephalon (telencephalon and diencephalon) is encountered (Photos 4.55-4.57). Also identify in these sections the metencephalon, optic vesicles, infundibulum (future posterior pituitary gland), skin ectoderm, and head mesenchyme. As sections are traced posteriorly, the metencephalon is replaced by the myelencephalon, the level of the brain identifiable by its thin roof plate (Photos 4.56-4.59). At about these levels, also identify the precardinal veins and neural crest cells, leaving the dorsal part of the myelencephalon. At about the level where the prosencephalon begins to fade out, identify the mandibular processes of the first branchial arches, stomodeum, and telencephalon

(Photos 4.57, 4.58). Note the presence of the **semilunar ganglia of the trigeminal (V) cranial nerves** alongside the **myelencephalon** (these ganglia receive major contribution from **neural crest cells**), as well as the **notochord** beneath the **myelencephalon** (Photo 4.58).

Continue tracing the myelencephalon caudally (that is, continue to trace sections posteriorly) until the **auditory vesicles** are encountered (Photo 4.59). At this level, identify the **notochord, dorsal aortae, pharynx, ventricle** of the **heart, second branchial grooves**, and **second pharyngeal pouches**.

More caudally, the **spinal cord** appears in place of the myelencephalon (Photos 4.60-4.64). As you trace sections posteriorly (Photos 4.60, 4.61), identify the **laryngotracheal groove, lung buds, pleural cavity, peritoneal cavity, somites, descending aorta, midgut, nephrotome**, and **somatic** and **splanchnic mesoderm**.

More caudally, toward the **tail**, the **hindleg buds** are encountered (Photo 4.62). Identify at this level the **spinal cord, dermomyotome** and **sclerotome** of the **somites, notochord, dorsal aortae, hindgut**, and **peritoneal cavity**. After examining the section through the hindleg buds, examine the section through the **foreleg buds** (Photo 4.63). Identify in this section the **spinal cord, dermomyotome** and **sclerotome** of the **somites, descending aorta**, and **apical ectodermal ridges** of the **foreleg buds**.

Finally, examine the section at which the **spinal cord** is cut frontally (Photo 4.64). Identify the **spinal cord** and **dermomyotome** and **sclerotome** of the **somites**.

Complete your study of the mouse embryo by examining scanning electron micrographs (Photos 4.65-4.69). In the whole mount shown in Photo 4.65, identify the **neural folds** of the **midbrain** and **hindbrain** levels, **skin ectoderm**, and **auditory pits**. In the transverse slices shown in Photos 4.66, 4.67, identify the **skin ectoderm, hindbrain, head mesenchyme, precardinal veins, dorsal aortae, notochord, third aortic arches, aortic sac, spinal cord, descending aorta, foreleg buds, peritoneal cavity**, and **amnion**. In the sagittal slices shown in Photos 4.68, 4.69, identify the **telencephalon** and **diencephalon** regions of the **prosencephalon, mesencephalon, metencephalon, myelencephalon, spinal cord, optic sulcus (cup), stomodeum, mouth opening, pharynx, pharyngeal pouches, infundibulum**, and **ventricle** and **conotruncus** of **heart**.

F. COMPARISON OF EARLY AVIAN AND MAMMALIAN DEVELOPMENT

The early development of chick and mouse embryos occurs similarly. Both types of embryos develop from a two-layered **blastoderm**, consisting of **epiblast** and **hypoblast**, which becomes three layered, consisting of **ectoderm, mesoderm**, and **endoderm**, with the formation of the **primitive streak** during the process of **gastrulation**. The **germ layers** form similarly in the two organisms during gastrulation, and **neurulation, cardiogenesis, somitogenesis**, and **body folding** are essentially identical. The **tube-within-a-tube body plan** of both chick and mouse embryos again is essentially identical. Early chick and mouse embryos differ from one another in only one significant way: the mouse blastoderm is cup-shaped, whereas the chick blastoderm is flat. Return to the clay models you constructed earlier, as described in the Introduction to this chapter. Again, compare the models for the chick and mouse, and note the similarities in the position of the three **germ layers** (cut sections through the clay models if not yet done so to help you visualize the order of the germ layers): the **ectoderm, mesoderm**, and **endoderm** have the same order regardless of whether the blastoderm is cup-shaped or flat. Thus the early development of chick and mouse embryos exhibits far more similarities than differences.

G. SUMMARY OF THE CONTRIBUTIONS OF THE GERM LAYERS TO THE STRUCTURES PRESENT IN THE EARLY MOUSE EMBRYO

Ectoderm

auditory pits and vesicles

branchial grooves

cranial ganglia

> V. semilunar

infundibulum

lens vesicles

neural crest cells

neural folds and groove

neural plate and tube

optic sulcus (cup) and vesicles

stomodeum

subcaudal pocket

subcephalic pocket

Mesoderm

aortic arches

aortic sac

dermomyotomes

descending aorta

dorsal aortae

heart

lateral plate mesoderm

Mesoderm continued

- nephrotomes
- notochord
- precardinal veins
- sclerotome
- somites
- ventral aortae

Endoderm

- foregut
- hindgut
- laryngotracheal groove
- lung buds
- midgut
- pharynx
- pharyngeal pouches

Ectoderm and mesoderm

- amnion
- chorion
- head mesenchyme
- limb buds

Ectoderm, mesoderm, and endoderm

- branchial arches

Endoderm and mesoderm

- allantois
- yolk sac

H. TERMS TO KNOW

You should know the meaning of the following terms, which appeared in boldface in the preceding discussion of mouse embryos.

- allantois
- amnion
- amniotic cavity
- aortic sac
- apical ectodermal ridges
- auditory pits
- auditory vesicles
- blastocoel
- blastocyst
- blastoderm
- blastomeres
- body axes
- body folding
- body folds
- brain
- branchial arches
- branchial grooves
- cardiogenesis
- caudal end of the embryo
- caudal intestinal portal
- caudal neuropore
- caudal spinal cord
- cavitation
- chorion

- chorionic cavity
- cleavage
- compaction
- conotruncus
- cranial intestinal portal
- dermomyotome
- descending aorta
- diencephalon
- distal level of the embryonic part of the egg cylinder
- dorsal aortae
- ectoderm
- ectoplacental cone
- egg
- egg cylinder
- eight-cell stage
- embryonic part of the egg cylinder
- endoderm
- endothelial cells
- epiblast/ectoderm
- epithelial-to-mesenchymal transformation
- estrus cycle
- extraembryonic coelom

- extraembryonic membranes
- extraembryonic mesoderm
- extraembryonic part of the egg cylinder
- eyes
- fertilization
- first aortic arches
- forebrain
- foregut
- foreleg buds
- four-cell stage
- gametogenesis
- gastrulation
- gene targeting
- germ layers
- gut
- head
- head fold of the body
- head mesenchyme
- head mesoderm
- head of sperm
- heart
- heart mesoderm
- hindbrain

hindgut

hindleg buds

homologous recombination

hypoblast

implantation

infundibulum

ingress

inner cell mass

intermediate primitive streak stage

intraembryonic coelom

intraembryonic mesoderm

isthmus

knocked-out genes

laryngotracheal groove

lateral body folds

lateral plate mesoderm

lens vesicles

looping of heart tube

lung buds

mandibular processes of the first branchial arches

meiotic divisions

mesencephalon

mesoderm

metencephalon

midbrain

midgut

morula stage

mouth opening

myelencephalon

nephrotome

neural crest cells

neural folds

neural groove

neural plate

neural tube

neurulation

node

notochord

oogenesis

optic sulcus (cup)

optic vesicles

organizer

organogenesis

oviduct

ovulation

pericardial cavity

peritoneal cavity

pharyngeal pouches

pharynx

placenta

pleural cavity

polar bodies

posterior pituitary gland

precardinal veins

presomitic mesoderm

primary oocytes

primitive streak

prosencephalon

prospective extraembryonic mesoderm

prospective fate maps

prospective head mesoderm

prospective heart

prospective lateral plate mesoderm

prospective notochord

prospective somitic mesoderm

proximal level of the embryonic part of the egg cylinder

roof plate

rostral end of the embryo

rostral level of the embryonic part of the egg cylinder

sclerotome

second branchial grooves

second pharyngeal pouches

secondary oocytes

semilunar ganglia of the trigeminal (V) cranial nerves

sexual maturity

sinoatrial region

skin ectoderm

somatic mesoderm

somites

somitic mesoderm

somitogenesis

sperm

spermatogenesis

spinal cord

splanchnic mesoderm

stomodeum

subcaudal pocket

subcephalic pocket

tail

tail fold of the body

tail of sperm

telencephalon

testes

third aortic arches

transgenic mice

trophoblast

trunk

tube-within-a-tube body plan

turning

two-cell stage

uterine cavity

uterine wall

ventral aortae

ventricle

yolk sac

zona pellucida

zygote

I. PHOTOS 4.1-4.69

Photos 4.1-4.69 depict the mouse embryos discussed in Chapter 4. These photos and their accompanying legends begin on the following page.

Photos 4.1-4.7

Mouse Embryos

Legend

1. Head of sperm
2. Tail of sperm
3. Fertilized egg (zygote)
4. Polar body

5. Zona pellucida
6. Blastomere
7. Morula
8. Inner cell mass

9. Trophoblast
10. Blastocoel

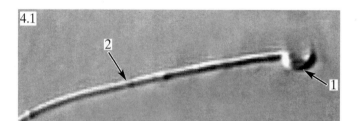

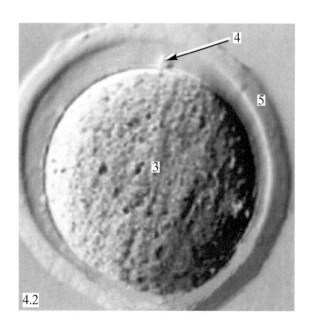

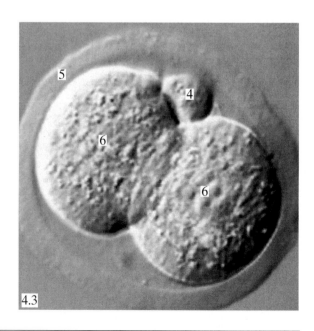

Photos 4.1-4.3. Early developmental stages of the mouse showing the sperm (Photo 4.1), fertilized egg (zygote; Photo 4.2), and two-cell stage (Photo 4.3).

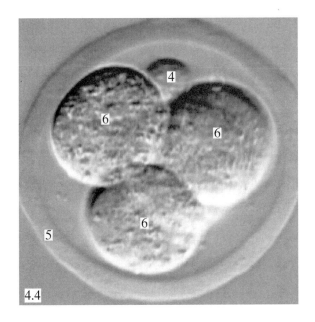

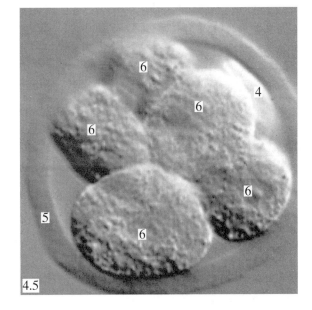

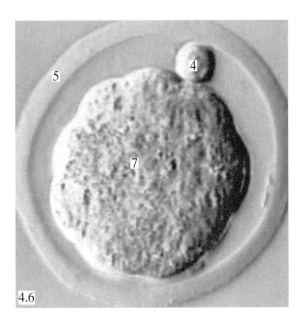

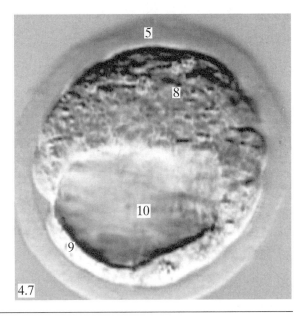

Photos 4.4-4.7. Early developmental stages of the mouse showing the four-cell stage (Photo 4.4), eight-cell stage (Photo 4.5), morula stage (Photo 4.6), and blastocyst stage (Photo 4.7).

Photos 4.8-4.13

Mouse Embryos

Legend

1. Extraembryonic part of the egg cylinder
2. Embryonic part of the egg cylinder
3. Proximal level of the embryonic part of the egg cylinder
4. Distal level of the embryonic part of the egg cylinder
5. Rostral end of blastoderm
6. Caudal end of blastoderm
7. Ectoplacental cone

8. Endoderm
9. Neural plate (seen through the endoderm in Photo 4.8)
10. Primitive streak (seen through the endoderm in Photo 4.8)
11. Node (organizer; seen through the endoderm in Photo 4.8)
12. Chorionic cavity (seen through the wall of the egg cylinder in Photo 4.8)

13. Amniotic cavity (seen through the wall of the egg cylinder)
14. Part of the chorion
15. Allantois
16. Wall of the extraembryonic part of the egg cylinder
17. Mesoderm
18. Epiblast

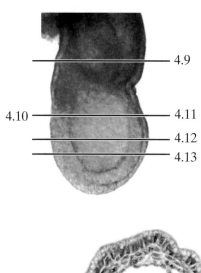

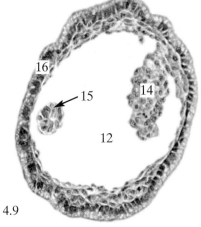

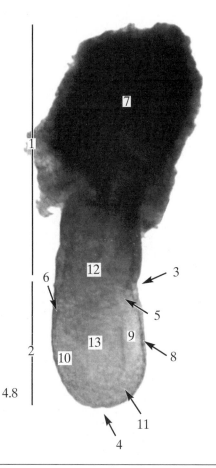

Photo 4.8. 7-day mouse embryo whole mount. Lateral view.

Photo 4.9. 7-day mouse embryo serial transverse section.

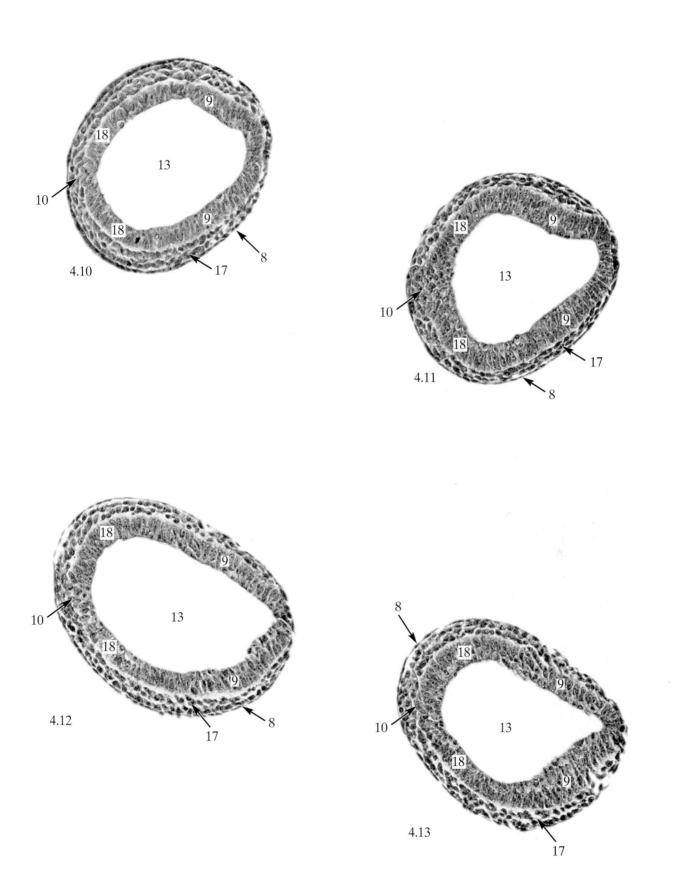

Photos 4.10-4.13. Continuation of 7-day mouse embryo serial transverse sections numbered in proximal to distal sequence.

Photos 4.14-4.18

Mouse Embryos

Legend

1. Extraembryonic part of the egg cylinder
2. Embryonic part of the egg cylinder
3. Caudal end of the embryo
4. Rostral end of the embryo
5. Ectoplacental cone
6. Endoderm
7. Node (organizer)
8. Chorion
9. Chorionic cavity
10. Amnion
11. Amniotic cavity
12. Neural plate
13. Primitive streak
14. Proximal level of the embryonic part of the egg cylinder
15. Distal level of the embryonic part of the egg cylinder
16. Intraembryonic mesoderm
17. Extraembryonic mesoderm
18. Epiblast

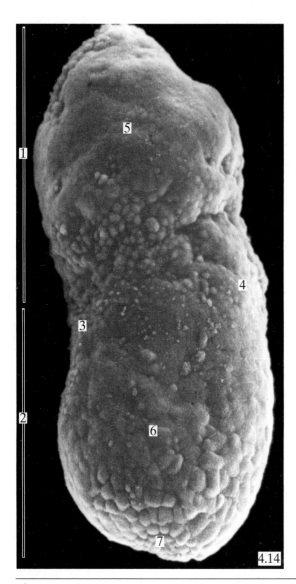
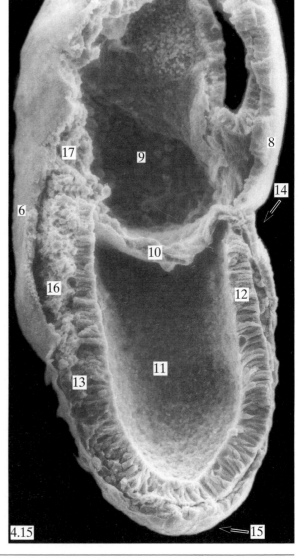

Photos 4.14, 4.15. Scanning electron micrographs of 7-day mouse embryos. Photo 4.14 shows an intact whole mount. Photo 4.15 shows an embryo sliced parasagittally.

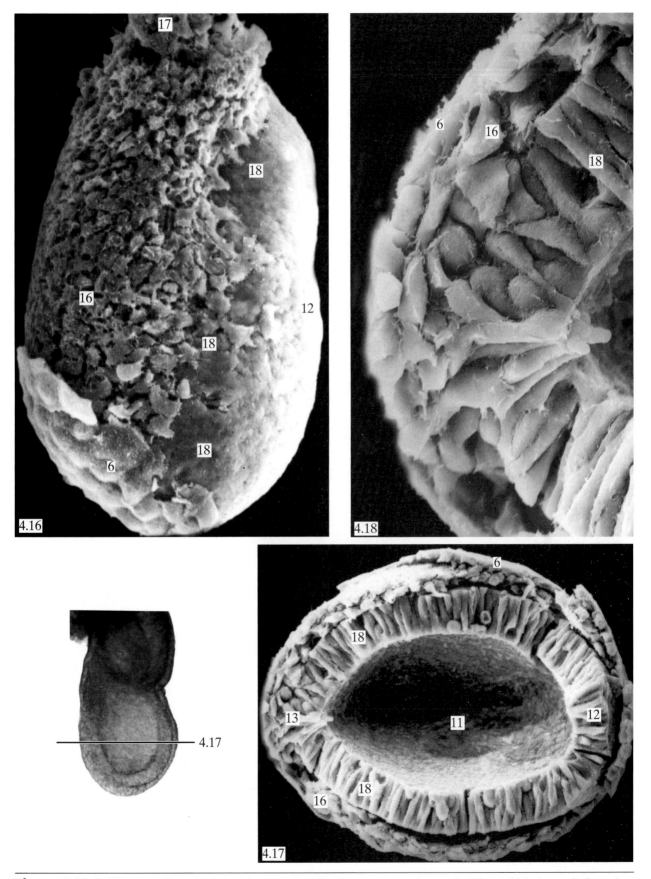

Photos 4.16-4.18. Scanning electron micrographs of 7-day mouse embryos. In Photo 4.16, the endoderm has been removed. In Photo 4.17, the embryo has been sliced transversely. Photo 4.18 shows an enlargement of the caudal side of Photo 4.17.

Photos 4.19-4.24

Mouse Embryos

Legend

1. Chorionic cavity
2. Amnion
3. Allantois
4. Endoderm
5. Amniotic cavity
6. Primitive streak

7. Neural plate
8. Node (organizer)
9. Neural folds
10. Head fold of the body
11. Foregut (opening ventrally as the cranial intestinal portal)

12. Incipient heart
13. Epiblast
14. Neural groove
15. Mesoderm

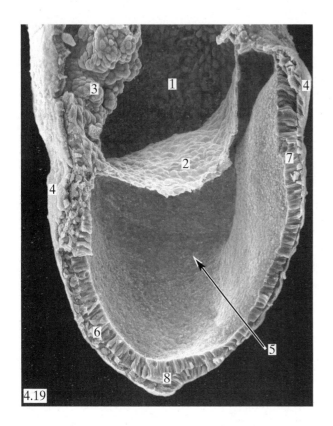

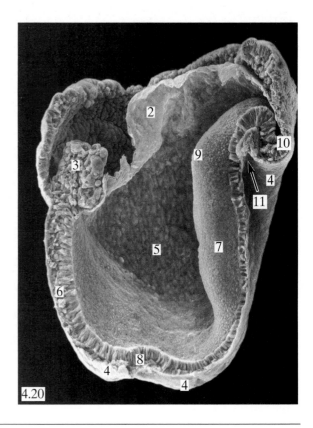

Photos 4.19, 4.20. Scanning electron micrographs of 7- to 7.5-day mouse embryos sliced in the midsagittal plane.

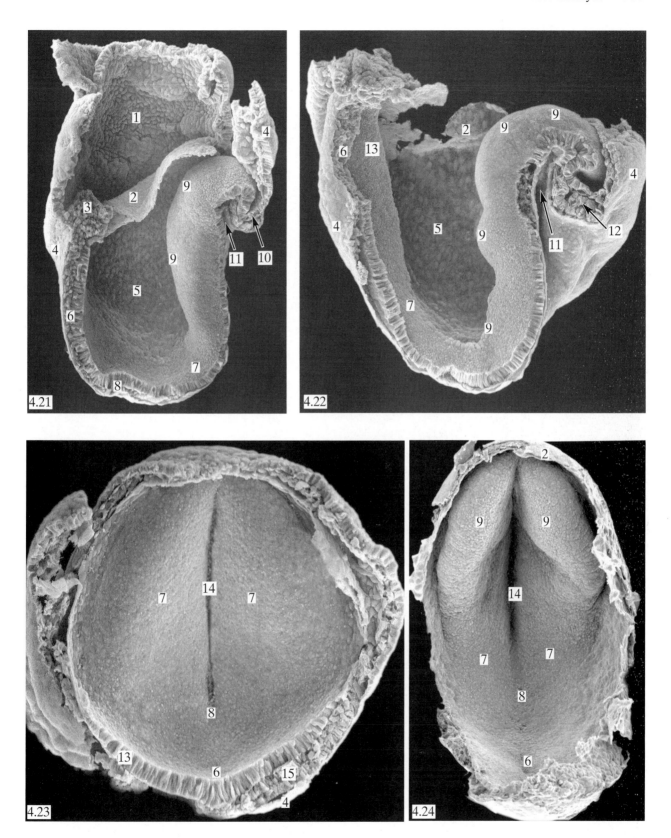

Photos 4.21, 4.22. Scanning electron micrographs of 8- to 8.5-day mouse embryos sliced in the midsagittal plane.

Photos 4.23, 4.24. Scanning electron micrographs of 7.5- 8-day mouse embryos with the epiblast viewed dorsally after removal of most of the amnion.

Photos 4.25-4.29

Mouse Embryos

Legend

1. Ectoplacental cone
2. Chorion
3. Chorionic cavity
4. Amnion (seen through the yolk sac in Photo 4.25)
5. Amniotic cavity (seen through the yolk sac and amnion in Photo 4.25)
6. Head (seen through the yolk sac and amnion in Photo 4.25)
7. Heart (seen through the yolk sac and amnion in Photos 4.25, 4.26)
8. Trunk (seen through the yolk sac and amnion in Photo 4.25)

9. Primitive streak
10. Allantois (seen through the chorion in Photo 4.25)
11. Neural plate (of forebrain) (seen through the yolk sac and amnion in Photos 4.25, 4.26)
12. Cranial intestinal portal
13. Hindgut
14. Neural plate (of midbrain)
15. Head mesenchyme
16. Yolk sac
17. Extraembryonic coelom
18. Dorsal aorta
19. Neural crest cells

20. Neural groove (forebrain level)
21. Neural groove (midbrain level)
22. Neural fold (forebrain level)
23. Neural fold (midbrain level)
24. Neural fold (caudal spinal cord level)
25. Optic sulcus (cup)
26. Foregut
27. First aortic arch
28. Neural groove (caudal spinal cord level)

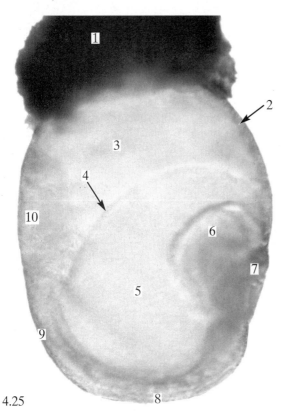

4.25

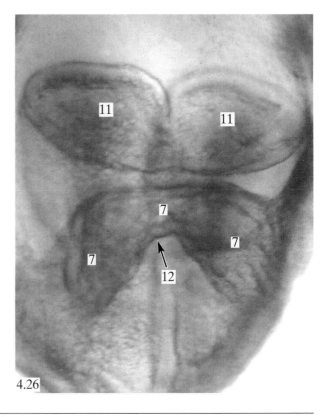

4.26

Photos 4.25, 4.26. 8.5-day mouse embryo whole mounts. Photo 4.20 is a lateral view. Photo 4.21 is a rostral view.

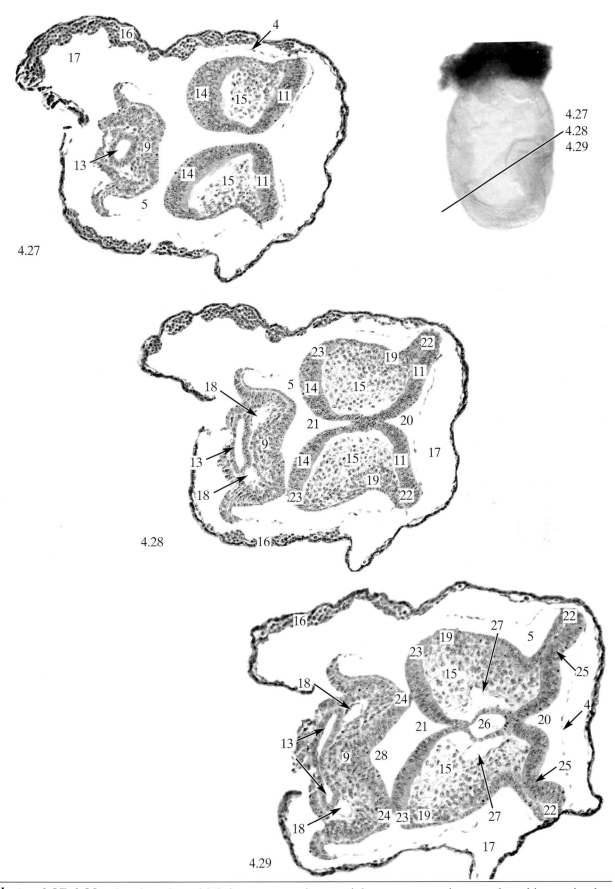

Photos 4.27-4.29. Continuation of 8.5-day mouse embryo serial transverse sections numbered in proximal to distal sequence.

Photos 4.30-4.33

Mouse Embryos

Legend

1. Yolk sac
2. Extraembryonic coelom
3. Amnion
4. Amniotic cavity
5. Primitive streak
6. Lateral body fold
7. Skin ectoderm
8. Somatic mesoderm
9. Splanchnic mesoderm

10. Endoderm
11. Neural groove (spinal cord level)
12. Neural groove (midbrain level)
13. Neural crest cells
14. Foregut
15. Notochord
16. Neural groove (forebrain level)

17. Ventral aorta
18. Dorsal aorta
19. Conotruncus region of heart
20. Pericardial cavity
21. Ventricle region of heart
22. Neural groove (hindbrain level)
23. Sinoatrial region of heart
24. Somite

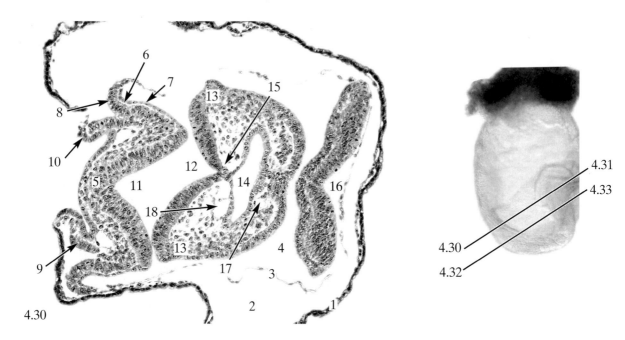

Photo 4.30. Continuation of 8.5-day mouse embryo serial transverse sections.

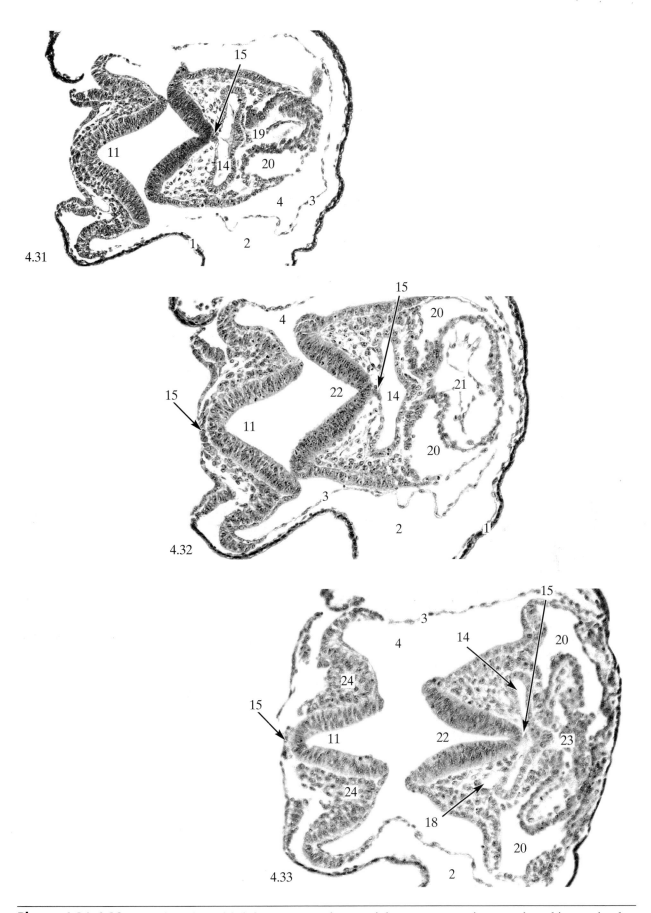

Photos 4.31-4.33. Continuation of 8.5-day mouse embryo serial transverse sections numbered in proximal to distal sequence.

Photos 4.34-4.37

Mouse Embryos

Legend

1. Notochord
2. Endoderm
3. Somite
4. Splanchnic mesoderm
5. Intraembryonic coelom

6. Somatic mesoderm
7. Amniotic cavity
8. Amnion
9. Neural groove (spinal cord level)
10. Skin ectoderm

11. Yolk sac
12. Extraembryonic coelom
13. Open midgut
14. Neural tube (spinal cord level)

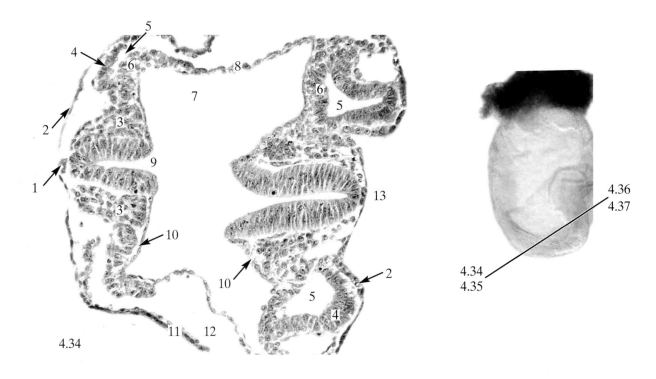

Photo 4.34. Continuation of 8.5-day mouse embryo serial transverse sections.

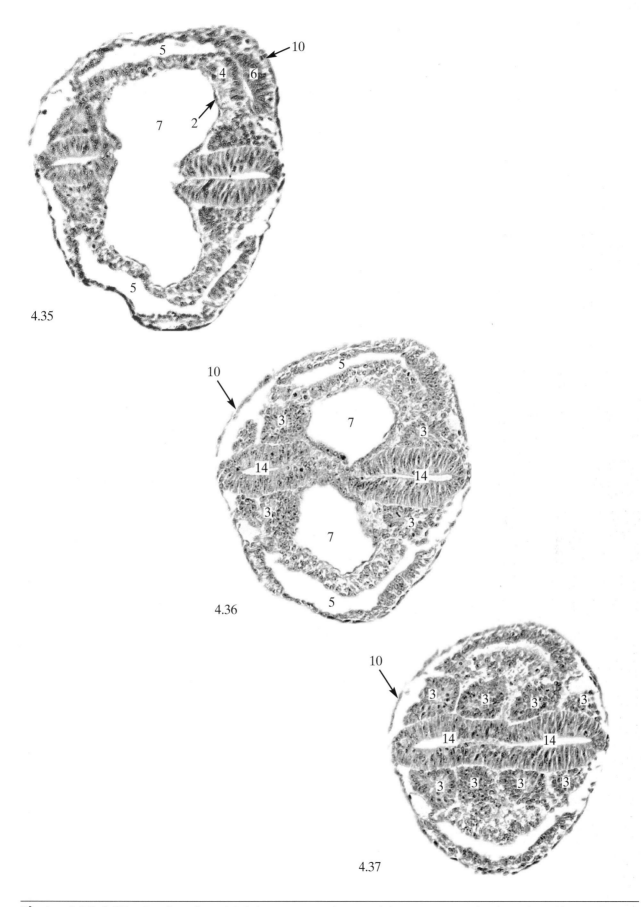

4.35

4.36

4.37

Photos 4.35-4.37. Continuation of 8.5-day mouse embryo serial transverse sections numbered in proximal to distal sequence.

Photos 4.38-4.41

Mouse Embryos

Legend

1. Allantois
2. Endoderm of yolk sac
3. Skin ectoderm
4. Neural groove
5. Neural fold (spinal cord level)
6. Neural fold (forebrain level)
7. Neural fold (midbrain level)
8. Neural plate
9. Optic sulcus (cup)
10. Heart
11. Neural tube (spinal cord level)
12. Notochord
13. Dorsal aorta
14. Endoderm
15. Presomitic mesoderm
16. Nephrotome
17. Lateral plate mesoderm

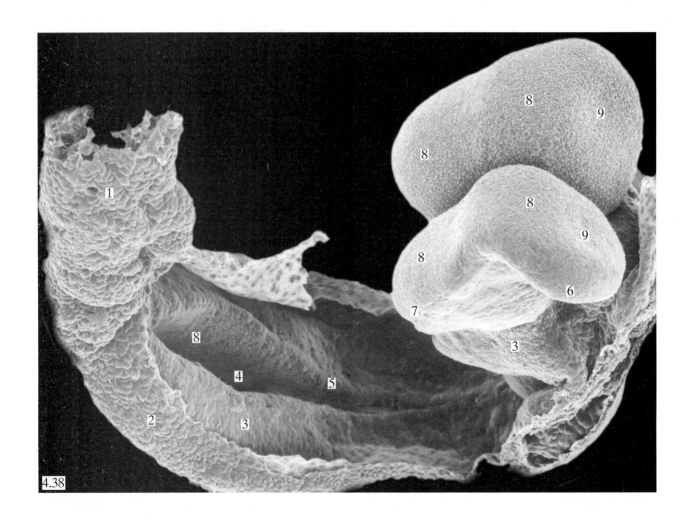

Photo 4.38. Scanning electron micrograph of an 8.5-day mouse embryo whole mount. Dorsolateral view.

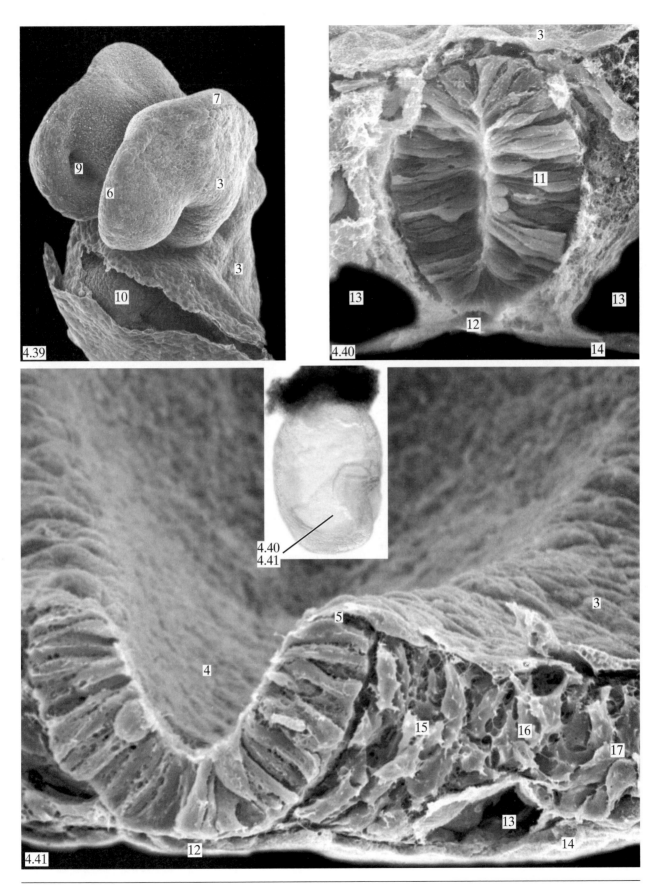

Photos 4.39-4.41. Scanning electron micrographs of 8.5-day mouse embryos. Photo 4.39 shows the rostral end of the embryo. Photos 4.40, 4.41 show transverse slices through the neural tube (Photo 4.40) and neural groove (Photo 4.41) at the leve of the spinal cord.

Photos 4.42-4.46

Mouse Embryos

Legend

1. Yolk sac
2. Amnion
3. Neural fold (forebrain level)
4. S-shaped (looping) heart
5. Cranial intestinal portal
6. Optic sulcus (cup)
7. Neural fold (midbrain level)
8. Neural groove

9. Neural fold (hindbrain level)
10. Head
11. Trunk
12. Allantois
13. Somites
14. Eye
15. Forebrain (seen through the skin ectoderm)

16. Midbrain (seen through the skin ectoderm)
17. Hindbrain (seen through the skin ectoderm)
18. Auditory vesicle (seen through the skin ectoderm)
19. Spinal cord
20. Tail

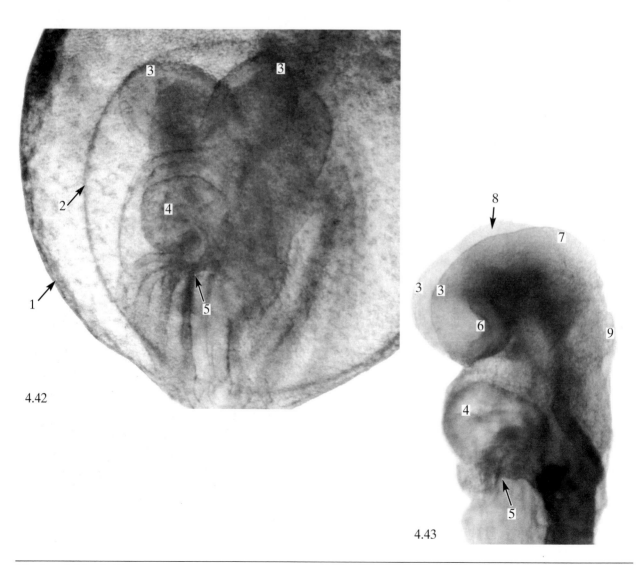

4.42

4.43

Photos 4.42, 4.43. 8.5- to 9-day mouse embryo whole mounts. Photo 4.33 is a ventral view of the rostral end of the embryo. Photo 4.34 is a ventrolateral view of the rostral end of the embryo.

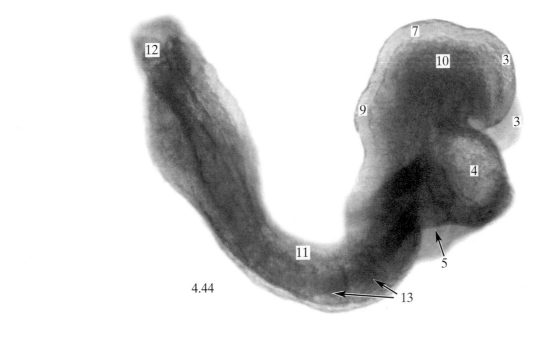

4.44

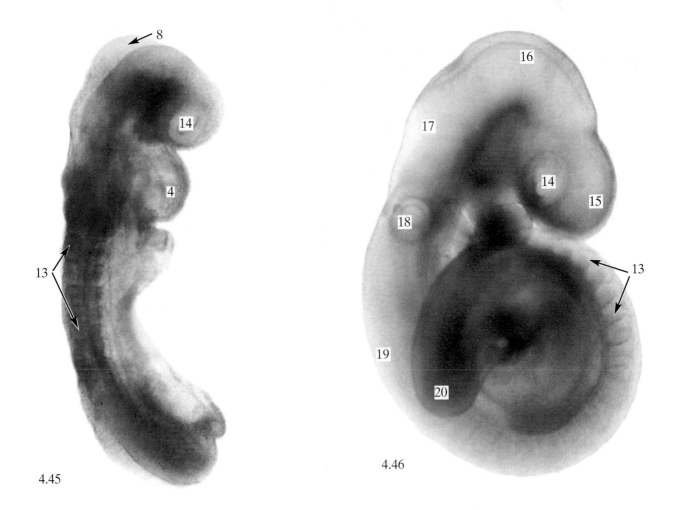

4.45

4.46

Photos 4.44-4.46. 8.5- to 9.5-day mouse embryo whole mounts. Lateral views.

Photos 4.47-4.51

Mouse Embryos

Legend

1. Neural fold (forebrain level)
2. Neural fold (midbrain level)
3. Skin ectoderm
4. Rostral spinal cord (beneath skin ectoderm)
5. Neural groove (spinal cord level)
6. Neural fold (hindbrain level)
7. Neural groove (brain level)
8. Allantois
9. Head
10. Trunk
11. Tail
12. Hindleg bud
13. Foreleg bud
14. Branchial arches
15. Branchial grooves
16. Somites
17. Neural groove (caudal neuropore; caudal spinal cord level)
18. Spinal cord
19. Neural crest cells
20. Lateral plate mesoderm
21. Stomodeum
22. Forebrain level of neural tube (beneath skin ectoderm)
23. Eye
24. Midbrain level of neural tube (beneath skin ectoderm)
25. Hindbrain level of neural tube (beneath skin ectoderm)
26. Trunk (spinal cord) level of neural tube (beneath skin ectoderm)
27. Tail level of neural tube (beneath skin ectoderm)

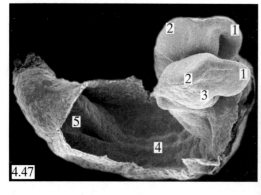

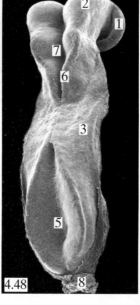

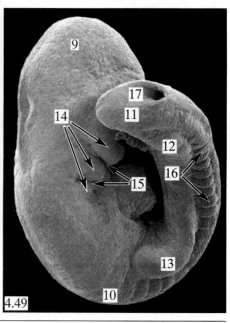

Photos 4.47-4.49. Scanning electron micrographs of 8.5- to 9.5-day mouse embryo whole mounts. Dorsolateral (Photos 4.47, 4.48) and lateral (Photos 4.49) views.

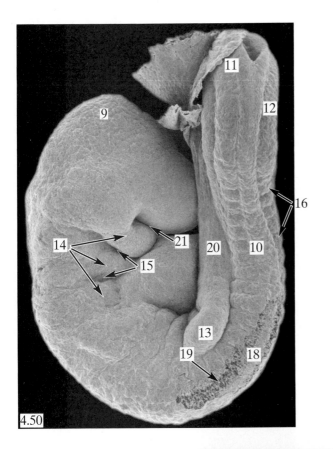

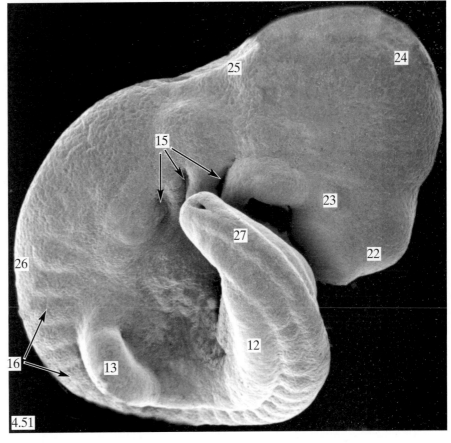

Photos 4.50, 4.51. Scanning electron micrographs of 9.5- to 10-day mouse embryo whole mounts.

Photos 4.52-4.54

Mouse Embryos

Legend

1. Prosencephalon
2. Eye
3. Mesencephalon
4. Isthmus
5. Metencephalon
6. Myelencephalon

7. Auditory vesicle
8. Spinal cord
9. Branchial grooves
10. Somites
11. Tail

12. Mandibular process of the first branchial arch
13. Stomodeum
14. Infundibulum
15. Head mesenchyme
16. Skin ectoderm

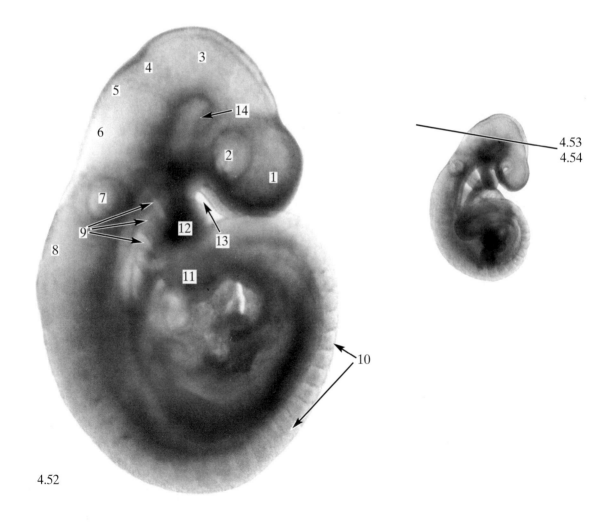

4.52

Photo 4.52. 9.5-day mouse embryo whole mount. Lateral view.

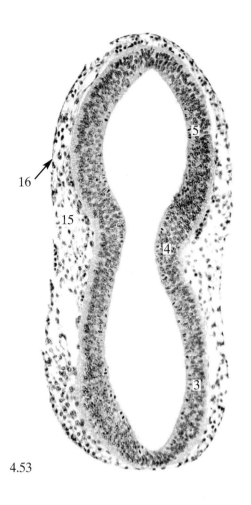

4.53

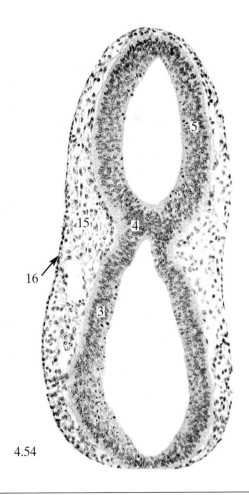

4.54

Photos 4.53, 4.54. 9.5-day mouse embryo serial transverse sections numbered in anterior to posterior sequence.

Photos 4.55-4.58

Mouse Embryos

Legend

1. Metencephalon
2. Head mesenchyme
3. Optic vesicle (connected to the diencephalon in Photo 4.55)
4. Skin ectoderm
5. Prosencephalon
6. Infundibulum
7. Myelencephalon
8. Precardinal vein
9. Neural crest cells
10. Thin roof plate of the myelencephalon
11. Mandibular process of the first branchial arch
12. Stomodeum
13. Telencephalon
14. Semilunar ganglion of the trigeminal (V) cranial nerve
15. Notochord

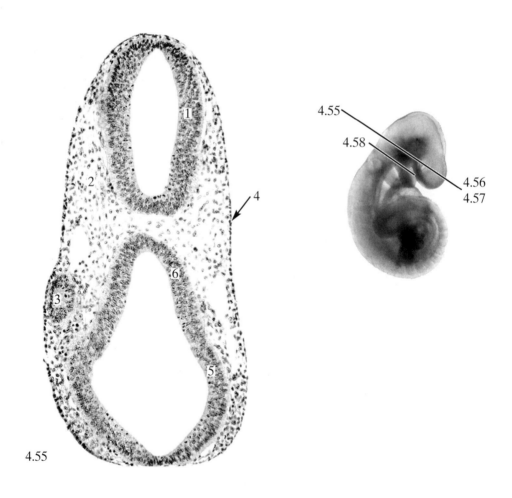

Photo 4.55. Continuation of 9.5-day mouse embryo serial transverse sections.

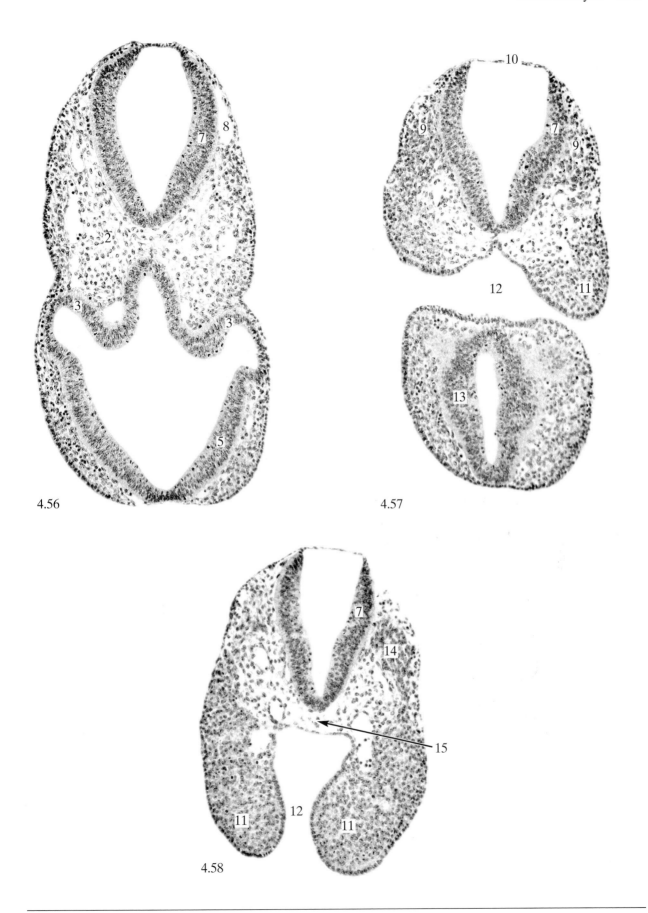

4.56

4.57

4.58

Photos 4.56-4.58.   Continuation of 9.5-day mouse embryo serial transverse sections numbered in anterior to posterior sequence.

Photos 4.59-4.61

Mouse Embryos

Legend

1. Myelencephalon
2. Auditory vesicle
3. Notochord
4. Dorsal aorta
5. Pharynx
6. Ventricle
7. Second branchial groove

8. Second pharyngeal pouch
9. Laryngotracheal groove
10. Lung buds
11. Pleural cavity
12. Peritoneal cavity
13. Spinal cord
14. Somites

15. Descending aorta
16. Midgut
17. Nephrotome (mesonephric duct)
18. Somatic mesoderm
19. Splanchnic mesoderm

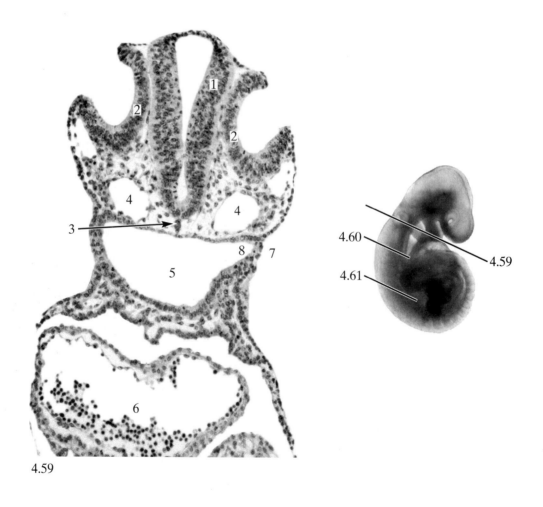

Photo 4.59. Continuation of 9.5-day mouse embryo serial transverse sections.

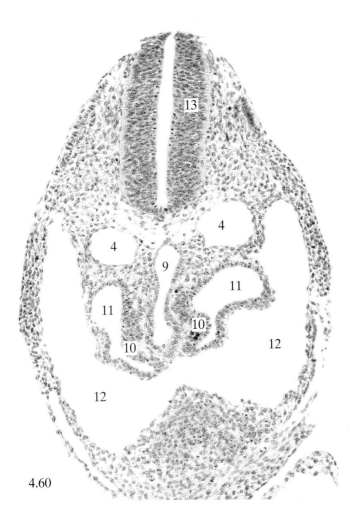

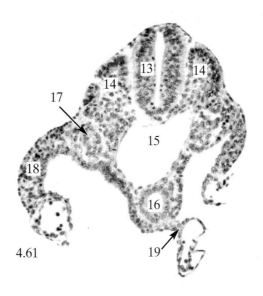

Photos 4.60, 4.61. Continuation of 9.5-day mouse embryo serial transverse sections numbered in anterior to posterior sequence.

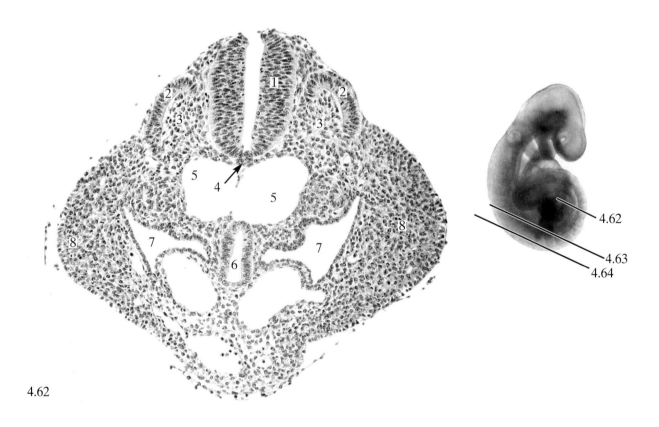

Photos 4.62-4.64
Mouse Embryos
Legend

1. Spinal cord
2. Dermomyotome of somite
3. Sclerotome of somite
4. Notochord

5. Dorsal aorta
6. Hindgut
7. Peritoneal cavity
8. Hindleg bud

9. Descending aorta
10. Foreleg bud
11. Apical ectodermal ridge of foreleg bud

4.62

Photo 4.62. Continuation of 9.5-day mouse embryo serial transverse sections.

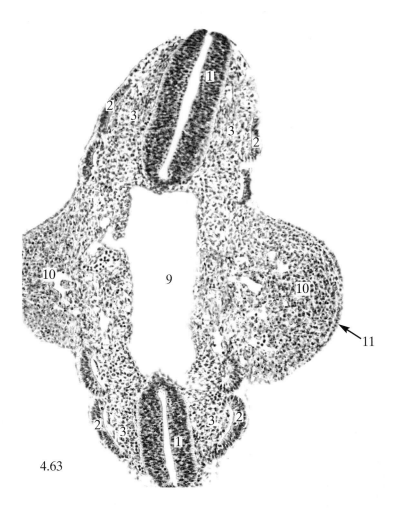

4.63

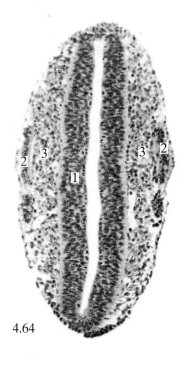

4.64

Photos 4.63, 4.64. Continuation of 9.5-day mouse embryo serial transverse sections numbered in anterior to posterior sequence.

Photos 4.65-4.67

Mouse Embryos

Legend

1. Neural folds (midbrain level)
2. Neural folds (hindbrain level)
3. Skin ectoderm
4. Auditory pits
5. Hindbrain
6. Head mesenchyme
7. Pharynx
8. Precardinal vein
9. Dorsal aorta
10. Notochord
11. Third aortic arch
12. Aortic sac
13. Spinal cord
14. Descending aorta
15. Foreleg bud
16. Peritoneal cavity
17. Amnion

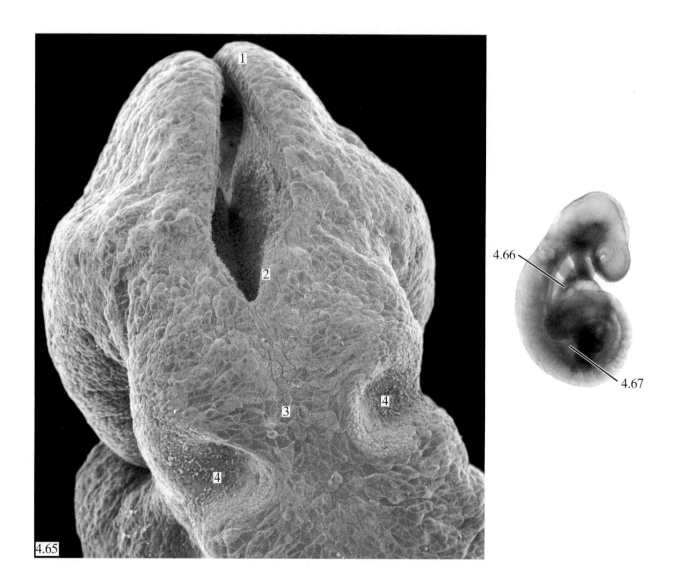

Photo 4.65. Scanning electron micrograph of the dorsal surface of the head of a 9.5-day mouse embryo.

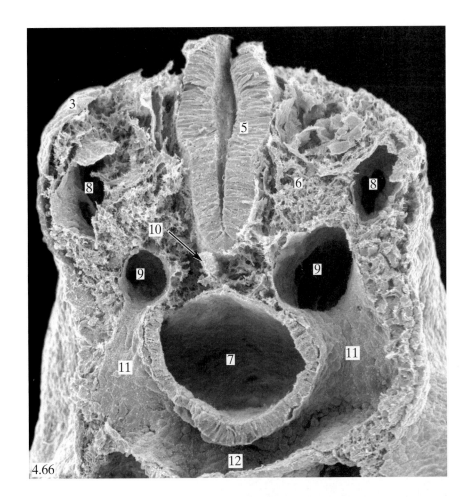

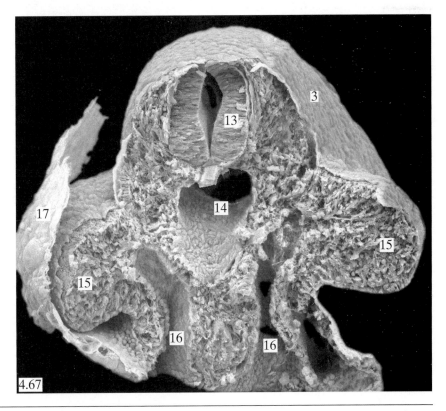

Photos 4.66, 4.67. Scanning electron micrographs of transverse slices through the hindbrain (Photo 4.66) and spinal cord (Photo 4.67) levels of 9.5-day mouse embryos.

Photos 4.68, 4.69
Mouse Embryos

Legend

1. Telencephalon region of the prosencephalon
2. Diencephalon region of the prosencephalon
3. Mesencephalon
4. Metencephalon
5. Myelencephalon
6. Spinal cord
7. Optic sulcus (cup)
8. Stomodeum
9. Mouth opening
10. Pharynx
11. Pharyngeal pouches
12. Infundibulum
13. Ventricle of heart
14. Conotruncus of heart

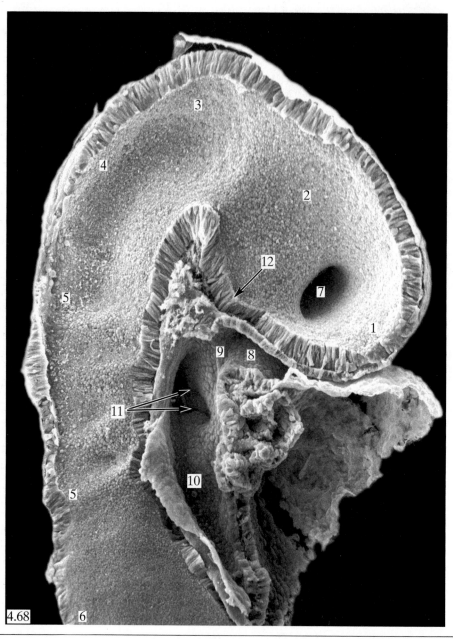

Photo 4.68.   Scanning electron micrograph of a sagittal slice through a 9.5-day mouse embryo.

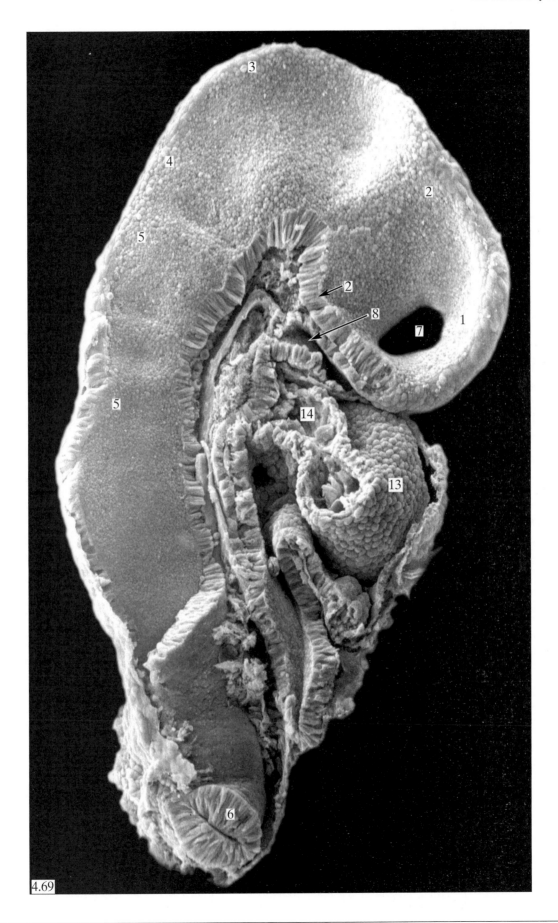

Photo 4.69. Scanning electron micrograph of a sagittal slice through a 9.5-day mouse embryo.

Chapter 5

Pig Embryos

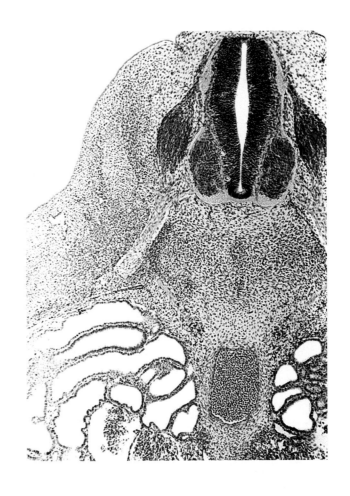

Chapter 5

Pig Embryos

A. INTRODUCTION

The development of chick embryos becomes increasingly less typical of vertebrates in general and more avian in character beyond the 72-hour stage. Consequently, for more advanced development, it is advisable to examine mammalian embryos. The 10-mm pig embryo serves as an excellent model for understanding early **organogenesis** of the human embryo. 10-mm pig embryos and equivalent stages of human embryos have virtually identical internal and external morphologies, with two major exceptions. First, the external morphology differs in that although both human and pig embryos develop **tails**, the length of the tail is much longer in the pig than in the human. Second, the internal morphology differs in that although both human and pig embryos develop mesonephric kidneys, the size of the **mesonephric kidneys** is much greater in the pig than in the human. The pig **placenta** is far less efficient in removing nitrogenous wastes than is the human placenta. Thus, the mesonephros must assume this role in the pig. Only the 10-mm pig embryo will be studied here. It has been developing for 20-21 days of a total gestation period of approximately 4 months and it contains the rudiments of essentially all adult structures. Therefore, examination of this stage is almost a study of adult anatomy in miniature.

B. OVERVIEW OF EARLY DEVELOPMENT

The total gestation period in the pig consists of approximately 4 months (usual range is 110-116 days). After **ovulation**, the **ovum** (**secondary oocyte**) enters the **oviduct** where it encounters the **sperm** and **fertilization** occurs. The **zygote** then initiates a series of **cleavage** divisions, forming a **morula** by about 3.5 days after fertilization and a **blastocyst** by about 5 days. The blastocyst contains a cavity, the **blastocoel**, and two groups of cells: an outer **trophoblast** and an inner cluster at one pole, the **inner cell mass**. It is the inner cell mass that under-

goes **gastrulation**, forming initially a **hypoblast** and **epiblast**, and then the three primary germ layers, the **ectoderm**, **mesoderm**, and **endoderm**. The early stages of development of the pig just described appear very similar to those occurring in the mouse and discussed in Chapter 4.

The pig blastocyst does not implant into the uterine wall. Rather, it undergoes a rapid elongation to form a thread-like structure up to a meter long. This elongation occurs about 8-9 days after fertilization, concomitant with the beginning of gastrulation. Elongation provides a greatly increased surface area for exchange between the **extraembryonic membranes** of the developing pig embryo and the **uterine horns**. A litter size of 8 or more is typical, with embryos generally spaced evenly between the two uterine horns.

Neurulation, **cardiogenesis**, segmentation to form **somites**, and formation of the **body folds** occur in a manner similar to that in the mouse, with one major exception: the pig blastoderm, unlike the mouse blastoderm, is a flat structure, more similar to that of the early chick. As a result of these early developmental processes, a **tube-within-a-tube body plan** is formed. **Organogenesis** then begins. Our study of organogenesis is restricted to the 10-mm stage.

C. 10-MM PIG EMBRYOS

1. Whole specimens

Examine models or, preferably, *whole specimens* preserved in 70% alcohol. Whole preserved specimens are far superior to any model and to specimens embedded in plastic. If preserved specimens are used, your instructor should remove the amnion and carefully cut the umbilical cord parallel to and as close to the body wall as possible (if these structures are still present). The preserved embryo should be examined in a small dish containing sufficient 70% alcohol to cover it; observations should be made with a dissecting microscope. Manipulate your embryo very carefully, preferably with

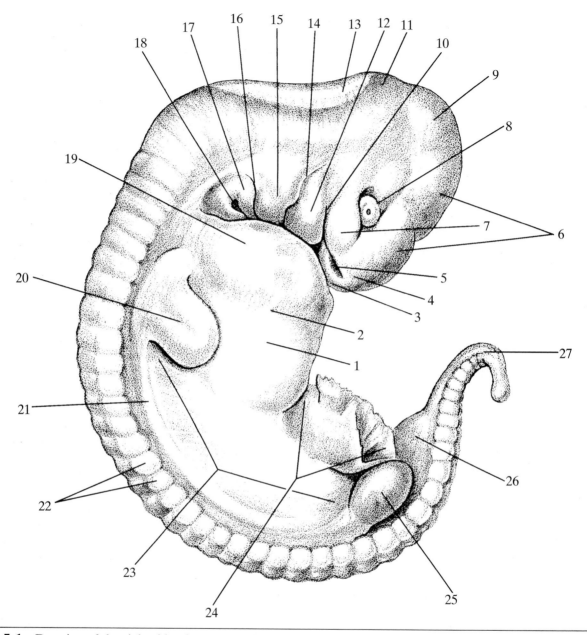

Fig. 5.1. Drawing of the right side of a preserved 10-mm pig embryo.

1. Liver region
2. Location of developing septum transversum
3. Medial nasal process
4. Lateral nasal process
5. Nasal pit
6. Prosencephalon region
7. Maxillary process
8. Eye
9. Mesencephalon region

10. Stomodeum
11. Metencephalon region
12. Mandibular process
13. Myelencephalon roof plate region
14. First branchial groove
15. Second branchial arch
16. Second branchial groove
17. Third branchial arch
18. Cervical sinus

19. Heart region
20. Foreleg bud
21. Mammary ridge
22. Somites
23. Mesonephric kidney region
24. Umbilical cord
25. Hindleg bud
26. Genital eminence
27. Tail

the enlarged rounded end of a small glass rod. Make every effort not to damage the embryo, so that it may be used repeatedly.

Note the general C-shape of the embryo (Fig. 5.1). This shape is due to the **cranial** (level of **mesencephalon**), **cervical** (level of **myelencephalon**), and **tail flexures**. Because of these pronounced flexures, the most *cranial* regions of your serially transverse sectioned embryo will not be seen in the most *anterior* sections of your set of slides. Similarly, the most *caudal* regions of the embryo will not be found in the most *posterior* sections.

Identify the **head, foreleg** and **hindleg buds, eyes** (note dark coloration due to **pigment granules** within the **pigmented retinas**), **nasal pits** (note the raised ridges bounding them, the **lateral** and **medial nasal processes**), and **somites**. Also identify the regions of the **prosencephalon, mesencephalon, metencephalon, myelencephalon** (note its thin **roof plate**), and **spinal cord**. The **stomodeum** partially subdivides the large **first branchial arch** on each side into **maxillary** and **mandibular processes**. The large **first branchial grooves** are immediately behind the mandibular processes, followed by the **second branchial arches**, and then by the **second branchial grooves**. The **third branchial arches** are partially hidden behind the second branchial arches and grooves. They are directly in front of the small, deep **cervical sinuses** (combined **third** and **fourth branchial grooves**).

The sites of three large developing organs can be seen within the relatively straight trunk. The most cranial organ, indicated by a large bulge, is the **heart**. The bulge immediately behind the heart, toward which the foreleg buds project, indicates the **liver**. The depression between the heart and liver bulges indicates the location of the developing **septum transversum** (source of part of the adult **diaphragm**). The very large bulges, one on each side, between foreleg and hindleg buds indicate the **mesonephric kidneys**. Identify the **mammary ridges** (small ridges, one on each side, lying parallel to the ventral edges of the somites between foreleg and hindleg buds), which form the **mammary glands**. The trunk ends with a ventral swelling, the **genital eminence** (rudiment of the **external genitalia**), which lies between the hindleg buds and the tapering tail. Any part of the **umbilical cord** still attached to the body is located between the liver region and hindleg buds.

2. Serial transverse sections

Position your slide on the microscope stage so that when viewed through the microscope, each section is oriented as in Photos 5.1-5.42. Examine sections in anteroposterior sequence under low magnification unless directed otherwise. Because the amount of flexion encountered is quite variable and the plane of section is not always exactly the same from embryo to embryo, the order of appearance of structures in *your* specimen will not always coincide *exactly* with the sequence described

here, but it should closely approximate it.

a. Nervous system

1. Brain. The first sections usually cut through the **myelencephalon**, but in some specimens the **metencephalon** appears first. The former is recognized by its thin **roof plate**, the latter by its thick walls. The myelencephalon and metencephalon are usually cut frontally when first encountered (Fig. 5.2). This is also true of many other cranial structures. The thin roof plate of the myelencephalon together with an outer adjacent vascular layer, the **pia mater**, are the rudiments of the **choroid plexus**. (The pia mater is just developing and can be identified under high magnification as a thin layer of head mesenchymal cells just outside the brain wall. See Photo 5.2.) The choroid plexus is formed by invagination of these rudiments into the cavity of the rhombencephalon, the **fourth ventricle**, and is the source of **cerebrospinal fluid** (that is, the fluid that fills the cavities of the brain and spinal cord). Continue tracing sections posteriorly and note that another section of the brain appears near the pointed apex of the metencephalon; this is the **mesencephalon**. The cavity of the mesencephalon, the **cerebral aqueduct**, soon becomes continuous with the fourth ventricle (Photo 5.1). The constricted region between mesencephalon and metencephalon is the **isthmus**.

The rhombencephalon and mesencephalon are the sources of several major adult brain components. The **myelencephalon** forms the **medulla**, the **metencephalon** forms the **cerebellum** and **pons**, and the **mesencephalon** forms the **corpora quadrigemina** (**superior** and **inferior colliculi**) and the **cerebral peduncles**.

Observe under high magnification that the walls of all regions of the brain are subdivided into three mediolateral regions (Photo 5.2). The innermost and darkest staining region (rich in nuclei) is the **ventricular zone**; the middle region containing fewer nuclei is the **intermediate (mantle) zone**; and the outer region, relatively free of nuclei and the lightest staining, is the **marginal zone**, consisting mainly of nerve fibers. The intermediate zone is formed by cells that migrate from the ventricular zone. Most of these migratory cells are **young neurons** at this stage. Some migratory cells may be **glioblasts** (**spongioblasts**), primitive nonnervous cells (that is, supporting cells) that remain within the wall of the brain.

Continue to trace sections posteriorly until the brain separates into two parts. The smaller oval-shaped (lower) section of the brain is still the mesencephalon; the larger (upper) section is still rhombencephalon (Photos 5.4, 5.5). The neural tube is separated into three parts in more posterior sections (Photo 5.6). The smaller (upper) section is either myelencephalon or **spinal cord** (the boundary between myelencephalon and spinal cord is indistinct), the middle section is still rhombencephalon,

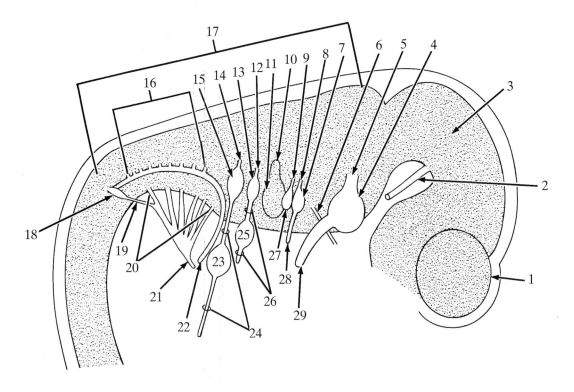

Fig. 5.2. Schematic drawing of the cranial nerves and ganglia in the 10-mm pig embryo.

1. Cerebral hemisphere
2. Oculomotor nerve
3. Mesencephalon
4. Semilunar ganglion
5. Sensory root of the trigeminal nerve
6. Abducens
7. Geniculate ganglion
8. Sensory root of the facial nerve
9. Auditory nerve
10. Endolymphatic duct
11. Auditory vesicle

12. Sensory root of the glossopharyngeal nerve
13. Superior ganglion
14. Sensory root of the vagus nerve
15. Jugular ganglion
16. Roots of the spinal accessory nerve
17. Rhombencephalon
18. Froriep's ganglion
19. Nerve fibers contributed by Froriep's ganglion to the hypoglossal nerve

20. Roots of the hypoglossal nerve
21. Hypoglossal nerve
22. Spinal accessory nerve
23. Nodose ganglion
24. Vagus nerve
25. Petrosal ganglion
26. Glossopharyngeal nerve
27. Acoustic ganglion
28. Facial nerve
29. Mandibular branch of the trigeminal nerve

and the remaining section is still mesencephalon. The floor of the rhombencephalon fades out in more posterior sections (Photo 5.9). The **infundibulum** appears at about this same level; it quickly becomes continuous with the **diencephalon**, which now occupies the position occupied by the mesencephalon in more anterior sections. The cavity of the diencephalon is the major part of the **third ventricle**.

As you continue tracing sections posteriorly, identify **Rathke's pouch**, a small flattened vesicle contacting the floor of the infundibulum (Photos 5.11, 5.12). It quickly becomes continuous with the **stomodeum** in more posterior sections. The **eyes** can be identified at about

this level. Quickly trace sections *anteriorly*. As the eyes fade out, identify dense masses of mesodermal cells, the **eye muscle rudiments**, lying lateral to Rathke's pouch (Photos 5.11, 5.12). These rudiments will later form the extrinsic **eye muscles** (**inferior** and **superior oblique eye muscles**; **inferior**, **superior**, **medial** [or **internal**] and **lateral** [or **external**] **rectus eye muscles**). Return to the level of the eyes and identify the following components (Photos 5.13-5.15): **sensory** and **pigmented retinas**, **lens vesicles**, **optic stalks**, and **optic fissures**. Observe under high magnification that the pigmented retinas now contain **pigment granules**. Each lens vesicle is completely separated from, and covered externally by, **skin ecto-**

derm, the future **corneal epithelium**. (The adult **cornea** is derived from both skin ectoderm and **head mesenchyme** [neural crest].) Note that the head mesenchyme (both neural crest and mesoderm) surrounding each pigmented retina is condensed, initiating formation of the **choroid** (from both neural crest and mesoderm) and **sclera** (from neural crest) of the eye. Trace sections posteriorly and note that the lateral walls of the brain begin to bulge outward to form the **cerebral hemispheres** of the **telencephalon** and that the diencephalon fades out (Photos 5.18, 5.19). The cavities of the cerebral hemispheres are the **lateral ventricles** (or **first** and **second ventricles**) and are broadly continuous with the cavity of the middle portion of the telencephalon (minor part of the **third ventricle**). The connections between the lateral and third ventricles later become greatly narrowed as the **interventricular foramina**. Continue tracing sections posteriorly and note that the two cerebral hemispheres separate from one another (Photo 5.23). The portion of the telencephalon wall lying between the two cerebral hemispheres is the **lamina terminalis**. **Commissures** (transverse bundles of nerve fibers) will *later* form within the lamina terminalis to interconnect the cerebral hemispheres.

The prosencephalon is the source of several major adult brain components. The **diencephalon** forms the **thalamus, epithalamus, hypothalamus, pineal gland**, and **posterior pituitary gland** (derivative of the infundibulum); the **telencephalon** forms the **cerebrum, olfactory lobes**, and **corpora striata**.

2. Cranial nerves. The cranial nerves will be described in order, beginning with the first pair. Refer periodically to Fig. 5.2 to help you understand the three-dimensional structure of the cranial nerves and their relationships to the various subdivisions of the brain. Return to the most anterior section through the cerebral hemispheres (Photo 5.18) and trace sections posteriorly. Identify the **nasal pits**. The cavity of each nasal pit is separated at first from the **amniotic cavity**, but these cavities become confluent in more posterior sections (Photo 5.21). Many of the cells within the walls of the nasal pits are **young neurons**. Their **axons** are growing toward the cerebral hemispheres and may be identified in some specimens. These axons constitute on each side the **olfactory (I) cranial nerve** (a **sensory nerve**). **Dendrites** will develop *later* from the young neurons within each nasal pit and differentiate as **olfactory receptors** (receptors for the sense of smell). Identify elevated regions of head mesenchyme covered by skin ectoderm to either side of each nasal pit, the **lateral** and **medial nasal processes**. Each medial nasal process will eventually fuse laterally with a maxillary process and medially with the other medial nasal process to form the **upper jaw**. The lateral nasal processes eventually form the sides of the **nose**.

Return to the level of the optic fissures (Photo 5.15) and

examine the optic stalks and fissures under high magnification. Recall that within the region of each optic fissure the ventral lip of the optic cup is absent. Therefore the **marginal zone** of each sensory retina is continuous at this level, via the marginal zone of the optic stalk, with the marginal zone of the floor of the diencephalon.

Many of the cells within each sensory retina are **young neurons**. Some of these young neurons form **ganglion cells**, whose **axons** grow into the marginal zone of the sensory retina. These axons remain within the marginal zone and grow to the floor of the diencephalon, constituting the **optic (II) cranial nerves (sensory nerves)**. The optic nerves cannot be identified at this stage. (Because the axons constituting each optic nerve remain within the marginal zone throughout their extent, the so-called optic nerves are not actually true nerves; rather, they are **fiber tracts** located within derivatives of the diencephalon wall.)

Return to the level where the mesencephalon begins to separate from the metencephalon (Photo 5.4) and trace sections posteriorly. Notice two slight thickenings of the floor of the mesencephalon, one on either side of the midline. Each thickening constitutes a **motor nucleus**; it is formed by an accumulation of **cell bodies** of **young neurons** within the **intermediate zone**. Identify **axons** growing out from these motor nuclei. These axons constitute on each side the **oculomotor (III) cranial nerve** (a **motor nerve**). Continue to trace sections posteriorly, noting that the oculomotor nerves terminate near the eye muscle rudiments. These nerves will later innervate four pairs of extrinsic **eye muscles (inferior oblique eye muscles; inferior, superior**, and **medial rectus eye muscles)**.

The fourth cranial nerves, the **trochlear (IV) cranial nerves (motor nerves)**, consist of **axons** that also grow out from **motor nuclei** developing within the floor of the mesencephalon. (The trochlear nerves grow dorsad within the wall of the brain, cross over to the opposite side, and emerge from the brain in the isthmus region.) These nerves and their motor nuclei cannot be identified at this stage. The trochlear nerves eventually innervate the **superior oblique eye muscles**.

Return again to the level where mesencephalon begins to separate from metencephalon (Photo 5.4). As sections are traced posteriorly, **axons** appear at the broadest part of the rhombencephalon, continuous with the latter (Photo 5.5). These axons constitute on each side the **sensory root** of the **trigeminal (V) cranial nerve** (a **sensory** and **motor**, that is, **mixed**, nerve). These sensory roots are continuous with the large **semilunar ganglia** a few sections more posteriorly (Photo 5.6). Continue to trace sections posteriorly. Each semilunar ganglion eventually gives rise to three branches of the **trigeminal nerve**, as in the 72-hour chick embryo: the **ophthalmic branch** (to eye region), **maxillary branch** (to maxillary process), and **mandibular branch** (to mandibular process). Usually only the mandibular branches can be readily identified in the 10-mm pig embryo (Photo 5.10; but also see Photo 5.54, which shows the ophthalmic branch). The ophthalmic and maxillary branches are usually very small and are entirely sensory, consisting only of **dendrites**. The mandibular branches are eventually mixed but consist only of **dendrites** at this stage. (**Axons** later will grow out from a pair of **motor nuclei** within the floor of the rhombencephalon as the motor component of the mandibular branches.)

Return to the level where the neural tube is separated into three parts (Photo 5.6) and trace sections posteriorly. Identify **axons** growing out on each side from the floor of the rhombencephalon as the latter begins to fade out (Photo 5.8). These axons constitute the **abducens (VI) cranial nerves (motor nerves)**. The abducens nerves are usually cut across frontally, are very small, and quickly disappear as sections are traced posteriorly. They will later innervate the **lateral rectus eye muscles**.

Return to a level similar to Photo 5.1 and trace sections posteriorly. Note that a small vesicle soon appears on each side of the myelencephalon; these vesicles are the **endolymphatic ducts**. Identify the **auditory vesicles**, in slightly more posterior sections, which lie lateral to the endolymphatic ducts (Photo 5.3). The dorsal portion of each auditory vesicle forms the **utriculus** of the **inner ear**. The endolymphatic duct and auditory vesicle on each side are continuous in more posterior sections (Photo 5.5). The projection nearer the top of the section is the rudiment of the **posterior semicircular canal**. The projection nearer the bottom is the rudiment of the **anterior semicircular canal**. The rudiments of the anterior and posterior semicircular canals are attached to the dorsal portion of the auditory vesicle (that is, the future utriculus). (The rudiment of the **lateral semicircular canal** will form *later* from the utriculus portion of each auditory vesicle.) The rudiments of the anterior and posterior semicircular canals disappear in more posterior sections; this portion of each auditory vesicle is the future **sacculus** of the **inner ear** (Photo 5.6). The utriculi, semicircular canals, and sacculi form receptors for maintaining equilibrium. The auditory vesicles narrow and fade out as sections are traced posteriorly. The ventral portion of each auditory vesicle becomes the **cochlea** of the **inner ear** (Photo 5.8), which forms the receptor organ for sound (that is, the **spiral organ of Corti**).

The next two cranial nerves are closely associated with each other and with the auditory vesicles and will therefore be described together. Return to a level similar to Photo 5.6 and note that two ganglia lie just below each future sacculus. The lowermost ganglion is the **geniculate ganglion** of the **facial (VII) cranial nerve** (a **mixed nerve**); the uppermost ganglion is the **acoustic ganglion** of the **auditory (VIII) cranial nerve** (a **senso-** ry nerve). At about this level, or in slightly more posterior sections, **axons** constituting the **sensory roots** of the **facial nerves** interconnect the geniculate ganglia and rhombencephalon (Photo 5.7). (**Axons** constituting the **auditory nerves** similarly interconnect the acoustic ganglia and rhombencephalon, but they are difficult to identify with certainty because the auditory nerves are very short and closely applied to the sensory roots of the facial nerves.) Continue to trace sections posteriorly and identify the **facial nerves** extending from the geniculate ganglia into the *second* branchial arches (Photo 5.8). The facial nerves consist only of **dendrites** at this stage. (**Axons** will later grow out from a pair of **motor nuclei** within the floor of the rhombencephalon as the motor component of the facial nerves.) Note that the acoustic ganglia and auditory vesicles fade out at about the same level (Photo 5.8). (Each acoustic ganglion eventually subdivides into two parts [usually at later stages], the **vestibular** and **spiral [cochlear] ganglia**. **Dendrites** from **cell bodies** of **young neurons** within the **vestibular ganglia** terminate in equilibrium receptors in the semicircular canals, utriculi, and sacculi; **dendrites** from **cell bodies** of **young neurons** within the **spiral ganglia** terminate in the **spiral organ of Corti** within each cochlea.)

Return to the level at which the auditory vesicle and endolymphatic duct on each side become continuous (Photo 5.4). Identify **axons** continuous with the wall of the myelencephalon just above each auditory vesicle. These axons constitute on each side the **sensory root** of the **glossopharyngeal (IX) cranial nerve** (a **mixed nerve**). The **superior ganglia** of the **glossopharyngeal nerves** appear in slightly more posterior sections (Photo 5.5). Continue tracing sections posteriorly. The dorsomost portions of the glossopharyngeal nerves soon appear in place of the superior ganglia (Photo 5.6). Another pair of ganglia (much larger than the superior ganglia) appear in place of the glossopharyngeal nerves in more posterior sections (Photo 5.5). These are the **petrosal ganglia** of the glossopharyngeal nerves. Finally, the ventromost portions of the **glossopharyngeal nerves** appear in place of the petrosal ganglia in slightly more posterior sections (Photo 5.10, 5.11). Try to trace the ventromost portion of the glossopharyngeal nerves into the *third* branchial arches, where they terminate. **Cell bodies** of **young neurons** within the superior and petrosal ganglia form both **axons** and **dendrites**. The sensory roots of the glossopharyngeal nerves consist of **axons** formed by cell bodies of young neurons within both the superior and petrosal ganglia. The portion of the glossopharyngeal nerve that interconnects the superior and petrosal ganglia on each side (that is, the dorsomost portion) consists of **axons** formed by cell bodies of young neurons within the petrosal ganglion and **dendrites** formed by cell bodies of young neurons within the superior ganglion. The portion of the glossopharyngeal nerve that extends from the petrosal ganglion to the third branchial arch on each side (that is, the ventromost portion) consists of only **dendrites** formed by cell bodies of young neurons within both the superior and petrosal ganglia. (**Axons** constituting the motor component of the glossopharyngeal nerves will later grow out from a pair of **motor nuclei** within the floor of the myelencephalon.)

Return to a section similar to Photo 5.1 and trace sections posteriorly. At about the level at which the endolymphatic ducts appear, identify **axons** continuous with the wall of the myelencephalon just caudal to the auditory vesicles (Photo 5.3). These axons constitute on each side the **sensory root** of the **vagus (X) cranial nerve** (a **mixed nerve**).

The large **jugular ganglia** of the **vagus nerves** appear a few sections more posteriorly (Photo 5.4). Note that the jugular ganglia lie just caudal to the superior ganglia. The jugular ganglia fade out in more posterior sections and the **vagus nerves** appear in their place (Photo 5.6). (These portions of the vagus nerves consist of **dendrites** derived from **cell bodies** of **young neurons** within the jugular ganglia and axons derived from **cell bodies** of **young neurons** within a second pair of ganglia, the **nodose ganglia**.) Continue tracing sections posteriorly and observe that the vagus nerves fade out and that the **nodose ganglia** (derived from a pair of **epibranchial placodes**) appear in their place (Photo 5.11). The nodose ganglia fade out in the area of the *fourth* branchial arches in more posterior sections. It is difficult to trace the **vagus nerves** caudal to this level. They have extended caudad in many embryos and lie ventrolateral to the esophagus (Photo 5.23). At these caudal levels the vagus nerves consist mainly of **preganglionic axons** of the **parasympathetic division** of the **autonomic nervous system**. These axons grow out from **motor nuclei** (impossible to identify) developing within the floor of the myelencephalon. (The vagus nerves also contain **dendrites** formed by **cell bodies** of **young neurons** within both the jugular and nodose ganglia.)

Return to a level similar to Photo 5.1 and trace sections posteriorly. Identify **axons** growing out from the myelencephalon (Photo 5.2). These axons constitute on each side the **roots** of the **spinal accessory (XI) cranial nerve** (a **motor nerve**). A few sections more posteriorly these roots merge together on each side and are cut frontally as the **spinal accessory nerve** (Photo 5.3). Each spinal accessory nerve is cut across twice, a few sections more posteriorly (Photo 5.4). The part of each spinal accessory nerve nearest the top of the section can be traced caudally where it eventually fades out adjacent to the spinal cord. Trace the lower part of each spinal accessory nerve caudally. It first lies adjacent to the jugular ganglion, then adjacent to the vagus nerve (Photo 5.6), and then adjacent to the nodose ganglion (Photo 5.12); finally, it fades out. The spinal accessory nerves eventually innervate mainly derivatives of the *fourth* branchial arches and certain **neck** and **shoulder muscles**

(that is, the **sternocleidomastoid** and **trapezius muscles**).

Return to the level where the spinal accessory nerves are cut frontally (Photo 5.3) and trace sections posteriorly. Identify several bundles of **axons** growing out from the myelencephalon (Photo 5.5). These bundles constitute on each side the **roots** of the **hypoglossal (XII) cranial nerve** (a **motor nerve**). Trace these roots as far caudally as possible. It may be possible to observe in some embryos that the roots merge together on each side and continue caudally as the **hypoglossal nerve**. The hypoglossal nerves eventually innervate the **tongue muscles**.

The chart on the following page summarizes the development of the cranial nerves through the 10-mm stage.

3. Spinal cord and spinal nerves. Return to the level where the spinal accessory nerves are cut frontally (Photo 5.3) and trace sections posteriorly, watching the uppermost part of the neural tube in your sections. Observe a rather small but prominent ganglion on each side just caudal to the roots of the spinal accessory nerve; this is **Froriep's ganglion** (Photo 5.5), which contributes nerve fibers to the hypoglossal nerve. The level of Froriep's ganglia is still the level of the myelencephalon, but the **spinal cord** gradually appears a few sections more posteriorly. At about this level, identify a pair of ganglia lying adjacent to the cranial end of the spinal cord, the most cranial **spinal ganglia**.

Quickly trace the spinal cord caudally and find one or more sections that show all the following structures (Photo 5.32): **dorsal root** (composed of **axons** entering the spinal cord, **spinal ganglion**, and **dendrites** entering the spinal ganglion), **ventral root** (composed of **axons** leaving the spinal cord), and **spinal nerve** with its three branches: the **dorsal ramus**, **ventral ramus**, and **visceral ramus** (**ramus communicans**). The visceral ramus extends ventromedially on each side toward an accumulation of **neural crest cells**, one of the **sympathetic chain ganglia** of the **autonomic nervous system**. (The **sympathetic collateral** and **parasympathetic terminal ganglia** of the **autonomic nervous system** will form *later* from other accumulations of **neural crest cells**.)

Identify the following components of the spinal cord: **ventricular**; **intermediate**, and **marginal zones**, **alar plate** (dorsal half of spinal cord); **basal plate** (ventral half of spinal cord); **sulcus limitans** (slight depression on each side partially separating alar and basal plates), **roof** and **floor plates**; and **motor horns** (paired enlargements of the ventral part of the intermediate zone). Note the small diameter of the **notochord**; it is enclosed within a mass of sclerotome cells that will later form the **centrum** of each **vertebra**. Most of the notochord ultimately degenerates.

At the level of the foreleg buds, the ventral rami of adjacent spinal nerves are interconnected in complex ways to form the **brachial plexus** (Photo 5.25). (A **lum-**

bosacral plexus will form *later* at the level of the hindleg buds.) Note that the **foreleg** and **hindleg buds** consist of an outer layer of ectoderm, part of which is thickened as the **apical ectodermal ridge**, and a core of somatic mesoderm. Note a small dorsolateral swelling on each side at section levels between the limb buds, the **mammary ridge** (Photo 5.27).

Trace the spinal cord caudally until it is cut frontally due to the tail flexure (Photo 5.42). Identify the **dermatome**, **myotome**, and **sclerotome** subdivisions of the **somites**. Note that each sclerotome has a dense *caudal* portion and a less dense *cranial* portion. Each **vertebra** is formed by the fusion of the caudal half of one sclerotome with the cranial half of the next caudal sclerotome. By examining sections anterior or posterior to this level, identify the **spinal ganglia** and the **ventral roots** of the **spinal nerves**, which leave the ventrolateral walls of the spinal cord.

Cranial Nerves	Cranial Ganglia Present	Origin of Cranial Cranial Ganglia	Type of Nerve Fibers Present	Regions Innervated
olfactory	———————	———————	sensory	lining of nasal cavities
optic (nerves cannot be identified)	———————	———————	sensory	retinas
oculomotor	———————	———————	motor	inferior oblique eye muscles; inferior, superior, and medial rectus eye muscles
trochlear (nerves cannot be identified)	———————	———————	motor	superior oblique eye muscles
trigeminal	semilunar	neural crest cells and epibranchial placodes	sensory (*later* also motor)	primarily derivatives of first branchial arches
abducens	———————	———————	motor	lateral rectus eye muscles
facial	geniculate	neural crest cells and epibranchial placodes	sensory (*later* also motor)	primarily derivatives of second branchial arches
auditory	acoustic	auditory placodes	sensory	inner ears
glosso-pharyngeal	superior and petrosal	neural crest cells and epibranchial, placodes, respectively	sensory (*later* also motor)	primarily derivatives of third branchial arches
vagus	jugular and nodose	neural crest cells and epibranchial placodes, respectively	sensory and motor	primarily derivatives of fourth branchial arches
spinal accessory	———————	———————	motor	primarily derivatives of fourth branchial arches and certain neck and shoulder muscles
hypoglossal	———————	———————	motor	tongue muscles

b. Respiratory and digestive systems

Find the most anterior section showing a round **esophagus** (Photo 5.20). The round **trachea** usually lies beneath the esophagus at this level. Trace sections posteriorly a considerable distance, watching both esophagus and trachea. Note a distinct asymmetrical evagination of the trachea toward the *right* (apparent left) side. This evagination is the **eparterial bronchus** (Photo 5.23); it will form the upper lobe of the right **lung**. Continue tracing sections posteriorly and observe that a vesicle appears on each side of the trachea and quickly becomes continuous with the latter. These vesicles are the **lung buds**. The trachea fades out in more posterior sections and the lung buds continue caudally for a short distance (Photo 5.25). (Each of the two lung buds bifurcates *later* to form two lobes of the lung. Therefore, in the adult the *right* lung contains three lobes, and the *left* lung, only two.) Note at this level that the **pleural cavities** lie lateral to the lung buds and that they are continuous laterally with the **peritoneal cavity**. A thin layer of splanchnic mesoderm, the **visceral pleura**, lies just medial to each pleural cavity (examine this layer under high magnification). The **parietal pleura** has not yet formed at this level. The pleural cavities are isolated from all other portions of the coelom more *cranially*, and visceral and parietal pleura can be identified (Photo 5.24).

Return to a level similar to Photo 5.25 and trace sections posteriorly, noting that the esophagus shifts ventrad and the lung buds fade out (Photo 5.26). Identify the **mesoesophagus**. The **stomach** appears in more posterior sections (Photo 5.27). It has undergone a rotation around its longitudinal axis of approximately 45°, so its original *dorsal* surface lies to the *left* (apparent right) and its *ventral* surface lies to the *right* (apparent left). The original dorsal surface forms the **greater curvature** of the adult stomach; the original ventral surface forms the **lesser curvature**. Identify the **dorsal mesogaster**, which enlarges vastly as the adult **greater omentum**. (Mesenchymal cells will aggregate *later* within the dorsal mesogaster and form the **spleen**.) Also identify the mesentery connecting stomach and **liver**, the **hepatogastric ligament**. Note that the liver is subdivided into paired **dorsal** and **ventral lobes**, and that it consists of irregularly shaped cords of cells, the **hepatic cords**; the hepatic cords are separated by irregular vascular spaces, the **hepatic sinusoids**. The liver at this stage is functioning as a **hematopoietic organ** (that is, it is temporarily producing blood cells). Identify the **omental bursa**, a closed cavity to the *right* (apparent left) of the stomach.

Trace sections posteriorly, noting that the omental bursa opens into the **peritoneal cavity** via the **epiploic foramen** (Photo 5.29). Lining the peritoneal cavity are the **visceral peritoneum** (a thin layer of splanchnic mesoderm on the surface of the abdominal organs) and the

parietal peritoneum (a thin layer of somatic mesoderm on the internal surface of the ventral and lateral body walls). Examine these layers under high magnification.

Return to the level of the stomach (Photo 5.27) and trace sections posteriorly, noting that the stomach fades out and the **duodenum** appears in its place (Photo 5.28). Sections at about this level also cut through the endodermal **hepatic duct**, located within the dorsal part of the liver. The mesentery connecting duodenum and liver is the **hepatoduodenal ligament**. (The hepatogastric and hepatoduodenal ligaments collectively constitute the adult **lesser omentum**.) The mesentery connecting the liver to the ventral body wall is the **falciform ligament**. In slightly more posterior sections, the hepatic duct constricts into an upper **common bile duct**, which quickly joins the duodenum, and a lower **cystic duct** (Photo 5.29). The cystic duct gradually expands as the **gallbladder** in more posterior sections. The gallbladder may be bilobed or trilobed in some specimens. Continue to trace sections posteriorly until the liver disappears. Then reverse direction and trace sections *anteriorly*, noting the reappearance of the liver and finally the disappearance of its cranial end. Note that the cranial end of the liver terminates in a transverse partition, the **septum transversum** (Photo 5.25).

Return to the level where the common bile duct joins the duodenum (Photo 5.29) and trace sections posteriorly. As the common bile duct disappears, a small, darkly stained solid mass of cells, the **ventral pancreatic rudiment**, appears in its place (Photo 5.30). The ventral pancreatic rudiment develops as an outgrowth from the common bile duct; its original connection to the common bile duct, the **ventral pancreatic duct (duct of Wirsung)**, is degenerating and probably cannot be identified. (In humans this connection usually persists.) Identify the **dorsal pancreatic rudiment** at about this same level. This rudiment develops as an outgrowth from the duodenum; its connection to the duodenum, the **dorsal pancreatic duct (duct of Santorini)**, persists in the adult pig to transport to the duodenum certain pancreatic secretions involved in digestion. (In humans this connection usually degenerates.) Dorsal and ventral pancreatic rudiments will fuse *later* to form the **pancreas**.

Trace sections posteriorly from the level at which the dorsal pancreatic duct connects to the duodenum (Photo 5.30) and note that the duodenum lies dorsal to another region of the gut, which is contained within the lightly stained tissue of the **umbilical cord** (that is, the latter region of the gut is contained within **Wharton's jelly**, a mesodermal derivative). This latter region of the gut is the **cranial limb of the intestinal loop** (Photo 5.31). The **intestinal loop** has herniated into the **extraembryonic coelom** of the umbilical cord, forming the temporary **umbilical hernia**. (The cranial limb of the intestinal loop later forms mainly the **jejunum** and most of the **ileum** of the adult **small intestine**.) The duodenum and cranial limb of the intestinal loop progressively approach one another and become continuous as sections are traced posteriorly. These regions of the gut are cut frontally at this level (Fig. 5.3, level a; Photo 5.34), and they quickly disappear as sections are traced posteriorly. Return to the level where the duodenum and cranial limb of the intestinal loop are cut frontally and trace sections *anteriorly,* observing the cranial limb of the intestinal loop. Note that it rotates somewhat toward the *right* (apparent left) to lie alongside another portion of the gut, the **caudal limb of the intestinal loop** (Photo 5.33). (The caudal limb of the intestinal loop later forms principally the **ascending colon** and most of the **transverse colon** of the adult **large intestine**.) Trace sections *anteriorly* a considerable distance and try to observe the continuity of the two limbs at the apex of the intestinal loop (Fig. 5.3, level b). The **yolk sac** normally attaches to the intestinal loop at its apex. If the yolk sac is present in your specimen, it will first appear as a slight circular expansion of the gut tube as the apex of the intestinal loop is traced *anteriorly.* The yolk sac appears as a large, irregularly shaped vesicle in more *anterior* sections (Photo 5.26). (Unfortunately, in some specimens the umbilical cord has been cut off so close to the body that the apex of the intestinal loop and the yolk sac are not present in sections.)

Return to the level of the apex of the intestinal loop, or as close to this level as possible, and trace sections *posteriorly*, observing the caudal limb of the intestinal loop. It rotates somewhat toward the *left* (apparent right) and is eventually suspended between the two large mesonephric kidneys by a long, thin mesentery (Photo 5.36). Because this portion of the caudal limb of the intestinal loop will form part of the **colon** of the adult **large intestine**, this mesentery is part of the **mesocolon**. The caudal limb of the intestinal loop approaches another region of the gut as you continue to trace sections posteriorly. This latter region is the caudal part of the **colon** (Photo 5.38); it quickly joins the caudal limb of the intestinal loop, and these two structures are cut frontally (Fig. 5.3, level c). They then quickly disappear as sections are traced posteriorly. Reverse direction and trace sections *anteriorly,* following the lower part of the gut in your sections (that is, the colon). Note that the gut becomes slightly elongated dorsoventrally, indicating the level of the **rectum**. The rectum then quickly joins a large cavity, the **cloaca** (Photo 5.36). The cloaca is subdividing at this stage into a dorsal (lower) **rectum** and a ventral (upper) **urogenital sinus**. The mesodermal ingrowth that separates the cloaca into these two chambers is the **cloacal septum**.

Trace sections *anteriorly* while watching the cloaca; note that as the cloaca fades out, a solid cord of cells, the **cloacal membrane**, can be identified (Photo 5.35). (This membrane looks different from the one seen in the 72-hour chick embryo because it is cut frontally rather than transversely.) Note that the cloacal membrane is contained within a large swelling, the **genital eminence**. A small degenerating **tail gut** can be identified at about this level in some specimens.

Trace sections *posteriorly* and try to observe that the tail gut opens cranially into the cloaca. Note that the cloaca becomes completely subdivided into the **urogenital sinus** and **rectum** as sections are traced posteriorly. The **colon** again appears in more posterior sections (note its rounded, rather than elongated, shape) and an endoderm-lined cavity, the **allantois**, emerges from the umbilical cord to join the urogenital sinus ventrally (Photo 5.37). Reverse direction and trace sections *anteriorly*, following the allantois out into the umbilical cord (Photos 5.36-5.31). Note that within the umbilical cord the allantois lies between two large arteries, the **umbilical (allantoic) arteries**.

Return to the most anterior section showing the round **trachea** beneath the round **esophagus** (Photo 5.20) and trace sections *anteriorly*. The trachea elongates dorsoventrally as the **larynx**, which lies beneath the somewhat triangularly shaped esophagus. (The walls of the larynx tend to be in contact and may therefore be difficult to identify.) In more *anterior* sections, the esophagus expands, indicating the level of the **pharynx**, and the latter immediately becomes continuous with the larynx (Photo 5.17). The opening of the larynx into the pharynx is the **glottis**. The **fourth pharyngeal pouches** can be identified at about this level. They usually first appear as isolated vesicles ventrolateral to the pharynx, but they then quickly join the pharynx as sections are traced *anteriorly*. The fourth pharyngeal pouches are the sources of the adult **superior parathyroid glands**. A small dorsolateral evagination of each fourth pharyngeal pouch can be identified in some specimens (Photo 5.16). These are the rudimentary **fifth pharyngeal pouches (ultimobranchial bodies)**. They are later incorporated into the thyroid gland, where they give rise to **C (parafollicular) cells**, which produce the hormone **calcitonin**.

Note a broad *ectodermal* invagination on each side, lateral to the pharynx, as you continue to trace sections *anteriorly* (Photo 5.14), the **cervical sinus** (combined **third** and **fourth branchial grooves**). The larynx usually opens broadly into the pharynx via the glottis at about this level. Note the prominent swellings on either side of the glottis, the **arytenoid swellings**. These will later form the **arytenoid cartilages** of the **larynx**. Identify the **thyroid gland**, lying beneath the floor of the pharynx at about this level. Its original connection to the floor of the pharynx, the **thyroglossal duct**, has degenerated. Such a ductless gland that drains its secretions only into the blood stream is an **endocrine gland**. Continue to trace sections *anteriorly* and identify a midline swelling of the floor of the pharynx just cranial to the glottis. This

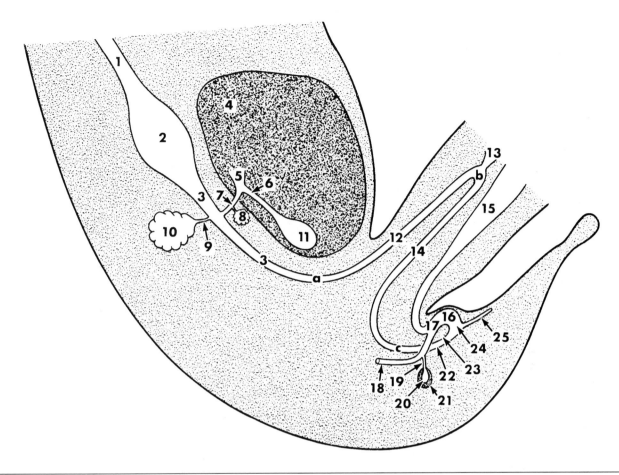

Fig. 5.3. Schematic drawing of the digestive and urogenital systems in the 10-mm pig embryo. Letters a–c indicate gut levels described in the text. The outer line indicates the boundaries of the caudal two-thirds of the embryo and of the umbilical cord.

1. Esophagus
2. Stomach
3. Duodenum
4. Liver
5. Hepatic duct
6. Cystic duct
7. Common bile duct
8. Ventral pancreatic rudiment
9. Dorsal pancreatic duct
10. Dorsal pancreatic rudiment
11. Gallbladder
12. Cranial limb of the intestinal loop
13. Yolk sac
14. Caudal limb of the intestinal loop
15. Allantois
16. Cloaca
17. Urogenital sinus
18. Mesonephric duct
19. Ureter
20. Renal pelvis
21. Nephrogenic tissue
22. Colon
23. Cloacal septum
24. Rectum
25. Tail gut

swelling is the **epiglottis** (Photo 5.13). It is derived from the third and fourth branchial arches.

Return again to the most posterior section through the glottis (Photo 5.17) and trace sections *anteriorly*. Identify a pair of flattened vesicles lying just above (dorsal to) the third branchial arches, the **third pharyngeal pouches** (Photo 5.15). The ventral portions of the two third-pharyngeal pouches (that is, the portions shown in Photo 5.15) *later* fuse in the midline to form the **thymus gland**. The third pharyngeal pouches move dorsad and become continuous with the pharynx as sections

are traced *anteriorly*. Identify the **third closing plates** at about this level. The dorsal portions of the third pharyngeal pouches (that is, the portions shown in Photos 5.13, 5.14) form the **inferior parathyroid glands**. The rudiments of these glands can be seen in most embryos under high magnification as a thickening on each side just beneath the cavity of the third pharyngeal pouch. (Each entire third pharyngeal pouch, including the rudiments of the inferior parathyroid and thymus glands, later migrates caudal to the fourth and fifth pharyngeal pouches. Thus, in the adult the inferior parathyroid glands [derived from the third pharyngeal pouches] lie *caudal* to the superior parathyroid glands [derived from the fourth pharyngeal pouches].)

Continue to trace sections *anteriorly* and identify the **second pharyngeal pouches** (Photos 5.12-5.10), the sources of the adult **palatine tonsils**. Also identify the **second branchial grooves** and **second closing plates**. The latter may be partially ruptured, forming the **second branchial clefts**. If so, the exact boundary between the second branchial groove and pharyngeal pouch on each side may be difficult to determine. Identify the **tongue rudiments** at about the level of the second pharyngeal pouches. They consist of a pair of **lateral lingual swellings**, a midline **tuberculum impar**, and a midline **copula** (Photo 5.11-5.12). The copula is derived from the second branchial arches; the lateral lingual swellings and tuberculum impar from the first branchial arches (although these three latter structures are now fused with the second branchial arches). Note that the tongue rudiments protrude into the pharynx.

Trace sections *anteriorly*. At about the level at which the tongue rudiments fade out, identify the **first branchial grooves**, **closing plates**, and **pharyngeal pouches** (Photos 5.10, 5.9). The first branchial groove on each side forms the adult **external auditory meatus** (**external ear canal**). Each first closing plate, after mesodermal cells penetrate between its ectoderm and endoderm, forms a **tympanic membrane**. Note that each first pharyngeal pouch consists of two portions (Photo 5.10). The upper portion in your sections connects to the pharynx and forms the adult **eustachian** (**pharyngotympanic**) **tube**; the lower portion forms the **tympanic** (**middle ear**) **cavity**. The tympanic cavity on each side will eventually surround the **ear ossicles** (that is, the **malleus** and **incus** [derivatives of the first branchial arch] and the **stapes** [derivative of the second branchial arch]). Elevations will *later* form from the first and second branchial arches to either side of the first branchial grooves. These elevations will subsequently fuse on each side to form the **pinna** (**auricle**) of the **external ear**.

c. Urogenital system

Return to the level of the cloacal membrane (Photo 5.35) and trace sections posteriorly until the cloaca is reached. Note that the ventral (upper) portion of the cloaca broadens transversely. This broadened portion is the **urogenital sinus**. The urogenital sinus is continuous with the **mesonephric ducts** in more posterior sections (Fig. 5.3; Photo 5.37). Continue tracing sections posteriorly, following the mesonephric ducts. Note that they lie lateral to the colon (Photo 5.38). They are quickly cut frontally, as they join another portion of the mesonephric ducts located in the ventral part of the large **mesonephric kidneys** (Photo 5.39). The mesonephric ducts disappear a few sections more posteriorly.

Return to the level where the mesonephric ducts are cut frontally (Photo 5.39) and trace that part of each mesonephric duct within the mesonephric kidney *anteriorly*. Note that numerous **mesonephric tubules** are continuous with it. Continue to trace sections *anteriorly* until the mesonephric ducts can no longer be identified. The mesonephric duct on each side will form the **epididymis**, **vas deferens**, and **ejaculatory duct** of the adult male reproductive system; it will also give rise to an evagination that forms the **seminal vesicle**. The mesonephric ducts mainly degenerate in the female.

Note the tremendous size of the mesonephric kidneys relative to that of other structures of the pig embryo. This large size probably counteracts the rather inefficient **placenta** of pig embryos (**diffuse, epitheliochorialis type** with a **placental membrane** [**barrier**] composed of many layers). The placenta apparently does not remove nitrogenous wastes from the bloodstream of pig embryos very readily; pig embryos therefore possess massive mesonephric kidneys to take care of this function. Examine a section cut through about the middle of the mesonephric kidneys (Photos 5.34-5.36). Note that this type of kidney consists mostly of **mesonephric tubules**. These tubules mainly degenerate in females. In males some mesonephric tubules persist as the **efferent ductules** (**vasa efferentia**) of the adult reproductive system; the remaining tubules degenerate. At the medial side of each mesonephric kidney, note several large spaces bounded by a very flat epithelium and filled with cells. These expanded spaces are the **glomerular** (**Bowman's**) **capsules** of the mesonephric tubules (Photo 5.34). The cells within the glomerular capsules are capillaries, constituting the **glomeruli**, and contained blood cells (Photo 5.35). Glomerular capsules and glomeruli of mesonephric kidneys later degenerate in both males and females.

The **gonad rudiments** are just forming at this stage as a thickening on the medial side of each mesonephric kidney (Photos 5.29, 5.33, 5.35). Each gonad rudiment consists of a localized thickening of visceral peritoneum (that is, a localized thickening of the splanchnic mesoderm that covers the organs of the peritoneal cavity), called the **germinal epithelium**, and a subjacent region of condensed mesenchyme (Photo 5. 30). The gonad rudiments contain **primordial germ cells** (not readily identifiable), which in mammals originate from the **endoderm** of the caudal portion of the **yolk sac**. These cells then undergo an extensive migration through the splanchnic mesoderm of the yolk sac and the dorsal gut mesentery to reach the gonad rudiments.

Return to the level where the mesonephric ducts connect to the urogenital sinus (Photo 5.37) and trace sections posteriorly. Note that a small duct emerges from the dorsal (lower) side of each mesonephric duct shortly after the urogenital sinus fades out (Photo 5.39). These ducts are the ureters. Continue tracing sections posteriorly, noting that the **ureters** separate from the mesonephric ducts and that each eventually expands as the **renal pelvis** (Photo 5.40). The distinct epithelial walls of these structures (that is, the layer next to the lumen) later bud repeatedly, forming the **major** and **minor calyces** and **collecting tubules** of the **metanephric kidneys**. The very dark mass of cells just outside each renal pelvis consists of condensed **nephrogenic tissue** (Photo 5.40), which forms the **secretory tubules** of the **metanephric kidneys**. Thus, development of the metanephric kidneys involves both epithelial and mesenchymal components. It has been demonstrated experimentally that neither of these components can develop in the absence of the other. Therefore, development of the metanephric kidneys depends on **epithelial-mesenchymal interactions** (**inductions**), which are similarly involved in the development of many other structures whose rudiments consist of epithelial and mesenchymal components.

d. Circulatory system

1. Arterial system. Begin with the most posterior section through the **conotruncus (bulbus cordis)** (Photo 5.22) and trace sections *anteriorly*. Observe that the cavity of the conotruncus becomes separated into two cavities by the approximation of two ingrowths of the conotruncus wall. These ingrowths are the **bulbar ridges** (Photo 5.20); they are fused in more *anterior* sections as the **bulbar septum** (Photo 5.19), which separates the conotruncus into two vessels. The vessel to the *right* (apparent left) is the **aortic trunk (ascending aorta)**; the vessel to the *left* is the **pulmonary trunk**. Note that the pulmonary trunk extends dorsad at about this level, so as to be cut frontally, and it is continuous with the two **sixth-aortic arches** (Photo 5.19). The *left* (apparent right) sixth aortic arch is usually larger in diameter than is the *right*. Continue tracing sections *anteriorly* and observe that the sixth aortic arches join the **dorsal aortae** (Photos 5.17, 5.16). The distal end of the *left* sixth aortic arch (that is, the portion that connects to the *left* dorsal aorta) is the **ductus arteriosus** (Fig. 5.4); it persists until birth as an important blood channel acting as a shunt, enabling much of the blood to bypass the functionless lungs. The ductus arteriosus is converted after birth into a fibrous ligament, the **ligamentum arteriosum**. The distal end of the *right* sixth aortic arch soon degenerates. The *right* (apparent left) dorsal aorta is usually smaller than the *left* at the level where the sixth aortic arches join the dorsal aortae. This level of the *right* dorsal aorta becomes part of the adult **right subclavian artery**, carrying blood mainly to the right foreleg, whereas this level of the *left* dorsal aorta becomes part of the adult **arch of the aorta**, carrying blood to most regions of the body (Fig. 5.4).

Return to the level where the pulmonary trunk gives rise to the sixth aortic arches (Photo 5.19) and trace sections *posteriorly*. Identify the **pulmonary arteries** as the sixth aortic arches fade out (Photo 5.20). These arter-

ies originated as caudal outgrowths from the sixth aortic arches. Try to trace the pulmonary arteries back toward the lung buds.

Return to the level of the bulbar septum (Photo 5.19) and trace sections *anteriorly*, noting the disappearance of the pulmonary trunk after it gives rise to the sixth aortic arches (Photo 5.18). (The **fifth aortic arches** are rudimentary and have degenerated by this stage.) Identify the **fourth aortic arches** (Photo 5.15), continuous with the aortic trunk, and note that they quickly join the dorsal aortae (Photos 5.15, 5.14). The *left* (apparent right) fourth aortic arch is typically larger than the *right*. The former is incorporated into the adult **arch of the aorta**, and the latter into the adult **right subclavian artery** (Fig. 5.4). Observe that the aortic trunk terminates cranially by dividing into the two **third-aortic arches** (Photo 5.14). (The **first** and **second aortic arches** have degenerated or are degenerating by this stage.) Try to identify a pair of small vessels extending ventrad from the third aortic arches at about this level. These vessels are the **external carotid arteries**; attempt to trace them into the developing lower jaw (that is, into the mandibular processes).

Return to the level where the third aortic arches join the dorsal aortae (Photo 5.14) and trace sections *anteriorly*. Note that the third aortic arches join the dorsal aortae (Photo 5.10). The third aortic arches form the adult **common carotid arteries** and part of the adult **internal carotid arteries**. The remainder of each internal carotid artery is formed from an extension of each dorsal aorta, cranial to the connection between the third aortic arch and dorsal aorta (Fig. 5.4). Observe that the portion of each dorsal aorta lying between its connection to the fourth and third aortic arches is considerably reduced in diameter (Fig. 5.4; Photos 5.10-5.12). This level of each dorsal aorta, the **ductus caroticus**, will soon completely degenerate. Following this degeneration, blood passing from the aortic trunk into the *third* aortic arches will flow cranially through the **internal carotid arteries**, whereas blood passing from the aortic trunk into the *fourth* aortic arches will flow caudally through the **right subclavian artery** and the **arch of the aorta**.

Continue tracing sections *anteriorly*, following the internal carotid arteries (that is, the cranial extensions of the dorsal aortae). They are quickly cut frontally (Fig. 5.5, level a; Photo 5.9) and disappear a few sections more *anteriorly*. Return to the level where the internal carotid arteries are cut frontally (Fig. 5.5, level a; Photo 5.9). Shift your attention to the lower end of each internal carotid artery (that is, the end nearest the diencephalon) and trace sections *posteriorly*. The internal carotid arteries gradually move downward toward the diencephalon (Fig. 5.5, level b; Photos 5.10-5.12), where each joins another vessel cut frontally. These latter vessels are the cranial ends of the internal carotid arteries (Fig. 5.5, level c). Each is continuous at its lower end with a small **anterior cerebral artery** (Photo 5.12). Continue tracing sections *posteriorly,* watching the anterior cerebral arteries. They are difficult to follow because they are just developing, but they can usually be identified in more *posterior* sections lateral to the diencephalon (Photo 5.15).

Return to the level where the internal carotid and anterior cerebral arteries on each side are continuous (Photo 5.12) and trace sections *anteriorly,* watching the cranial end of each internal carotid artery (that is, the portion that is cut frontally). The **posterior communicating arteries** quickly appear in place of these latter frontally cut vessels (Fig. 5.5, level d; Photo 5.10). (The anterior cerebral and posterior communicating arteries will later become components of the **arterial circle of Willis,** a complex of blood vessels delivering blood to

the ventral surface of the brain.) Continue tracing sections *anteriorly,* following the posterior communicating arteries. They first lie lateral to the infundibulum (Photo 5.9) and then ventral to the mesencephalon (Photos 5.8-5.6). Note that they approach a vessel lying in the midline, just beneath the rhombencephalon, the **basilar artery,** and that the three vessels become continuous (Fig. 5.5, level e). (The exact region where the two posterior communicating arteries and basilar artery become continuous is difficult to identify.)

Reverse direction and trace sections *posteriorly,* following the basilar artery. As the rhombencephalon begins to fade out, identify a single vessel lying in the midline, just above the rhombencephalon (Photos 5.7, 5.8). This vessel is part of the basilar artery; in more

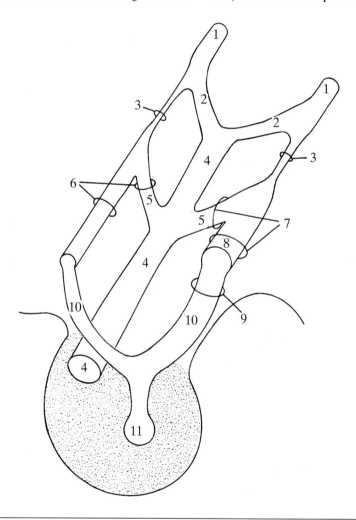

Fig. 5.4. Drawing of a reconstruction of the aortic arches of the 10-mm pig embryo. (Because this reconstruction was made directly from transverse sections, the *right* side of the embryo is drawn on the apparent left side of the reconstruction and the *left* side of the embryo is drawn on the apparent right side of the reconstruction.)

1. Internal carotid artery
2. Third aortic arch
3. Ductus caroticus
4. Aortic trunk
5. Fourth aortic arch
6. Two components of adult right subclavian artery
7. Two components of adult arch of the aorta
8. Dorsal aorta
9. Ductus arteriosus
10. Sixth aortic arch
11. Pulmonary trunk

posterior sections, the rhombencephalon fades out and the two parts of the basilar artery join and are cut frontally (Fig. 5.5, level f; Photo 5.10). The basilar artery fades out in more posterior sections.

Return to the level where the basilar artery is cut frontally (Fig. 5.5, level f; Photo 5.10) and trace sections *anteriorly*, following the upper portion of the basilar artery. Note that it thins dorsoventrally and is continuous with two vessels, the **vertebral arteries** (Fig. 5.5, level g; Photo 5.7), which extend dorsad. The vertebral arteries are irregularly shaped in this region. Trace sections *posteriorly*, following the vertebral arteries beneath the spinal cord. (In the adult, each vertebral artery receives its blood directly from a **subclavian artery**.)

Return to the level of the stomach (Photo 5.27) and identify the **descending aorta**. Trace sections posteriorly. At about the level where the stomach fades out, or, more commonly, in slightly more posterior sections, identify a single midline vessel continuous with the descending aorta ventrally, the **celiac artery**. It originated by fusion of a pair of **vitelline arteries** at an earlier stage. Continue tracing sections posteriorly and identify paired vessels continuous with the descending aorta dorsolaterally; these are the **intersegmental arteries**. (The adult **left subclavian artery** is derived exclusively from a left intersegmental artery at the level of the foreleg buds. The adult **right subclavian artery** is derived from a right intersegmental artery at the level of the foreleg buds [Photo 5.24], part of the right dorsal aorta, and part of the right fourth aortic arch; see Fig. 5.4.)

Continue to trace sections posteriorly and identify small ventrolateral branches of the descending aorta. These are the **mesonephric (lateral) arteries**, which carry blood to glomeruli of the mesonephric kidneys. Identify a large midline vessel continuous with the descending aorta ventrally in sections at about the level of the caudal end of the liver (Photo 5.33). This vessel is the **superior mesenteric artery**; it also originated by fusion of a pair of **vitelline arteries**. The distal end of the superior mesenteric artery can be identified within the umbilical cord at about this level (Photo 5.33). Trace sections *posteriorly* until the two portions of the superior mesenteric artery join and are cut frontally (Photo 5.35).

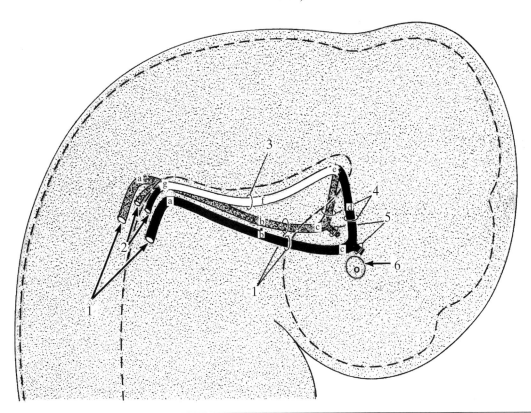

Fig. 5.5. Schematic drawing of the arteries in the head of the 10-mm pig embryo. Arteries on the right side of the embryo are indicated by heavy black lines and on the left side by stippling. The basilar artery (3) is in the midline. Letters a–g indicate specific levels of arteries described in the text. The dashed line indicates the boundaries of the brain and spinal cord, and the outer line, the boundaries of the head.

1. Internal carotid arteries
2. Vertebral arteries
3. Basilar artery
4. Posterior communicating arteries
5. Anterior cerebral arteries
6. Eye

Quickly trace sections posteriorly until the descending aorta is cut frontally due to the tail flexure (Photo 5.41). The descending aorta disappears in more posterior sections. Return to the level at which the descending aorta is cut frontally (Photo 5.41) and trace sections *anteriorly,* following the lower part of this vessel. Identify the right and left **umbilical arteries** continuous with the descending aorta (Photo 5.41). A smaller midline portion of the descending aorta can be traced into the tail as the **caudal artery** (Photos 5.41, 5.40). Try to identify a small lateral branch from each umbilical artery; these branches are the **external iliac arteries** (Photo 5.40). Continue tracing sections *anteriorly.* The umbilical arteries first lie at the base of the hindleg buds (Photo 5.39, 5.38). This portion of the umbilical arteries forms part of the adult **internal iliac arteries**. The umbilical arteries can be identified within the umbilical cord in more *anterior* sections (Photos 5.35-5.33, 5.31).

2. *Heart.* Return to the most posterior section through the bulbar septum (Photo 5.19) and trace sections posteriorly. Identify the **right** and **left atria** and the muscular **left ventricle** (Photo 5.20, 5.21). Continue tracing sections posteriorly and watch the two atria approach each other and become continuous (Photo 5.22). The conotruncus and muscular **right ventricle** are also continuous at about this level. The cavities of the right and left atria are partially separated by a partition, the **septum primum (septum I)**. This septum is incomplete dorsally as the **foramen secundum (foramen II)**. (Earlier there existed a more ventral opening in the septum primum, the **foramen primum [foramen I]**, which is now closed.) In slightly more posterior sections, the septum primum seems to separate completely the two atria (Photo 5.23). Continue tracing sections posteriorly and observe that the cavities of the atrium and ventricle on each side are continuous via an **atrioventricular**

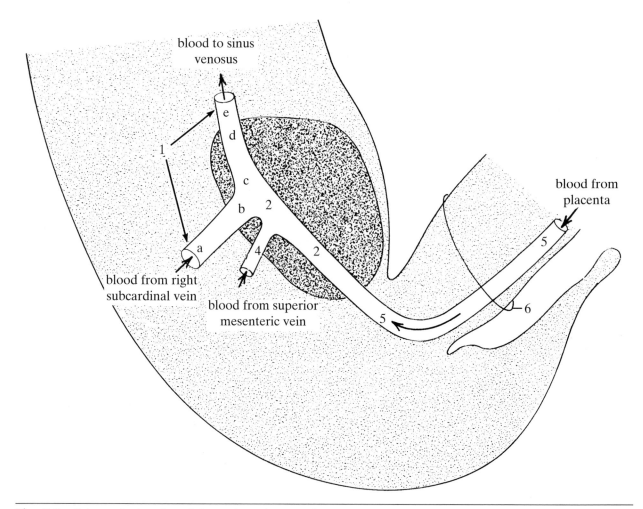

Fig. 5.6. Schematic drawing of the major veins associated with the liver in the 10-mm pig embryo. The *right* umbilical vein is not shown. Letters a-e indicate specific levels of the inferior vena cava (1) described in the text. The outer line indicates the boundaries of the caudal two-thirds of the embryo and of the umbilical cord. Arrows indicate direction of blood flow.

1. Inferior vena cava	3. Liver	5. Left umbilical vein
2. Ductus venosus	4. Portal vein	6. Umbilical cord

canal (Photo 5.24). (The **tricuspid valve** *later* develops within the *right* atrioventricular canal; the **bicuspid [mitral] valve** develops within the *left*.) The atrioventricular canals lie on either side of a midline mass of lightly stained tissue, the **endocardial cushion**. Identify the prominent **interventricular septum**, partially separating the cavity of the ventricle into right and left sides. At this stage the interventricular septum is incomplete dorsally as the **interventricular foramen**. (This foramen will be closed later by the **septum membranaceum**, derived from outgrowths from the endocardial cushion and part of the bulbar septum.)

Return to the level of the foramen secundum (Photo 5.22) and trace sections posteriorly. Identify the **sinus venosus** lying above the *right* atrium (Photos 5.23, 5.24) and the **sinoatrial valve** projecting into the cavity of the *right* atrium (Photo 5.23). The sinus venosus is eventually absorbed by the wall of the *right* atrium as the **pacemaker** of the adult heart. The sinoatrial valve consists of right and left **valve flaps**, which are usually in contact with one another. The *left* (apparent right) valve flap forms part of the **septum secundum (septum II)**, which will therefore lie to the *right* (apparent left) of the septum primum. An opening, the **foramen ovale**, persists between the septum secundum and endocardial cushion (Photo 5.23). The foramen ovale is staggered with respect to the opening in the septum primum (that is, the **foramen secundum**). The positioning of these two foramina and atrial septa is such that the septum primum acts as a flutter valve allowing blood to pass only *from the right to the left* atrium.

The wall of the heart consists of three layers (which are difficult to distinguish from one another). The inner layer of the heart (the one that immediately encloses blood cells) is the **endocardium**. The middle (and thickest) layer is the **myocardium**. The outer layer is the **epicardium (visceral pericardium)**. This layer originates from the region of the **sinus venosus**, near the **dorsal mesocardium**. The epicardium grows downward from this region as a cellular sheet, eventually covering the outer surface of the entire myocardium.

3. Venous system. Return again to the level of the foramen secundum (Photo 5.22) and trace sections posteriorly. Identify a single **pulmonary vein** continuous with the *left* atrium shortly before it fades out (Photo 5.24). (This vein will later branch to form four **pulmonary veins**, which will open directly into the *left* atrium in the adult.)

Quickly trace sections posteriorly until the umbilical cord can be identified (Photo 5.31). Identify the right and left **umbilical (allantoic) veins** within the umbilical cord (Photos 5.31, 5.33-5.35). The *right* (apparent left) umbilical vein is smaller than the *left* because it is degenerating. After this degeneration occurs, all oxy-

genated blood will flow from the **placenta** to the embryo via the *left* umbilical vein. Each umbilical vein is cut across twice. The two portions of each vein become continuous and are cut frontally as sections are traced posteriorly (Photos 5.36, 5.37). The umbilical veins fade out in more posterior sections.

Return to the level where the umbilical veins are cut

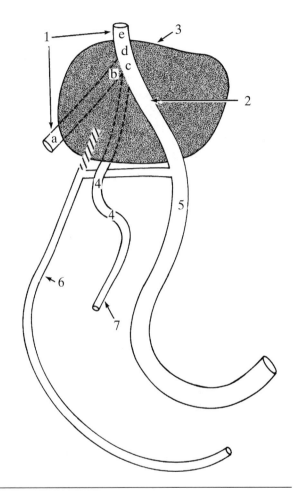

Fig. 5.7. Drawing of a reconstruction of the major veins associated with the liver in the 10-mm pig embryo. This drawing is similar to Fig. 5.6, but the vessels are viewed from the front rather than the side. Letters a-e indicate specific levels of the inferior vena cava (1) described in the text.

1. Inferior vena cava
2. Ductus venosus
3. Liver
4. Portal vein
5. Left umbilical vein
6. Right umbilical vein
7. Superior mesenteric vein

frontally (Photos 5.37, 5.36) and trace sections *anteriorly,* following the upper portion of the umbilical veins. In addition, observe the region between the two mesonephric kidneys, just beneath the descending aorta. Right and left **subcardinal veins** may be located there (Photos 5.33, 5.31), or the subcardinal veins may have fused to form a **subcardinal anastomosis** (Photos 5.34-5.36). (The subcardinal veins are fused at several localized levels in some embryos, forming multiple subcardinal anastomoses.) Continue tracing sections *anteriorly* and note that the *left* (apparent right) umbilical vein enters the liver (Photo 5.31). (The right umbilical vein usually joins the left shortly before the latter enters the liver, but the right umbilical vein also enters the liver in younger embryos.) Also note that as sections are traced *anteriorly,* the *left* (apparent right) subcardinal vein fades out, and the *right* continues as part of the **inferior vena cava** (Figs. 5.6 and 5.7 level a; Photo 5.30).

Return to the level where the left umbilical vein enters the liver (Photo 5.31). Identify a vessel lying above and to the *left* (apparent right) of the duodenum at about this level. This vessel shifts to the *right* of the duodenum in slightly more *anterior* sections (Photo 5.30). This vessel constitutes the **superior mesenteric vein** when it lies to the *left* of the duodenum (Photo 5.31) and the **portal vein** when it lies in the midline and to the *right* of the duodenum (Photo 5.30). Continue to trace sections *anteriorly* and note that the portal vein enters the right dorsal lobe of the liver and that the inferior vena cava begins to shift ventrad toward the liver (Photos 5.29, 5.28). The inferior vena cava enters the right dorsal lobe of the liver in more *anterior* sections (Figs. 5.6 and 5.7, level b; Photo 5.27). The **ductus venosus** can be identified at about this level. It is formed by the enlargement of hepatic sinusoids.

Return again to the level where the left umbilical vein enters the liver (Photo 5.31) and trace sections *anteriorly.* Try to trace the left umbilical vein (and possibly also the right) until it joins the ductus venosus. (The exact boundary between the left umbilical vein and ductus venosus is impossible to distinguish. The latter is irregularly shaped; the former is circular in transverse section.) Also try to trace the portal vein until it joins the ductus venosus (this happens at a section level between Photos 5.28 and 5.27). The inferior vena cava and ductus venosus become continuous in slightly more *anterior* sections (Figs. 5.6 and 5.7, level c; Photo 5.26). This fused vessel narrows in more anterior sections, indicating the level of the inferior vena cava (Figs. 5.6 and 5.7, level d; Photo 5.25). Continue tracing sections *anteriorly* and note that the inferior vena cava exits from the liver (Figs. 5.6 and 5.7, level e) and opens into the sinus venosus.

Return to a level where the portal vein can be identified within the right dorsal lobe of the liver (Photos 5.28, 5.29) and trace sections *posteriorly.* It quickly leaves the liver and lies between the dorsal and ventral pancreatic rudiments to the *right* (apparent left) of the duodenum (Photo 5.30). Note that it then shifts to the left of the duodenum to become the superior mesenteric vein (Photo 5.31), one of the main branches of the portal vein in the adult. Trace sections posteriorly, following the superior mesenteric vein. It soon joins a prominent **common vitelline vein** (Photo 5.33). (The superior mesenteric, portal, and common vitelline veins are derived from a pair of **vitelline veins** present earlier. The superior mesenteric vein is derived from the proximal part of the *left* vitelline vein; the portal vein is derived from the proximal part of the *right* vitelline vein and from a cross connection between right and left vitelline veins; and the common vitelline vein is derived from an anastomosis between the distal ends of the right and left vitelline veins.) In more posterior sections, the superior mesenteric and common vitelline veins can be traced into the umbilical cord and are cut frontally—first the common vitelline vein and then the superior mesenteric vein (Photo 5.35). These veins then disappear. Return to the levels at which the superior mesenteric (Photo 5.35) and the common vitelline veins are cut frontally and trace sections *anteriorly,* following the lower portion of each vein within the umbilical cord. Trace these vessels as far as possible.

Return to a section where the **precardinal veins** can first be identified (Photo 5.6). These veins are cut frontally at about this level. Trace sections *posteriorly,* following the lower portion of each precardinal vein until it fades out ventrolateral to the diencephalon. (The adult **superior vena cava** is derived in part from a portion of the *right* precardinal vein. The adult **internal jugular veins** are derived from the cranial ends of both precardinal veins.)

Return again to a section where the precardinal veins can first be identified (Photo 5.6) and quickly trace sections posteriorly, following the *upper* portion of each precardinal vein. Note that the precardinal veins are closely approximated to the nodose ganglia (Photo 5.12). The precardinal veins gradually move ventrad in more posterior sections (Photos 5.15, 5.18, 5.21). They elongate dorsoventrally at about the level of the foreleg buds (Photos 5.22, 5.23), indicating that sections are now cutting through the **common cardinal veins.** (The adult **superior vena cava** is derived from the *right* common cardinal vein, as well as from the *right* precardinal vein. The left common cardinal vein forms the adult **coronary sinus** and **oblique vein of the left atrium.**) The *right* common cardinal vein quickly enters the sinus venosus (Photo 5.23). Similarly, the *left* common cardinal vein enters the sinus venosus, but in more posterior sections, by means of a narrow connection beneath the trachea (use high magnification and carefully trace sections to observe this connection). Try to identify the irregularly shaped **subclavian veins** just developing at the base of the foreleg buds (Photos 5.21, 5.22).

Return to the level where the *right* common cardinal vein enters the sinus venosus (Photo 5.23) and trace sections posteriorly. Identify the **postcardinal veins** quickly appearing in place of the common cardinal veins (Photo 5.24). Continue tracing sections posteriorly, following the postcardinal veins as far as possible. Note that they first lie dorsolateral to the mesonephric kidneys (Photo 5.25). They are cut across frontally in more posterior sections (Photo 5.41). Reverse direction and trace sections *anteriorly,* following the lower portion of each postcardinal vein located at the base of a hind-leg bud (Photo 5.41) into the tail (Photo 5.39). (The *left* postcardinal vein later mainly degenerates. The proximal part of the *right* postcardinal vein forms part of the adult **azygos vein**)

3. Serial sagittal sections

The study of serial sagittal sections of the 10-mm pig has been included to help you assess your overall understanding of embryonic morphology at this stage. Do not begin your study until you fully understand whole mounts and serial transverse sections. Even then, do not try to identify every structure present in all sections. Rather, concentrate on only those structures identified in Photos 5.43-5.54, beginning with sections similar to those illustrated by Photos 5.50, 5.51; that is, sections that cut through approximately the midsagittal plane in the cranial third of the embryo. Then move to more peripheral structures on each side of the midline (that is, structures identified in principally in Photos 5. 43-5.49, 5.52-5.54). In your study, frequently refer to whole mounts and serial transverse sections to gain a deeper appreciation of the three-dimensional spatial relationships among adjacent structures.

4. Summary of the contributions of the germ layers to structures present in the 10-mm pig embryo but not present in the 72-hour chick embryo

Ectoderm

autonomic ganglia

brachial plexus

cornea

cranial ganglia:

> VII. geniculate
>
> VIII. acoustic
>
> IX. petrosal
>
> X. nodose

cranial nerves:

> VI. abducens
>
> VIII. auditory
>
> X. vagus
>
> XI. spinal accessory
>
> XII. hypoglossal

Froriep's ganglia

glioblasts

lamina terminalis

pigment granules (within pigmented retinas)

rudiments of the semicircular canals

sclera

spinal nerves

Mesoderm

cloacal septum

eye muscle rudiments

gonad rudiments

heart partitions:

> bulbar ridges and septum
>
> endocardial cushion
>
> interventricular septum
>
> septum primum
>
> septum secundum

mesenteries:

> falciform ligament
>
> mesocolon

mesonephric kidneys

pericardium

peritoneum

pleura

renal pelvis

septum transversum

sinoatrial valve

ureters

Wharton's jelly

Endoderm

caudal limb of the intestinal loop

colon

common bile duct

cranial limb of the intestinal loop

cystic duct

eparterial bronchus

gallbladder

hepatic cords

hepatic duct

larynx

pancreatic ducts and rudiments

rectum

thyroid gland

trachea

urogenital sinus

Ectoderm and Mesoderm

choroid

choroid plexus

genital eminence

mammary ridges

nasal processes

umbilical cord

Endoderm and Mesoderm

arytenoid swellings

epiglottis

tongue rudiments

D. TERMS TO KNOW

You should know the meaning of the following terms, which appeared in boldface in the preceding discussion of 10-mm pig embryos.

abducens (VI) cranial nerves

acoustic ganglion

alar plate

allantoic artery

allantoic vein

allantois

amniotic cavity

anterior cerebral artery

anterior semicircular canal

aortic trunk

apical ectodermal ridge

arch of the aorta

arterial circle of Willis

arytenoid cartilages

arytenoid swellings

ascending aorta

ascending colon

atrioventricular canal

auditory (VIII) cranial nerves

auditory vesicles

auricle

autonomic nervous system

axons

azygos vein

basal plate

basilar artery

bicuspid valve

blastocoel

blastocyst

body folds

Bowman's capsules

brachial plexus

bulbar ridges

bulbar septum

C cells

calcitonin

cardiogenesis

caudal artery

caudal limb of the intestinal loop

celiac artery

cell bodies

centrum

cerebellum

cerebral aqueduct

cerebral hemispheres

cerebral peduncles

cerebrospinal fluid

cerebrum

cervical flexure

cervical sinuses

choroid

choroid plexus

cleavage

cloaca

cloacal membrane

cloacal septum

closing plates

cochlea

cochlear ganglia

collecting tubules

colon

commissures

common bile duct

common cardinal veins

common carotid arteries

common vitelline vein

conotruncus

copula

cornea

corneal epithelium

coronary sinus

corpora quadrigemina

corpora striata

cranial flexure

cranial limb of the intestinal loop

cranial nerves

cystic duct

dendrites

dermatome

descending aorta

diaphragm

diencephalon

diffuse type placenta

dorsal aortae

dorsal lobe of liver

dorsal mesocardium

dorsal mesogaster

dorsal pancreatic duct

dorsal pancreatic rudiment

dorsal ramus of spinal nerve

dorsal root of spinal nerve

duct of Santorini

duct of Wirsung

ductus arteriosus

ductus caroticus

ductus venosus

duodenum

ear ossicles

ectoderm

efferent ductules

ejaculatory duct

endocardial cushion

endocardium

endocrine gland

endoderm

endolymphatic ducts

eparterial bronchus

epiblast

epibranchial placodes

epicardium

epididymis

epiglottis

epiploic foramen

epithalamus

epithelial-mesenchymal inductions

epithelial-mesenchymal interactions

epitheliochorialis type placenta

esophagus

eustacian tube

external auditory meatus

external carotid arteries

external ear

external ear canal

external genitalia

external iliac arteries

external rectus eye muscle

extraembryonic coelom

extraembryonic membranes

eye muscle rudiments

eye muscles

eyes

facial (vii) cranial nerves

falciform ligament

fertilization

fiber tracts

fifth aortic arches

fifth pharyngeal pouches

first aortic arch

first branchial arch

first branchial grooves

first pharyngeal pouches

first ventricle

floor plates

foramen I

foramen II

foramen ovale

foramen primum

foramen secundum

foreleg buds

fourth aortic arches

fourth branchial grooves

fourth pharyngeal pouches

fourth ventricle

Froriep's ganglion

gallbladder

ganglion cells

gastrulation

geniculate ganglion

genital eminence

germinal epithelium

glioblasts

glomerular capsules

glomeruli

glossopharyngeal (IX) cranial nerve

glottis

gonad rudiments

greater curvature of stomach

greater omentum

head

head mesenchyme

heart

hemopoietic organ

hepatic cords

hepatic duct

hepatic sinusoids

hepatoduodenal ligament

hepatogastric ligament

hindleg buds

hypoblast

hypoglossal (XII) cranial nerve

hypothalamus

ileum

incus

inductions

inferior colliculi

inferior oblique eye muscles

inferior parathyroid glands

inferior rectus eye muscles

inferior vena cava

infundibulum

inner cell mass

inner ear

intermediate zone

internal rectus eye muscles

internal carotid arteries

internal iliac arteries

internal jugular veins

intersegmental arteries

interventricular foramina

interventricular septum

intestinal loop

isthmus

jejunum

jugular ganglia

lamina terminalis

large intestine

larynx

lateral arteries

lateral lingual swellings

lateral nasal processes

lateral rectus eye muscles

lateral semicircular canal

lateral ventricles

left atria

left subclavian artery

left ventricle of heart

lens vesicles

lesser curvature of stomach

lesser omentum

ligamentum arteriosum

liver

lumbosacral plexus

lung

lung buds

major calyces

malleus

mammary glands

mammary ridges

mandibular branch of the
    trigeminal nerve

mandibular processes

mantle zone

marginal zone

maxillary branch of the
    trigeminal nerve

maxillary processes

medial nasal processes

medial rectus eye muscles

medulla

mesencephalon

mesocolon

mesoderm

mesoesophagus

mesonephric arteries

mesonephric ducts

mesonephric kidneys

mesonephric tubules

metanephric kidneys

metencephalon

middle ear cavity

minor calyces

mitral valve

mixed nerve

motor horns

motor nerves

motor nuclei

myelencephalon

myocardium

myotome

nasal pits

neck

nephrogenic tissue

neural crest cells

neurulation

nodose ganglia

nose

notochord

oblique vein of the left atrium

oculomotor (III) cranial nerve

olfactory (II) cranial nerve

olfactory lobes

olfactory receptors

omental bursa

ophthalmic branch of the
    trigeminal nerve

optic (II) cranial nerves

optic fissures

optic stalks

organogenesis

oviduct

ovulation

ovum

pacemaker of heart

palatine tonsils

pancreas

parafollicular cells

parasympathetic division of
    the autonomic nervous
    system

parasympathetic terminal
    ganglia

parietal peritoneum

parietal pleura

peritoneal cavity

petrosal ganglia

pharyngotympanic tube

pharynx

pia mater

pigment granules

pigmented retinas

pineal gland

pinna

placenta

placental barrier

placental membrane

pleural cavities

pons

portal vein

postcardinal veins

posterior communicating
    arteries

posterior pituitary gland

posterior semicircular canal

precardinal veins

preganglionic axons

primordial germ cells

prosencephalon

pulmonary arteries

pulmonary trunk

pulmonary veins

ramus communicans of spinal
    nerve

Rathke's pouch

rectum

renal pelvis

right atria

right subclavian artery

right ventricle of heart

roof plate

roots of the hypoglossal (XII) cranial nerve

roots of the spinal accessory (XI) cranial nerve

sacculus

sclera

sclerotome

second aortic arches

second branchial arches

second branchial clefts

second branchial grooves

second closing plates

second pharyngeal pouches

second ventricles

secondary oocyte

secretory tubules

semilunar ganglia

seminal vesicle

sensory nerves

sensory retina

sensory roots

septum I

septum II

septum membranaceum

septum primum

septum secundum

septum transversum

shoulder muscles

sinoatrial valve

sinus venosus

sixth aortic arches

skin ectoderm

small intestine

somites

sperm

spinal accessory (XI) cranial nerve

spinal cord

spinal ganglia

spinal nerves

spiral ganglia

spiral organ of Corti

spleen

spongioblasts

stapes

sternocleidomastoid

stomach

stomodeum

subcardinal anastomosis

subcardinal veins

subclavian artery

subclavian veins

sulcus limitans

superior colliculi

superior ganglia

superior mesenteric artery

superior mesenteric vein

superior oblique eye muscles

superior parathyroid glands

superior vena cava

sympathetic chain ganglia

sympathetic collateral ganglia

tail

tail flexure

tail gut

telencephalon

thalamus

third aortic arches

third branchial arches

third branchial grooves

third closing plates

third pharyngeal pouches

third ventricle

thymus gland

thyroglossal duct

thyroid gland

tongue muscles

tongue rudiments

trachea

transverse colon

trapezius muscles

tricuspid valve

trigeminal (V) cranial nerve

trochlear (IV) cranial nerve

trophoblast

trunk

tuberculum impar

tube-within-a-tube body plan

tympanic cavity

tympanic membrane

ultimobranchial bodies

umbilical arteries

umbilical cord

umbilical hernia

umbilical veins

upper jaw

ureters

urogenital sinus

uterine horns

utriculus

vagus (X) cranial nerves

valve flaps

vas deferens

vasa efferentia

ventral lobe of liver

ventral pancreatic duct

ventral pancreatic rudiment

ventral ramus of spinal nerve

ventral root of spinal nerve

ventral roots

ventricular zone

vertebra

vertebral arteries

vestibular ganglia

visceral pericardium

visceral peritoneum

visceral pleura

visceral ramus of spinal nerve

vitelline arteries

vitelline veins

Wharton's jelly

yolk sac

young neurons

zygote

E. PHOTOS 5.1-5.54

Photos 5.1-5.54 depict the pig embryos discussed in Chapter 5. These photos and their accompanying legends begin after the following page.

Photos begin on the following page. Use the space below for notes.

Photos 5.1-5.3

Pig Embryos

Legend

1. Myelencephalon
2. Fourth ventricle
3 Metencephalon
4. Isthmus
5. Mesencephalon
6. Cerebral aqueduct

7. Ventricular zone
8. Intermediate zone
9. Marginal zone
10. Roots of the spinal accessory nerve
11. Pia mater

12. Endolymphatic duct
13. Spinal accessory nerve
14. Sensory root of the vagus nerve
15. Auditory vesicle (future utriculus)

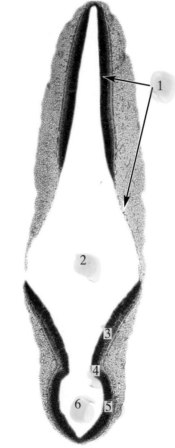

5.1

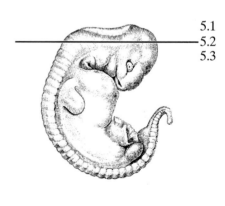

5.1
5.2
5.3

Photo 5.1. 10-mm pig embryo serial transverse sections numbered in anterior to posterior sequence.

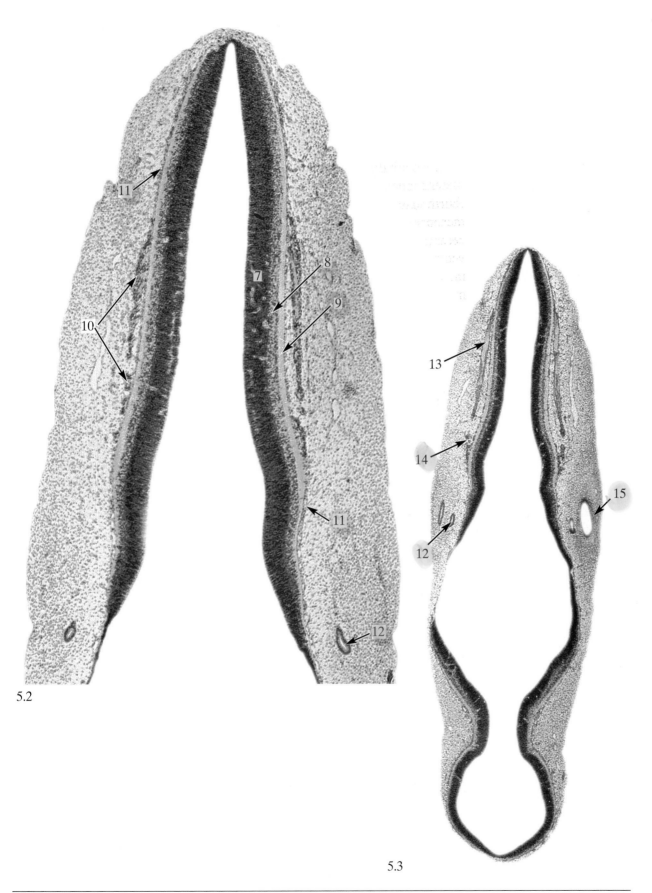

5.2

5.3

Photos 5.2, 5.3. Continuation of 10-mm pig embryo serial transverse sections numbered in anterior to posterior sequence. Photo 5.2 is an enlargement of the dorsal part of a transverse section.

Photos 5.4-5.6

Pig Embryos

Legend

1. Spinal accessory nerve
2. Myelencephalon
3. Jugular ganglion
4. Sensory root of the glossopharyngeal nerve
5. Auditory vesicle (future utriculus)
6. Endolymphatic duct
7. Neuromeres
8. Metencephalon
9. Mesencephalon
10. Froriep's ganglion
11. Roots of the hypoglossal nerve
12. Superior ganglion
13. Rudiment of the posterior semicircular canal
14. Rudiment of the anterior semicircular canal
15. Sensory root of the trigeminal nerve
16. Fourth ventricle
17. Cerebral aqueduct
18. Vagus nerve
19. Glossopharyngeal nerve
20. Precardinal vein
21. Auditory vesicle (future sacculus)
22. Acoustic ganglion
23. Geniculate ganglion
24. Semilunar ganglion
25. Rhombencephalon
26. Basilar artery
27. Oculomotor nerve
28. Posterior communicating artery

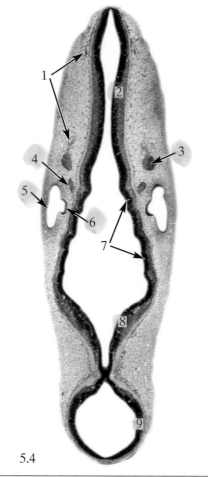

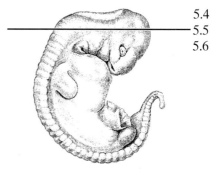

5.4
5.5
5.6

5.4

Photo 5.4. Continuation of 10-mm pig embryo serial transverse sections numbered in anterior to posterior sequence.

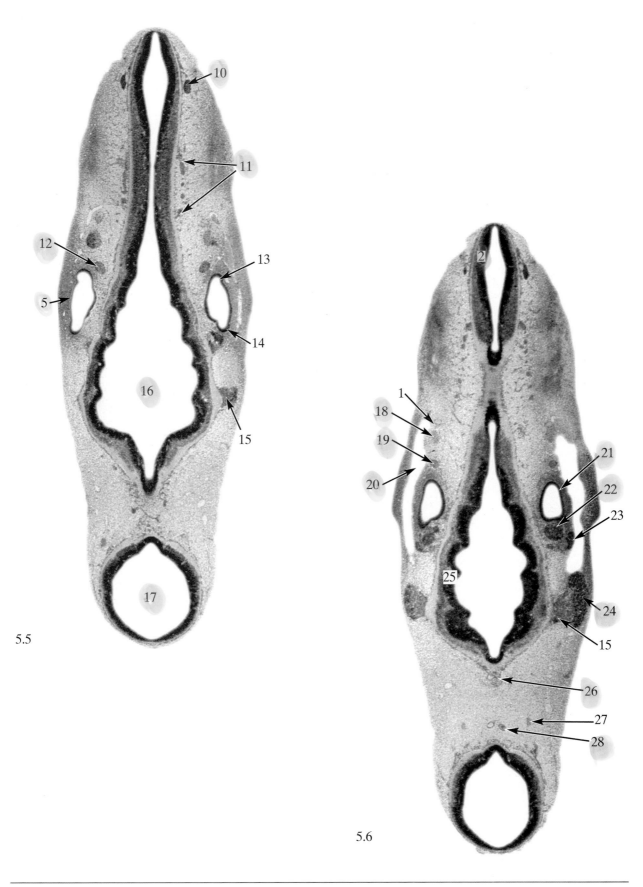

5.5

5.6

Photos 5.5, 5.6. Continuation of 10-mm pig embryo serial transverse sections numbered in anterior to posterior sequence.

Photos 5.7-5.9

Pig Embryos

Legend

1. Roots of the hypoglossal nerve
2. Vertebral artery
3. Basilar artery
4. Precardinal vein
5. Acoustic ganglion
6. Sensory root of the facial nerve
7. Rhombencephalon
8. Semilunar ganglion
9. Oculomotor nerve
10. Posterior communicating artery
11. Mesencephalon
12. Spinal ganglion
13. Petrosal ganglion
14. Auditory vesicle (future cochlea)
15. Facial nerve
16. Abducens nerve
17. First closing plate
18. First pharyngeal pouch (future tympanic cavity)
19. Internal carotid artery (near level a, Fig. 5.5)
20. First branchial groove
21. Infundibulum
22. Diencephalon
23. Third ventricle (major part)

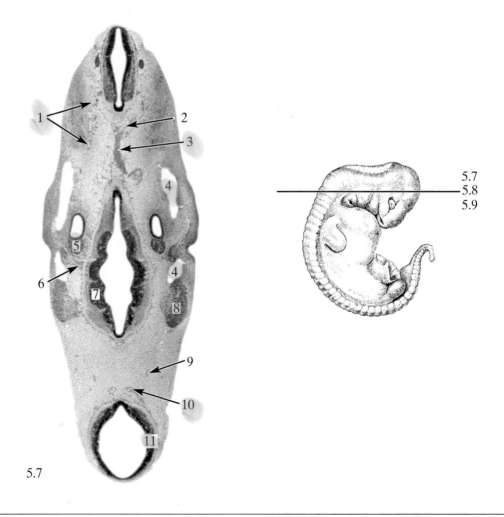

Photo 5.7. Continuation of 10-mm pig embryo serial transverse sections numbered in anterior to posterior sequence.

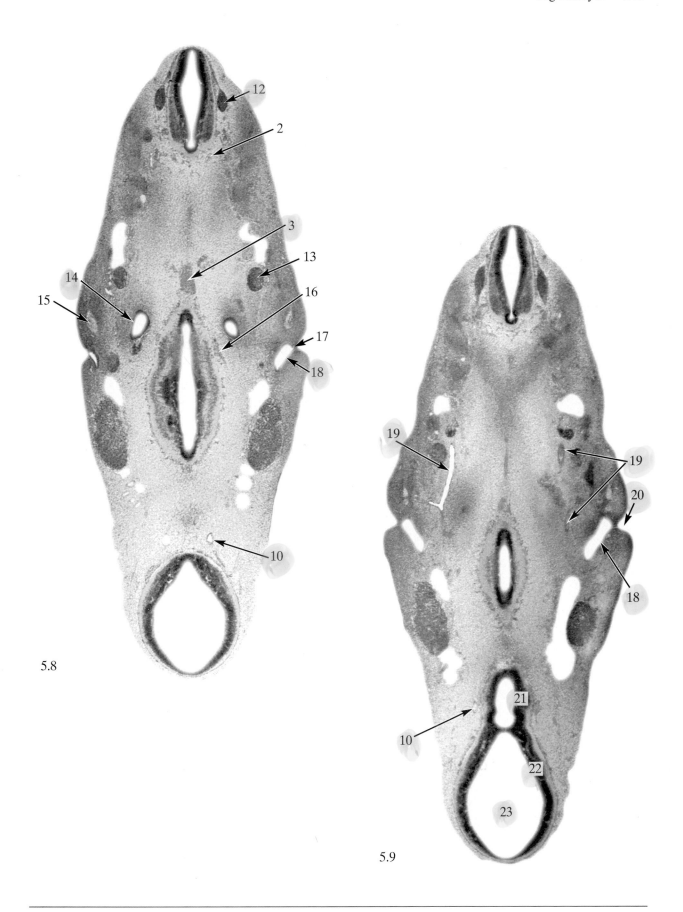

5.8

5.9

Photos 5.8, 5.9. Continuation of 10-mm pig embryo serial transverse sections numbered in anterior to posterior sequence.

Photos 5.10-5.12

Pig Embryos

Legend

1. Ductus caroticus
2. Third aortic arch
3. Second pharyngeal pouch
4. Pharynx
5. Second branchial arch
6. Facial nerve
7. First pharyngeal pouch (future eustachian tube, continuous below with future tympanic cavity)
8. Basilar artery (near level f, Fig. 5.5)

9. Internal carotid artery (near level b, Fig. 5.5)
10. Mandibular branch of the trigeminal nerve
11. Semilunar ganglion
12. Posterior communicating artery
13. Glossopharyngeal nerve
14. Nodose ganglion
15. Second branchial groove
16. Copula
17. Tuberculum impar

18. Eye muscle rudiments
19. Rathke's pouch
20. Infundibulum
21. Diencephalon
22. Precardinal vein
23. Spinal accessory nerve
24. Lateral lingual swelling
25. Internal carotid artery (near level c, Fig. 5.5)
26. Anterior cerebral artery
27. Third ventricle (major part)

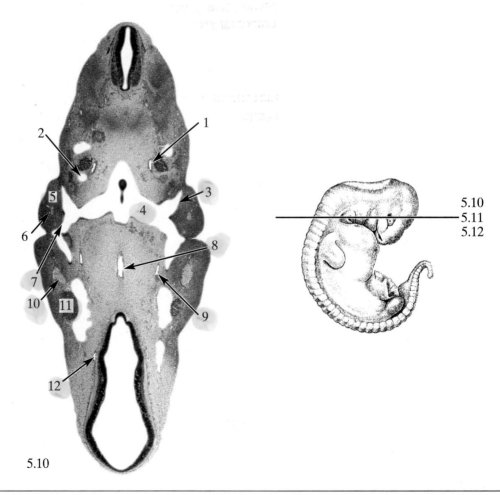

5.10

Photo 5.10. Continuation of 10-mm pig embryo serial transverse sections numbered in anterior to posterior sequence.

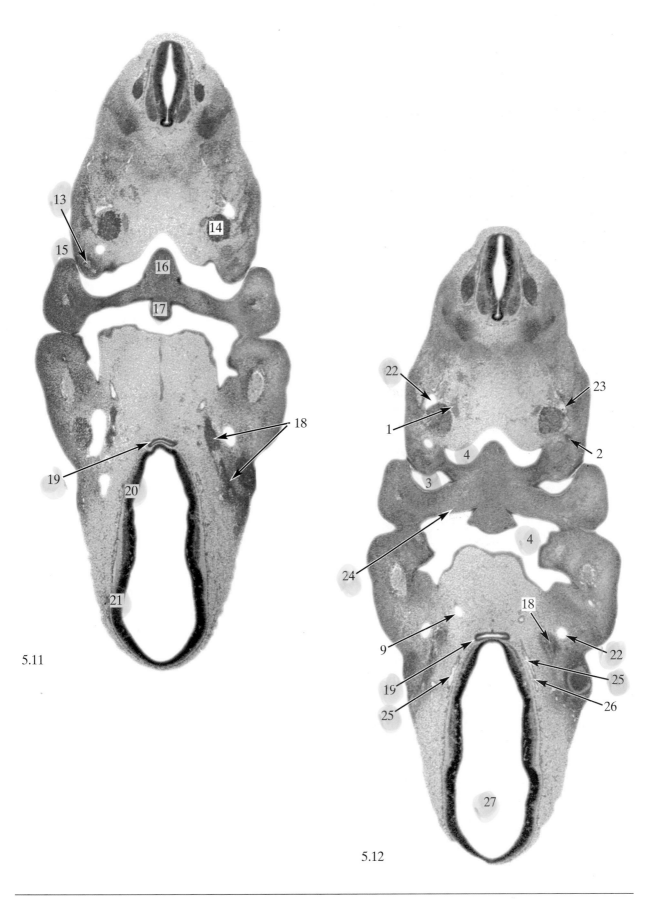

5.11

5.12

Photos 5.11, 5.12. Continuation of 10-mm pig embryo serial transverse sections numbered in anterior to posterior sequence.

Photos 5.13–5.15

Pig Embryos

Legend

1. Epiglottis
2. Third pharyngeal pouch (inferior parathyroid rudiment)
3. Third aortic arch
4. Third branchial arch
5. Second branchial arch
6. Mandibular branch of the trigeminal nerve
7. Stomodeum
8. Sensory retina
9. Pigmented retina

10. Diencephalon
11. Third ventricle (major part)
12. Spinal cord
13. Spinal ganglion
14. Dorsal aorta
15. Fourth aortic arch
16. Cervical sinus
17. Third closing plate
18. Thyroid gland
19. Mandibular process
20. Maxillary process
21. Lens vesicle

22. Glottis
23. Arytenoid swelling
24. Precardinal vein
25. Fourth branchial arch
26. Third pharyngeal pouch (thymus rudiment)
27. Aortic trunk
28. Optic stalk
29. Optic fissure
30. Future corneal epithelium
31. Anterior cerebral artery
32. Pharynx

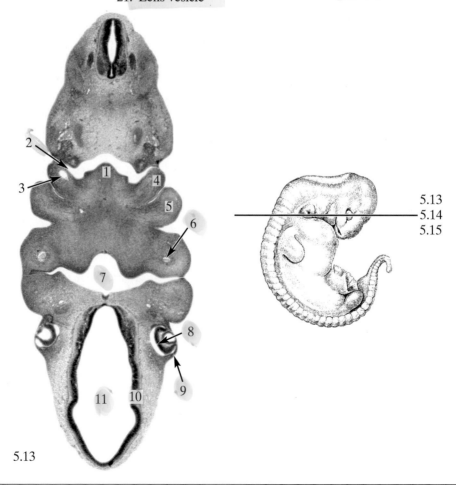

5.13

Photo 5.13. Continuation of 10-mm pig embryo serial transverse sections numbered in anterior to posterior sequence.

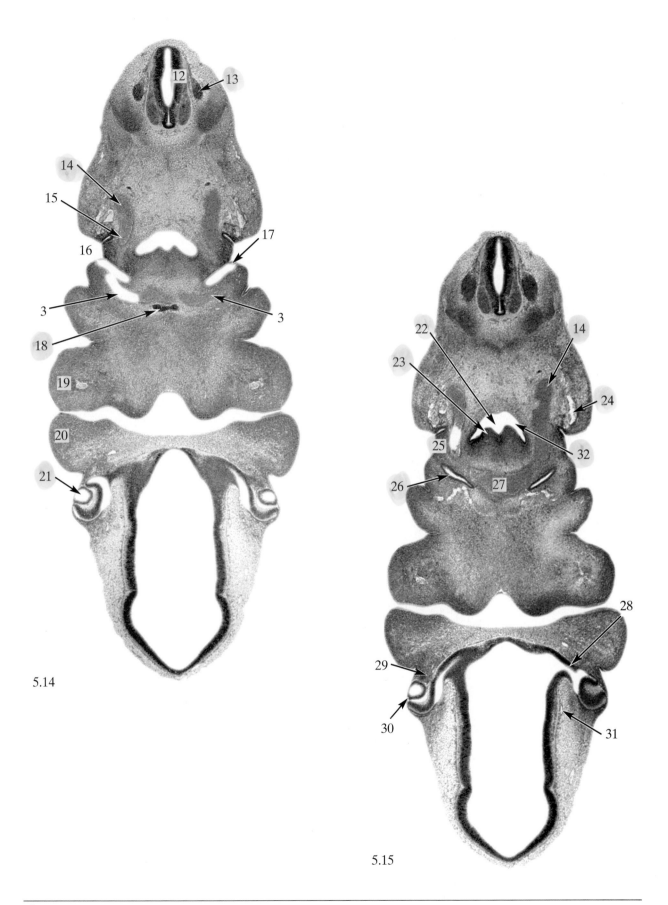

5.14

5.15

Photos 5.14, 5.15. Continuation of 10-mm pig embryo serial transverse sections numbered in anterior to posterior sequence.

Photos 5.16-5.18

Pig Embryos

Legend

1. Dorsal aorta
2. Fifth pharyngeal pouch
3. Fourth pharyngeal pouch
4. Larynx
5. Aortic trunk
6. Stomodeum

7. Diencephalon
8. Third ventricle (major part)
9. Vertebral artery
10. Glottis
11. Pharynx
12. Sixth aortic arch

13. Esophagus
14. Precardinal vein
15. Nasal pit
16. Cerebral hemisphere of telencephalon
17. Third ventricle (minor part)

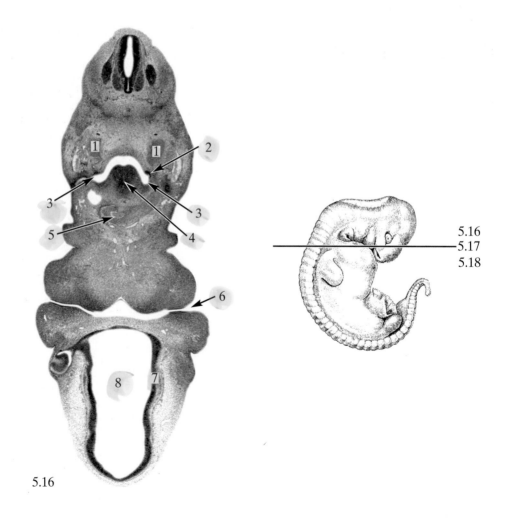

5.16

Photo 5.16. Continuation of 10-mm pig embryo serial transverse sections numbered in anterior to posterior sequence.

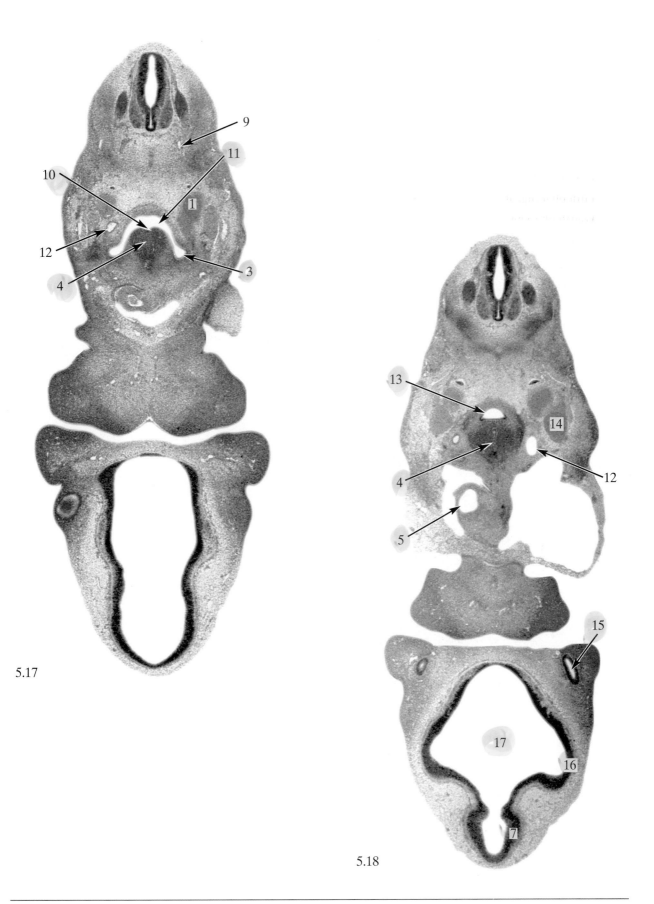

5.17

5.18

Photos 5.17, 5.18. Continuation of 10-mm pig embryo serial transverse sections numbered in anterior to posterior sequence.

Photos 5.19-5.21

Pig Embryos

Legend

1. Spinal cord
2. Spinal ganglion
3. Dorsal aorta
4. Precardinal vein
5. Sixth aortic arch
6. Amniotic cavity
7. Aortic trunk
8. Pulmonary trunk
9. Conotruncus
10. Bulbar septum

11. Cerebral hemisphere of telencephalon
12. Third ventricle (minor part)
13. Diencephalon
14. Sympathetic chain ganglion
15. Esophagus
16. Trachea
17. Pulmonary artery
18. Left atrium
19. Bulbar ridges

20. Nasal pit
21. Subclavian vein
22. Right atrium
23. Right ventricle
24. Left ventricle
25. Pericardial cavity
26. Medial nasal process
27. Lateral nasal process
28. Lateral ventricles

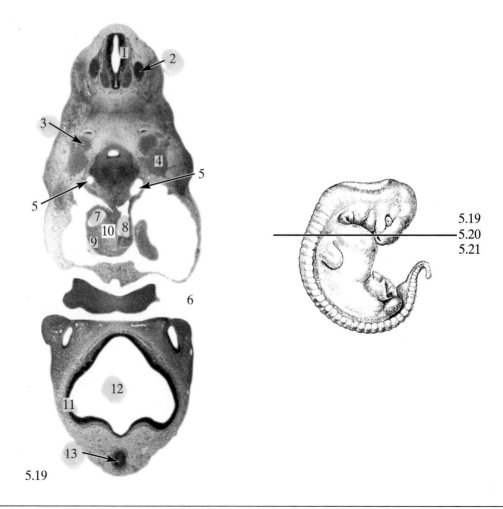

5.19

Photo 5.19. Continuation of 10-mm pig embryo serial transverse sections numbered in anterior to posterior sequence.

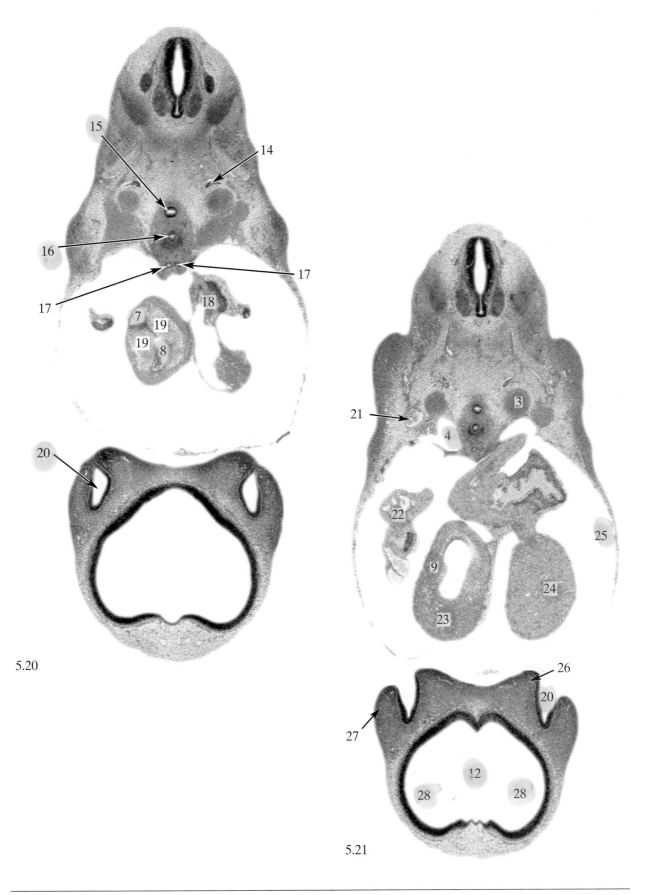

5.20

5.21

Photos 5.20, 5.21. Continuation of 10-mm pig embryo serial transverse sections numbered in anterior to posterior sequence.

Photos 5.22-5.24

Pig Embryos

Legend

1. Subclavian vein
2. Dorsal aorta
3. Common cardinal vein
4. Trachea
5. Foramen secundum
6. Septum primum
7. Left atrium
8. Right atrium
9. Conotruncus
10. Right ventricle
11. Nasal pit

12. Cerebral hemisphere of telencephalon
13. Esophagus
14. Vagus nerve
15. Eparterial bronchus
16. Sinus venosus
17. Sinoatrial valve
18. Foramen ovale
19. Endocardial cushion
20. Left ventricle
21. Lamina terminalis
22. Lateral ventricles

23. Intersegmental artery (forms part of adult right subclavian artery)
24. Postcardinal vein
25. Pleural cavity
26. Pulmonary vein continuous with left atrium
27. Atrioventricular canal
28. Interventricular foramen
29. Interventricular septum
30. Pericardial cavity
31. Parietal pleura
32. Visceral pleura

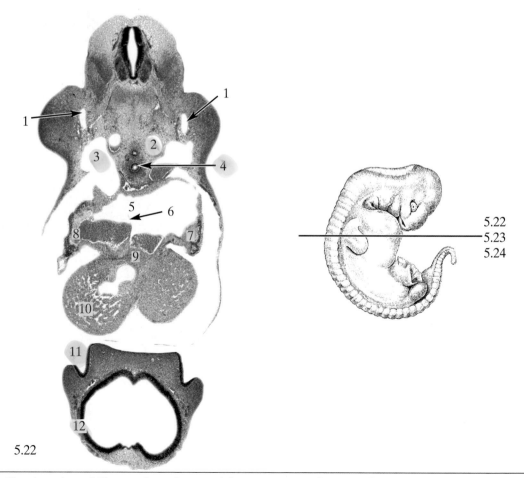

5.22

Photo 5.22. Continuation of 10-mm pig embryo serial transverse sections numbered in anterior to posterior sequence.

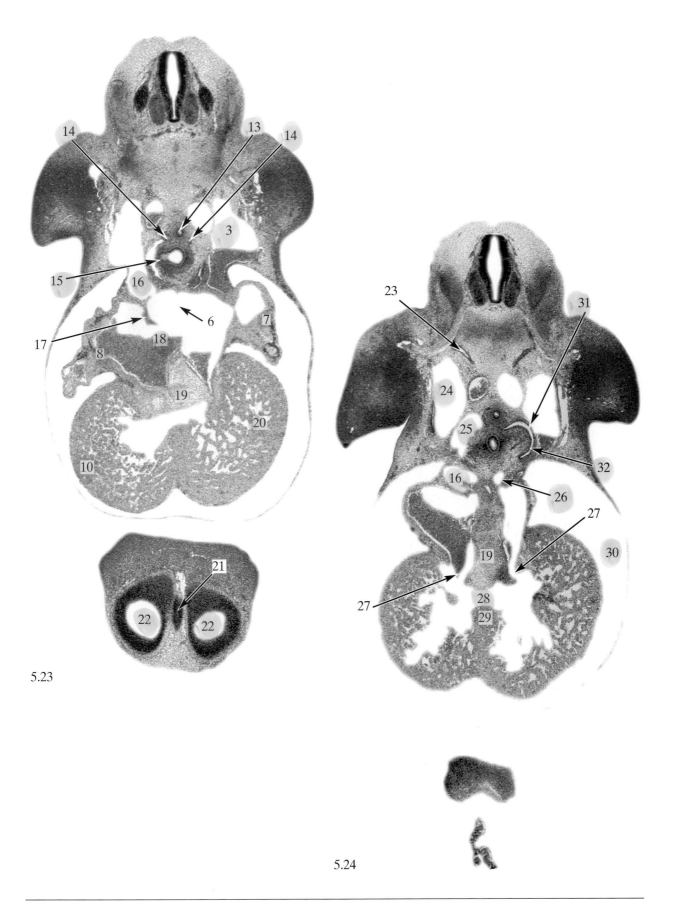

5.23

5.24

Photos 5.23, 5.24. Continuation of 10-mm pig embryo serial transverse sections numbered in anterior to posterior sequence.

Photos 5.25–5.27

Pig Embryos

Legend

1. Spinal cord
2. Spinal ganglion
3. Brachial plexus
4. Postcardinal vein
5. Somatic mesoderm of foreleg bud
6. Esophagus
7. Lung bud
8. Peritoneal cavity (continuous medially with pleural cavity)
9. Apical ectodermal ridge of foreleg bud

10. Pleural cavity
11. Visceral pleura
12. Septum transversum
13. Inferior vena cava
14. Ventricle
15. Mesoesophagus
16. Descending aorta
17. Mesonephric kidney
18. Left dorsal lobe of liver
19. Left ventral lobe of liver

20. Ductus venosus
21. Yolk sac
22. Dorsal mesogaster
23. Stomach
24. Mammary ridge
25. Omental bursa
26. Hepatogastric ligament
27. Peritoneal cavity
28. Hepatic cords
29. Hepatic sinusoids

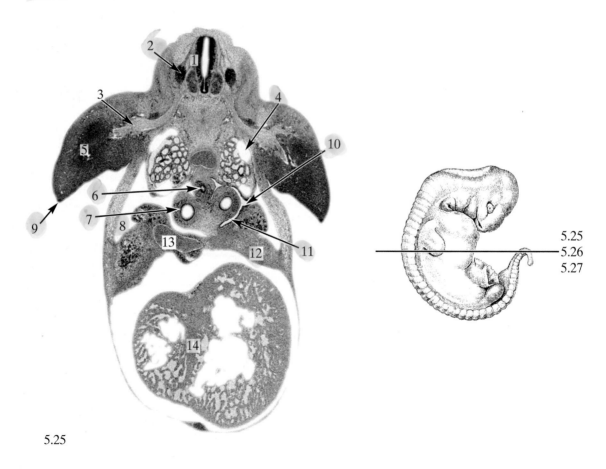

5.25

5.25
5.26
5.27

Photo 5.25. Continuation of 10-mm pig embryo serial transverse sections numbered in anterior to posterior sequence.

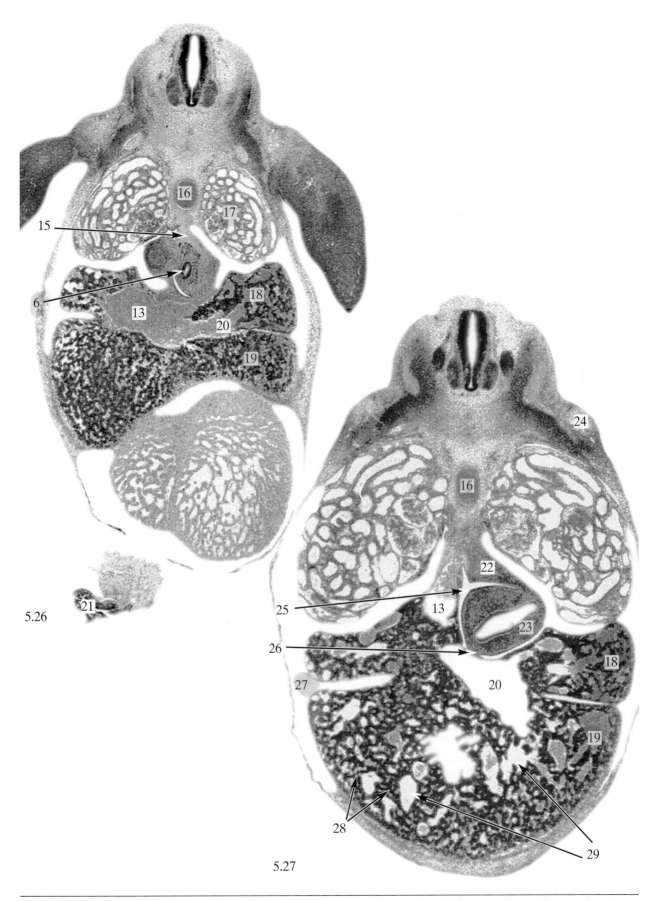

5.26

5.27

Photos 5.26, 5.27. Continuation of 10-mm pig embryo serial transverse sections numbered in anterior to posterior sequence.

Photos 5.28-5.30

Pig Embryos

Legend

1. Descending aorta
2. Mesonephric kidney
3. Inferior vena cava
4. Omental bursa
5. Portal vein
6. Duodenum
7. Hepatoduodenal ligament
8. Right dorsal lobe of liver
9. Hepatic duct
10. Right ventral lobe of liver
11. Falciform ligament

12. Temporary umbilical hernia
13. Gonad rudiment
14. Epiploic foramen
15. Parietal peritoneum
16. Visceral peritoneum
17. Common bile duct
18. Cystic duct
19. Right umbilical vein
20. Left umbilical vein
21. Common vitelline vein
22. Cranial limb of the intestinal loop

23. Caudal limb of the intestinal loop
24. Mesonephric tubules
25. Glomeruli
26. Dorsal pancreatic rudiment
27. Germinal epithelium
28. Ventral pancreatic duct
29. Dorsal pancreatic duct
30. Gallbladder
31. Hepatic cords
32. Hepatic sinusoids
33. Glomerular capsule

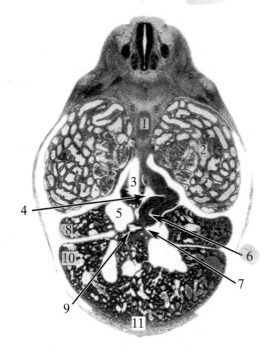

5.28

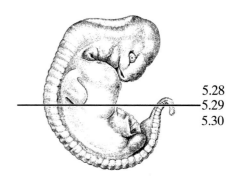

5.28
5.29
5.30

Photo 5.28. Continuation of 10-mm pig embryo serial transverse sections numbered in anterior to posterior sequence.

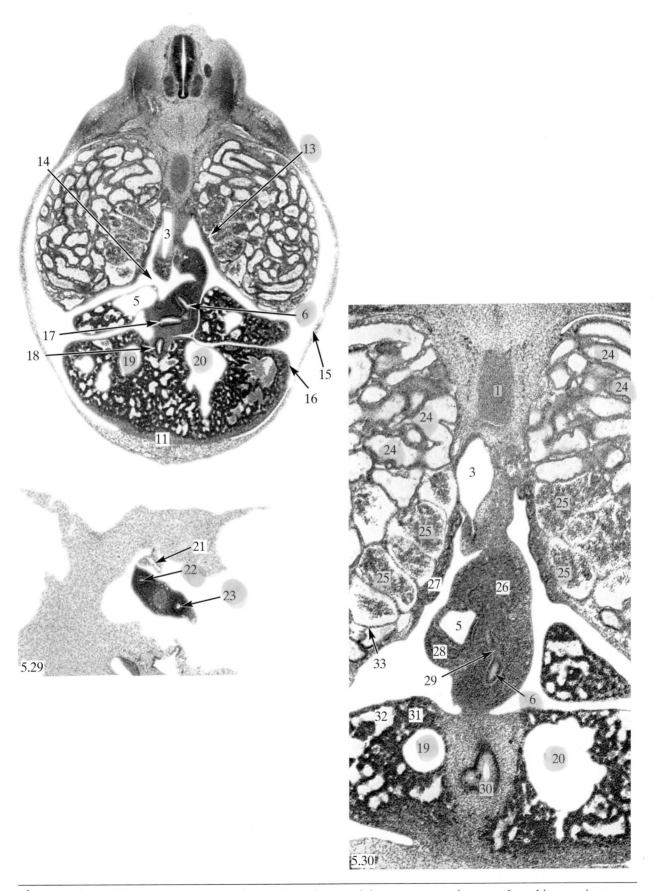

Photos 5.29, 5.30. Continuation of 10-mm pig embryo serial transverse sections numbered in anterior to posterior sequence. Photo 5.30 is an enlargement of the central part of a transverse section.

Photos 5.31-5.33

Pig Embryos

Legend

1. Descending aorta
2. Mesonephric kidney
3. Subcardinal vein
4. Superior mesenteric vein
5. Duodenum
6. Mesonephric duct
7. Gallbladder
8. Wharton's jelly of umbilical cord
9. Common vitelline vein
10. Umbilical vein
11. Cranial limb of the intestinal loop

12. Superior mesenteric artery
13. Umbilical artery
14. Tail
15. Marginal zone
16. Intermediate zone
17. Ventricular zone
18. Roof plate
19. Alar plate
20. Dorsal root axons
21. Spinal ganglion
22. Basal plate
23. Motor horn
24. Dorsal root dendrites

25. Spinal nerve—dorsal ramus
26. Ventral root
27. Floor plate
28. Notochord
29. Sclerotome
30. Spinal nerve—ventral ramus
31. Spinal nerve—visceral ramus
32. Sympathetic chain ganglion
33. Mesonephric tubules
34. Gonad rudiment
35. Caudal limb of the intestinal loop
36. Allantois

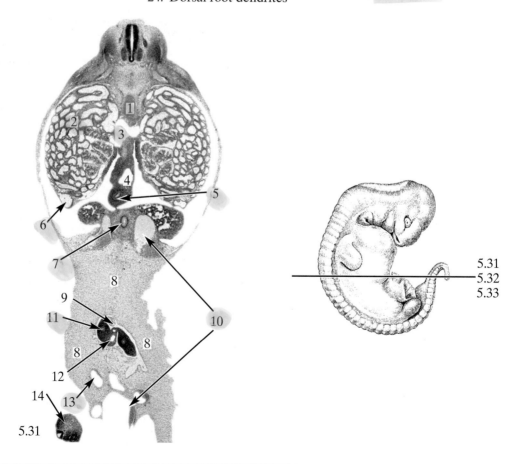

Photo 5.31. Continuation of 10-mm pig embryo serial transverse sections numbered in anterior to posterior

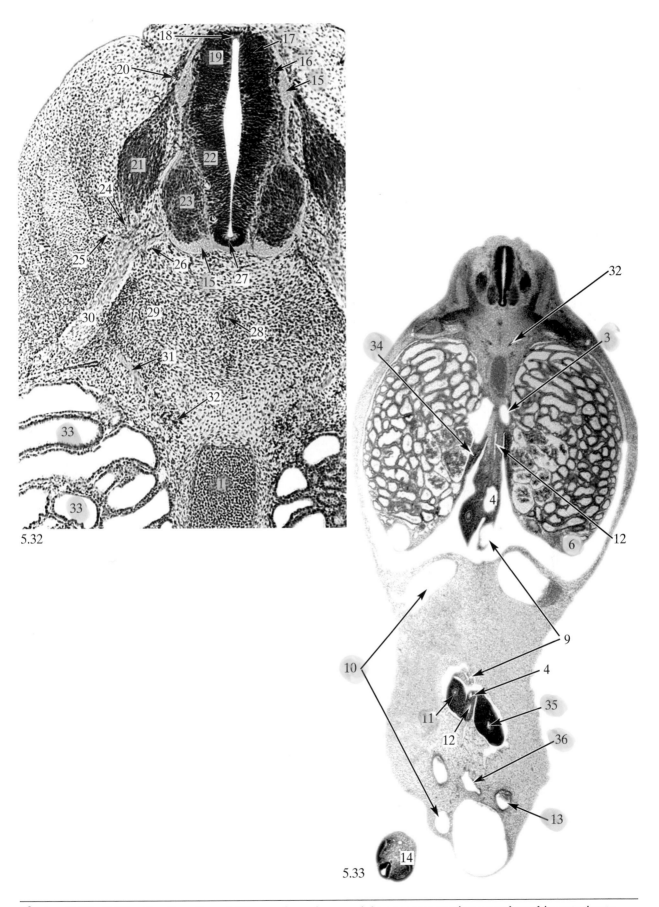

Photos 5.32, 5.33. Continuation of 10-mm pig embryo serial transverse sections numbered in anterior to posterior sequence. Photo 5.32 is an enlargement of the central part of a transverse section.

Photos 5.34-5.36

Pig Embryos

Legend

1. Mesonephric tubules
2. Subcardinal anastomosis
3. Glomerular capsule
4. Mesonephric duct
5. Continuity between the duodenum (above arrow) and cranial limb of the intestinal loop (below arrow)
6. Superior mesenteric vein
7. Caudal limb of the intestinal loop
8. Umbilical vein
9. Umbilical artery
10. Tail
11. Extraembryonic coelom
12. Descending aorta
13. Glomeruli
14. Gonad rudiment
15. Superior mesenteric artery
16. Mammary ridge
17. Genital eminence
18. Cloacal membrane
19. Tail gut
20. Mesocolon
21. Caudal limb of the intestinal loop (forms part of adult colon)
22. Allantois
23. Peritoneal cavity
24. Cloaca (future urogenital sinus)
25. Cloacal septum
26. Cloaca (future rectum)
27. Spinal cord
28. Mesonephric kidney

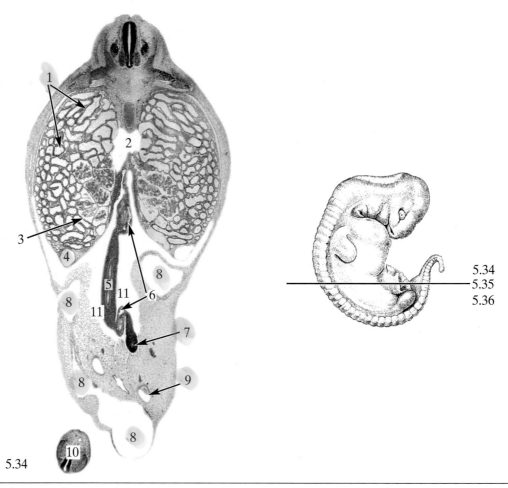

Photo 5.34. Continuation of 10-mm pig embryo serial transverse sections numbered in anterior to posterior sequence.

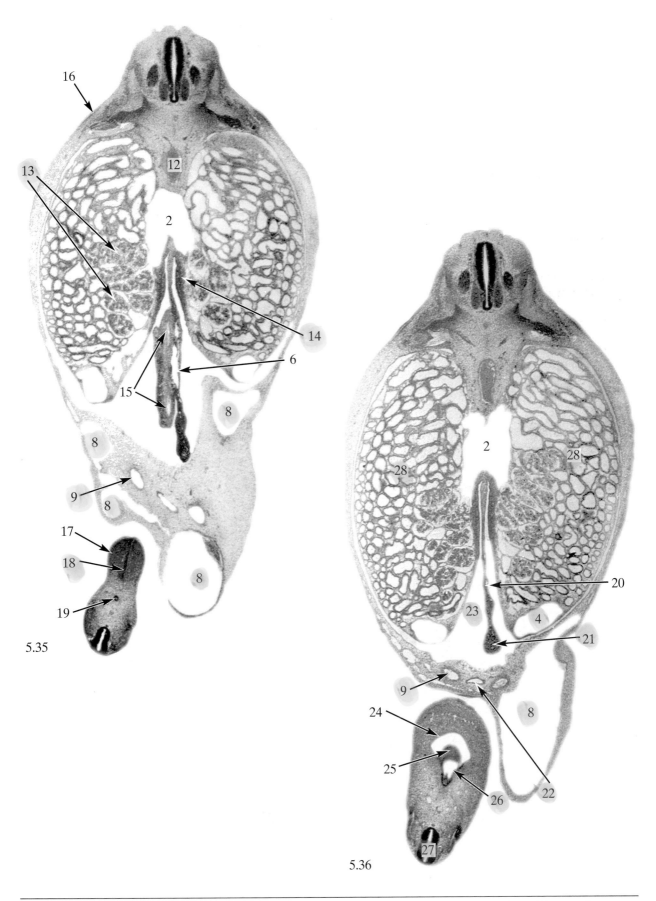

5.35

5.36

Photos 5.35, 5.36. Continuation of 10-mm pig embryo serial transverse sections numbered in anterior to posterior sequence.

Photos 5.37-5.39

Pig Embryos

Legend

1. Subcardinal anastomosis
2. Mesocolon
3. Umbilical artery
4. Allantois (continuous below with urogenital sinus)
5. Umbilical vein
6. Mesonephric duct
7. Urogenital sinus
8. Colon
9. Descending aorta
10. Mesonephric kidney
11. Caudal limb of the intestinal loop
12. Peritoneal cavity
13. Spinal cord
14. Motor horn
15. Spinal ganglion
16. Mammary ridge
17. Apical ectodermal ridge of hindleg bud
18. Somatic mesoderm of hindleg bud
19. Ureters
20. Postcardinal vein
21. Caudal artery
22. Notochord

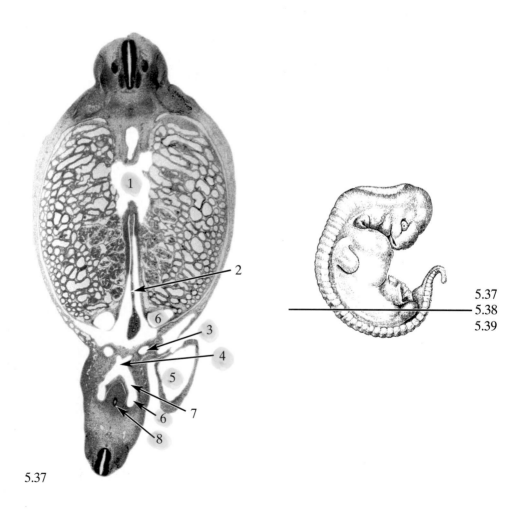

5.37

Photo 5.37. Continuation of 10-mm pig embryo serial transverse sections numbered in anterior to posterior sequence.

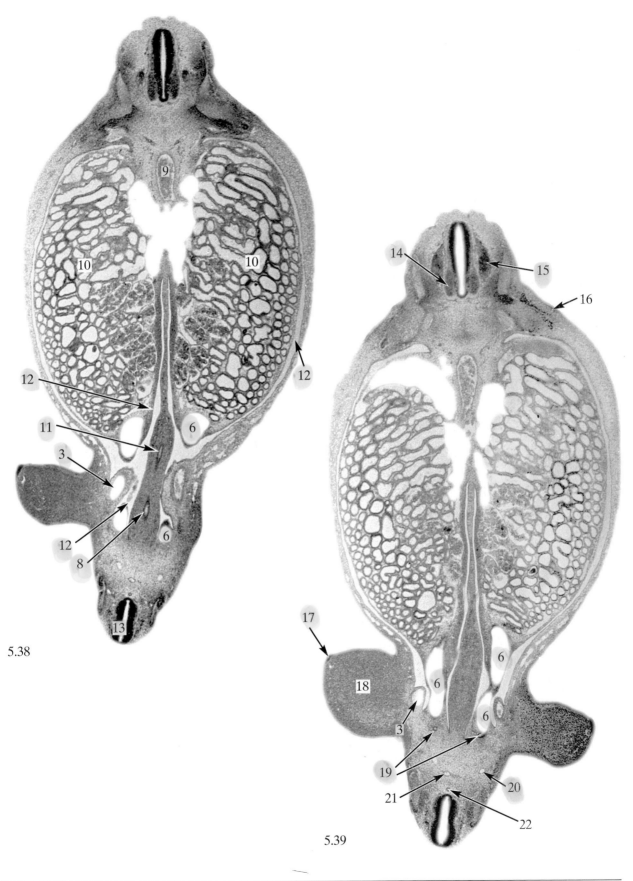

Photos 5.38, 5.39. Continuation of 10-mm pig embryo serial transverse sections numbered in anterior to posterior sequence.

Photos 5.40–5.42

Pig Embryos

Legend

1. Umbilical artery
2. External iliac artery
3. Renal pelvis
4. Caudal artery
5. Nephrogenic tissue
6. Descending aorta
7. Mesonephric kidney

8. Peritoneal cavity
9. Postcardinal vein
10. Notochord
11. Spinal ganglion
12. Spinal cord
13. Skin ectoderm
14. Myotome

15. Dermatome
16. Ventral root
17. Sclerotome—dense caudal portion
18. Sclerotome—less dense cranial portion
19. Cavity of the spinal cord

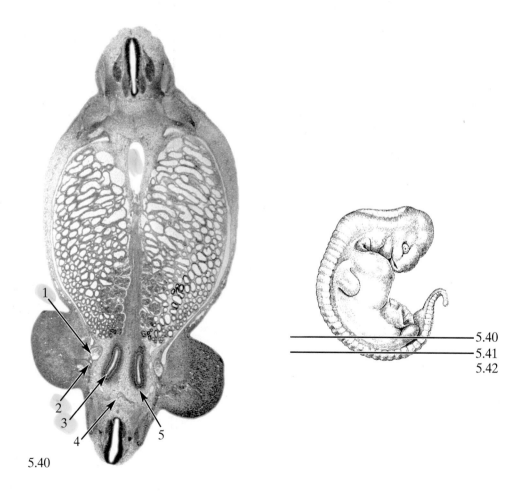

5.40

5.40
5.41
5.42

Photo 5.40.   Continuation of 10-mm pig embryo serial transverse sections numbered in anterior to posterior sequence.

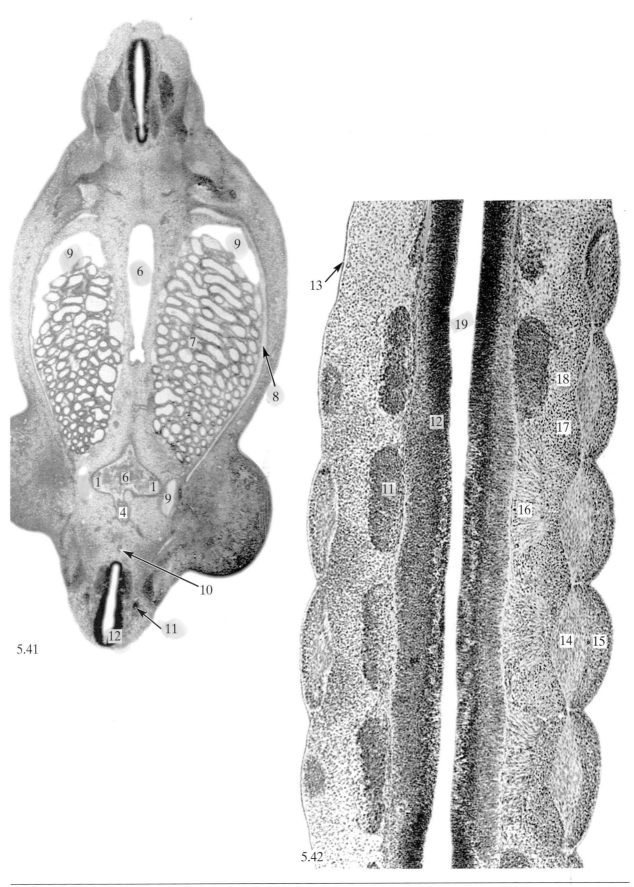

5.41

5.42

Photos 5.41, 5.42. Continuation of 10-mm pig embryo serial transverse sections numbered in anterior to posterior sequence.

Photos 5.43-5.45

Pig Embryos

Legend

1. Lens vesicle
2. Pigmented retina
3. Sensory retina
4. Optic fissure
5. Semilunar ganglion
6. Mandibular branch of the trigeminal nerve
7. Precardinal vein
8. Nasal pit
9. Maxillary process
10. Stomodeum
11. Mandibular process
12. First pharyngeal pouch
13. Second branchial arch
14. Second pharyngeal pouch

15. Second closing plate
16. Third branchial arch
17. Ventricle
18. Atrium
19. Pericardial cavity
20. Common cardinal vein
21. Ventral lobe of liver
22. Dorsal lobe of liver
23. Falciform ligament
24. Umbilical cord
25. Left umbilical vein
26. Ductus venosus
27. Mesonephric kidney
28. Glomerular capsules

29. Glomeruli
30. Mesonephric tubules
31. Mesonephric duct
32. Spinal ganglia
33. Umbilical artery
34. Auditory vesicle
35. Geniculate ganglion
36. Pharynx
37. Petrosal ganglion
38. Glossopharyngeal nerve
39. Third pharyngeal pouch
40. Cerebral hemisphere of telencephalon
41. Stomach
42. Spinal cord

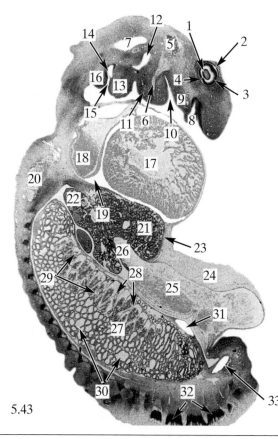

5.43

Photo 5.43. 10-mm pig embryo serial sagittal sections numbered in right to left sequence.

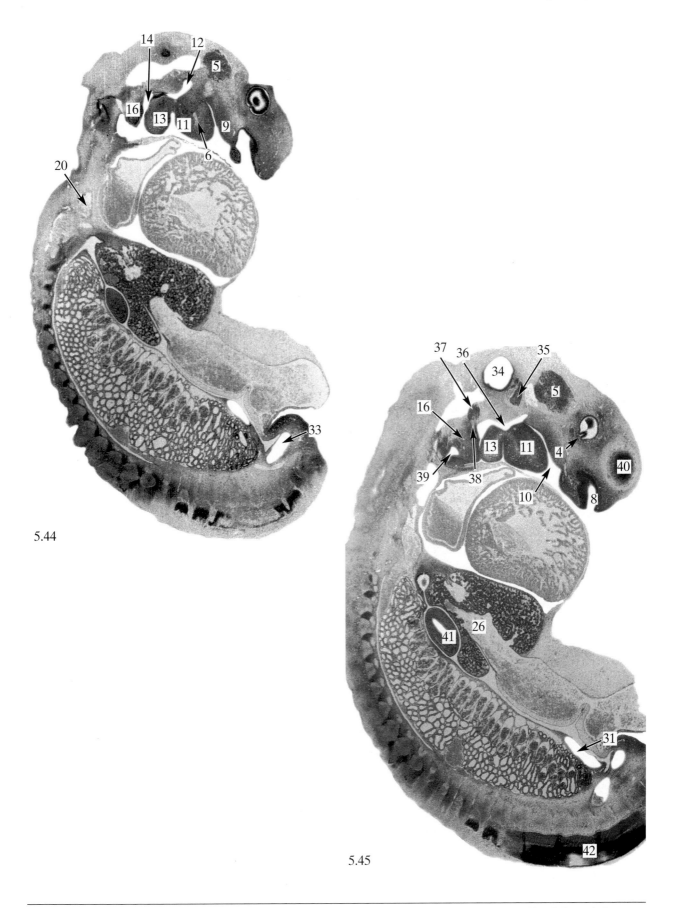

5.44

5.45

Photos 5.44, 5.45. Continuation of 10-mm pig embryo serial sagittal sections numbered in right to left sequence.

Photos 5.46-5.48

Pig Embryos

Legend

1. Lateral ventricle
2. Optic stalk
3. Eye muscle rudiments
4. Precardinal vein
5. Semilunar ganglion
6. Auditory vesicle
7. Endolymphatic duct
8. Superior ganglion
9. Jugular ganglion
10. Vagus nerve
11. Nodose ganglion
12. Internal carotid artery (near level a, Fig. 5.5)
13. Internal carotid artery (near level b, Fig. 5.5)

14. Copula
15. Pharynx
16. Maxillary process
17. Stomodeum
18. Mandibular process
19. Dorsal aorta (cranially)/descending aorta caudally)
20. Trachea/lung buds
21. Ductus venosus
22. Stomach
23. Omental bursa
24. Germinal epithelium
25. Notochord
26. Renal pelvis

27. Nephrogenic tissue
28. Mesonephric duct
29. Ureter
30. Umbilical cord
31. Acoustic ganglion
32. Second branchial arch
33. Third branchial arch
34. Fourth aortic arch
35. Umbilical artery
36. Epiglottis
37. Endocardial cushion
38. Sixth aortic arch
39. Caudal artery
40. Third pharyngeal pouch

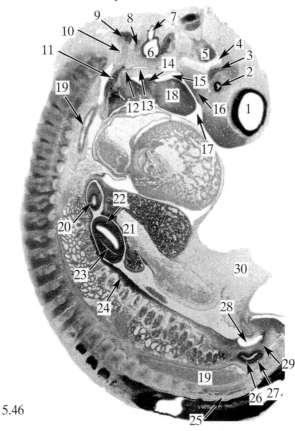

5.46

Photo 5.46. Continuation of 10-mm pig embryo serial sagittal sections numbered in right to left sequence.

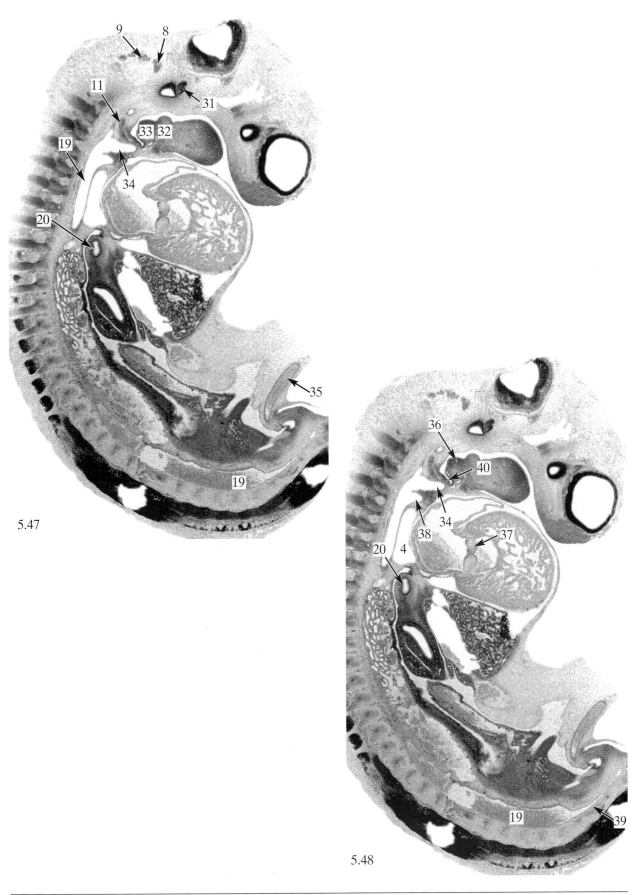

Photos 5.47, 5.48. Continuation of 10-mm pig embryo serial sagittal sections numbered in right to left sequence.

Photos 5.49-5.51

Pig Embryos

Legend

1. Lateral ventricle
2. Fourth ventricle
3. Mesencephalon
4. Metencephalon
5. Myelencephalon
6. Spinal cord
7. Infundibulum
8. Pharynx
9. Aortic trunk
10. Trachea/lung buds
11. Esophagus
12. Dorsal aorta
13. Duodenum (near level a, Fig. 5.3)
14. Cranial limb of the intestinal loop

15. Superior mesenteric artery
16. Superior mesenteric vein
17. Caudal limb of the intestinal loop
18. Allantois
19. Descending aorta
20. Subcardinal vein/anastomosis
21. Umbilical artery
22. Ureter
23. Third ventricle
24. Cerebral aqueduct
25. Isthmus
26. Choroid plexus
27. Basilar artery (near level e, Fig. 5.5)
28. Rathke's pouch

29. Glottis
30. Inferior vena cava
31. Gallbladder
32. Tail
33. Hindleg bud
34. Posterior communicating artery (near level d, Fig. 5.5)
35. Sixth aortic arch
36. Spinal ganglion
37. Sclerotome—less dense cranial portion
38. Sclerotome—dense caudal portion
39. Right subcardinal vein
40. Fourth and fifth pharyngeal pouches

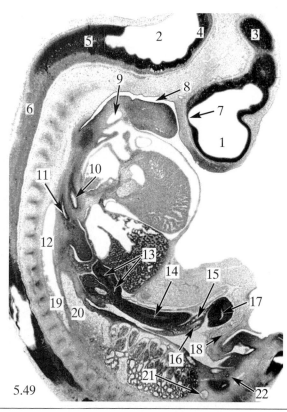

5.49

Photo 5.49. Continuation of 10-mm pig embryo serial sagittal sections numbered in right to left sequence.

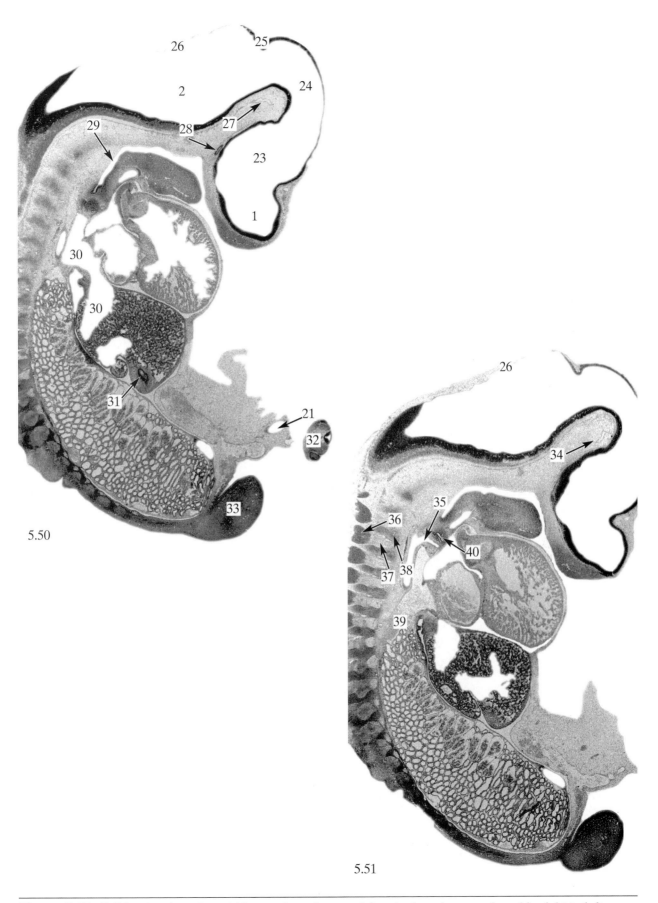

Photos 5.50, 5.51. Continuation of 10-mm pig embryo serial sagittal sections numbered in right to left sequence.

Photos 5.52-5.54

Pig Embryos

Legend

1. Fourth ventricle
2. Endolymphatic duct
3. Auditory vesicle
4. Semilunar ganglion
5. Optic stalk
6. Eye muscle rudiments
7. Precardinal vein
8. Nasal pit
9. Maxillary process

10. Mandibular process
11. Stomodeum
12. Second branchial arch
13. Third branchial arch
14. Jugular ganglion
15. Petrosal ganglion
16. Glossopharyngeal nerve
17. Pharynx
18. Hindleg bud

19. Foreleg bud
20. Acoustic ganglion
21. Sensory root of the facial nerve
22. Geniculate ganglion
23. Optic fissure
24. Sensory root of the trigeminal nerve
25. Ophthalmic branch of the trigeminal nerve

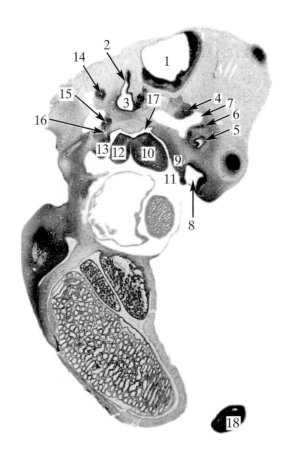

5.52

Photo 5.52. Continuation of 10-mm pig embryo serial sagittal sections numbered in right to left sequence.

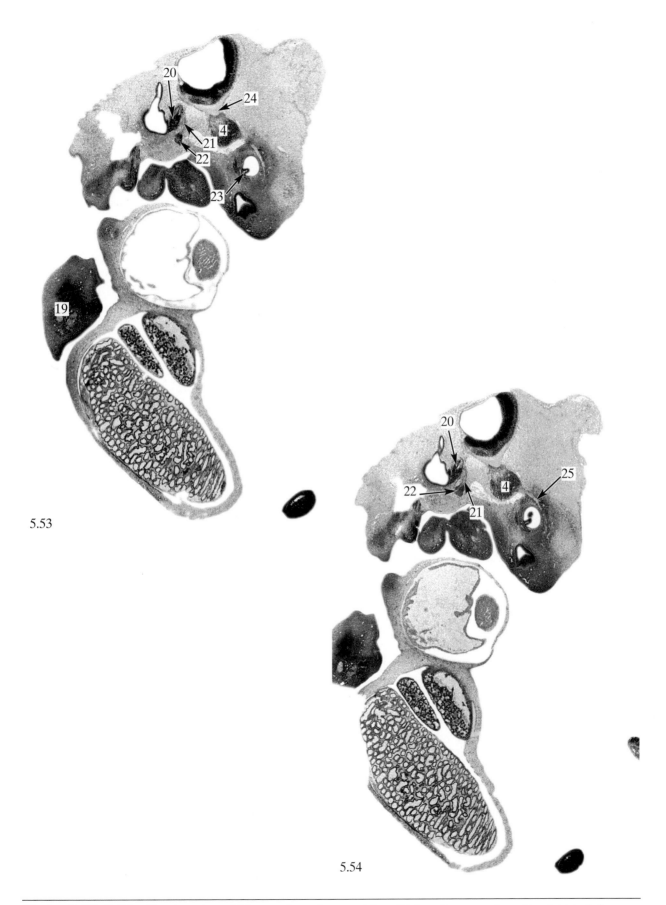

5.53

5.54

Photos 5.53, 5.54. Continuation of 10-mm pig embryo serial sagittal sections numbered in right to left sequence.

Chapter 6

Hands-On Studies

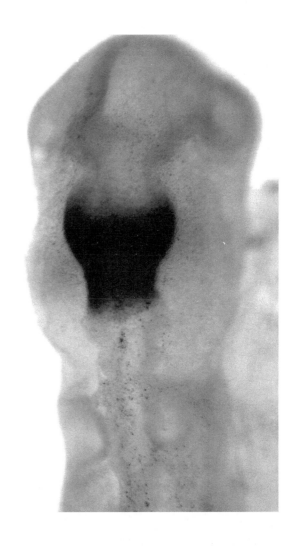

Chapter 6

Hands-On Studies

A. INTRODUCTION

Hands-on studies are an important part of your laboratory experience. Below are several exercises that will provide you with direct experience working with living embryos as well as with techniques used for analyzing embryos in advanced studies.

B. EXERCISES 1.1-1.3: SEA URCHIN EMBRYOS

1. Exercise 1.1: Spawning of gametes and observations of living embryos

a. Materials

Adult sea urchins

Natural seawater (filtered), artificial seawater (reagent grade NaCl, KCl, MgCl$_2$.6H$_2$O, MgSO$_4$.7H$_2$O, CaCl$_2$, NaHCO$_3$), or packaged sea salts (always use the entire package immediately after opening to avoid cytotoxicity; available from Ward's Natural Science Establishment, Inc., or Carolina Biological Supply Company)

0.5 M KCl in distilled water

Syringes (2-5 ml) and needles (20-26 gauge)

Glass microscope slides

Coverslips (22 mm square)

100- and 200-ml clean glass beakers (if possible, use new beakers to ensure that no contaminants are present) or disposable plastic beakers

200-ml graduated cylinders

Finger bowls

Disposable (plastic) petri dishes (35 mm and 100 mm)

Plasticine (or modeling clay)

Compound microscope, preferably with phase contrast optics, Hoffman optics, or differential interference contrast optics (Nomarski optics) and a mechanical stage

Polarizing film (available as an inexpensive Polarizing Effects Kit from Ward's Natural Science Establishment, Inc.)

India ink

Stock solution of Nile blue sulfate (dissolve 1 mg in 20 ml seawater, mix thoroughly, and filter to remove particulates)

Small vials (about 2-ml capacity) and caps

b. Procedures

In this exercise, gametes will be spawned and examined with a compound microscope, eggs and sperm will be mixed together, and the process of fertilization will be examined with a compound microscope; embryos will be studied as they undergo cleavage, blastulation, gastrulation, and formation of prism and pluteus larvae. (Several suppliers can provide living adult sea urchins for the "inland" colleges and universities. These can often be obtained as kits, including seawater, plasticware, KCl, etc. *Strongylocentrotus droehbachiensis* or *purpuratus* can be obtained year-round from Ward's Natural Science Establishment, Inc. [P.O. Box 92912, Rochester, NY 14692-9012; phone: 800-962-2660]). *Eucidaris tribuloides* or *Lytechinus variegatus* are available March through October from Carolina Biological Supply Company [2700 York Road, Burlington, NC 27215-3398; phone: 800-334-5551]. *Lytechinus variegatus* is available May through September and *Arbacia punculata* is available October through May from Gulf Specimen Co., Inc. [P.O. Box 237, Panacea, FL 32346; phone: 904-984-5297].) To make artificial seawater dissolve 28.32 g NaCl, 0.77 g KCl, 5.4 1g MgCl$_2$.6H$_2$O, 7.13 g MgSO$_4$.7H$_2$O, and 1.18 g CaCl$_2$ in 1l distilled water. Use reagent grade chemicals, which seem to have sufficient impurities to supply the necessary trace elements. After the salts are dissolved, add 0.2g NaHCO$_3$ and then adjust the pH to 8.2 if necessary. If possible, check the salinity of the artificial seawater (with a refractometer, a calibrated hydrometer, or a conductivity meter, or by titrating for chloride ions with silver nitrate) and adjust it to 28-34 parts per thousand.

To spawn gametes, the perivisceral coelom of adult animals will be injected with a 0.5 M solution of KCl. Randomly select a few urchins and inject each one with 1-2 ml KCl by inserting the injection needle into the membrane surrounding the mouth opening on the oral or ventral side of the animal. The perivisceral coelom is a large space, and no particular care must be taken in placement of the needle after the membrane is penetrated. Each animal should be placed mouth-side down on a clean sheet of paper and examined a few minutes later, at which time shedding of a creamy mass of eggs or a white, milky mass of sperm should be observed. Note that gametes are shed from the **gonopores** located on the dorsal or aboral surface of the animals. Each female will spawn as many as 10^7 eggs and each male will spawn as many as 10^{12} sperm. When spawning is evident, quickly collect a sample of the spawned material with a Pasteur pipette and place a very small drop of it on a glass microscope slide—use a new pipette and different glass slide for each sample to avoid mixing eggs and sperm. Invert each of the male sea urchins over a clean, dry 100-ml petri dish to collect the sperm. Also take the female sea urchins and invert each of them over a beaker filled with seawater to collect the eggs. As spawning is occurring, examine your samples of egg and sperm. To examine eggs, take a square coverslip and scrape each corner against plasticine or modeling clay to collect a small ball, which will be used as a foot to support the coverslip and prevent the eggs (and later embryos) from being crushed. Carefully place a coverslip over each sample on the slide, gently push down to spread the fluid meniscus beneath the coverslip, and examine the sample first with low magnification (4-10X) and then with the high, dry lens (40-60X). Identify the **mature egg**, its **pronucleus**, the surrounding **vitelline envelope**, and the **jelly layer.** (The latter is not directly visible but its presence can be inferred from the fact that eggs do not immediately abut one another; its presence can be better detected by adding india ink to the seawater, or by staining with Nile blue sulfate in the following way. Fill a small vial [about 2-ml capacity] with sea water, and add 1 drop of the Nile blue sulfate stock solution. Mix by inverting the vial [with its cap on] several times. After 30 seconds to 1 minute, remove the colored seawater and replace it with fresh seawater. Mix by inversion; after 30 seconds remove the seawater and replace it again with fresh seawater. This procedure should readily stain the jelly layer.) Also try to identify *immature* eggs, the **primary oocytes**. Primary oocytes can be identified by their large nucleus, the **germinal vesicle**. They cannot be fertilized, so they will not undergo further development. Some immature eggs are typically shed (and consequently

wasted) along with mature eggs during artificially induced spawning.

To examine sperm, place a coverslip (without "feet") over each sample on the slide. Then identify the **sperm**, with its **head**, **midpiece**, and **tail**.

To study fertilization and further development, prepare an egg dilution in the following way. Use a pipette to take a 1-ml aliquot of eggs from the collection beaker, and add these to another beaker containing 150 ml fresh seawater (Place the original suspension in a refrigerator for later use; this suspension can be stored for up to 6 hours). The egg dilution is now ready for insemination with sperm. To prepare a sperm dilution, take 2-4 drops of "dry" sperm (using a fresh pipette) from the petri dishes and add them to a 200-ml graduated cylinder containing 100 ml seawater (Place the *covered* petri dishes of dry sperm in a refrigerator for later use; dry sperm will keep for up to 24 hours). The sperm dilution is now ready for use. To inseminate eggs, add 1-2 drops of the diluted sperm to each beaker of diluted eggs (using a fresh pipette). Add 1 drop at a time and swirl the beaker between drops for 1-2 minutes to minimize the chance of polyspermy. The diluted sperm can be used for only about 15 to 30 minutes before they become inactive.

After insemination, let eggs settle in the beaker; remove the covering seawater and replace it with fresh seawater to eliminate excess sperm. Eggs (actually zygotes) should form a monolayer on the bottom of the beaker (if too many eggs are present, place some of them in another beaker; also, if desired, transfer eggs to petri dishes [35 or 100 ml] or finger bowls to facilitate subsequent study). After hatching, transfer swimming larvae to fresh seawater in a new finger bowl to optimize development.

Observe **fertilization** by placing a sample of the inseminated eggs (using a fresh pipette) on a glass microscope slide and attaching a supported coverslip. Observe the formation of the **fertilization envelope** and **fertilization cone**. Fertilization occurs very quickly, within the first 1-2 minutes after insemination, so work quickly. Be aware that in observing fertilization, you are seeing a process that began at least 2 billion years ago, a process that few people are fortunate enough actually to see.

If preferred, an alternative procedure can be used. Transfer a sample of eggs to 2-3 ml seawater in a 35-mm petri dish. Place the dish on the stage of a compound microscope and focus on the eggs. Then add a drop of freshly diluted sperm suspension and observe fertilization events (especially elevation of the **fertilization envelope**).

Over the next day or two, periodically collect embryos to observe the various stages of **cleavage**, **blastulation**, **gastrulation**, and formation of the **prism** and **pluteus larvae**. Allow embryos to develop in beakers at room temperature (about 23°C). Refer to the timetable given in the text to plan your schedule for observation. To facilitate observation, embryos can be deciliated. To do this, place them in "2X seawater" for 30 seconds to 1 minute. Make 2X seawater by adding an additional 27-29 g of NaCl to 1l natural or artificial seawater. Cilia regenerate rather quickly, that is, within 20 to 40 minutes, after embryos are returned to standard seawater.

At larval stages, examine embryos through crossed polarizers to better view the development of the **spicules**, which are birefringent (light passing through the crystalline spicules is split into two components). Lay a sheet of polarizing film (or a lens from an old pair of polarized sunglasses) below the substage condenser of a compound microscope and place another sheet over the microscope ocular (eyepiece; if your microscope has two oculars, two sheets can be used, one to cover each ocular; open the substage condenser aperture fully). Rotate the ocular until the spicules shine brightly (if two oculars are used, close the left eye and rotate the right ocular with the right eye open; then repeat the process for the left side).

Draw examples of gametes and embryos at all major stages of development. Label your drawings with as many structures identified in the text as possible. Also list the number of minutes or hours postfertilization at which each stage was obtained.

2. Exercise 1.2: Activation of cortical granules

a. Materials

Materials from Exercise 1.1

Glass microscope slides coated with poly-L-lysine (available from American Histology Reagent Co., Inc. [P.O. Box 7334, Stockton, CA 95267; phone: 209-477-5109])

Calcium-free isolation medium (make up a solution in distilled water consisting of 0.3 M KCl, 0.35 M glycine, 2 mM EGTA, and 2 mM $MgCl_2$; adjust the pH to 7.5)

Laboratory squirt bottle

b. Procedures

In this exercise, cortical "lawns" will be prepared as described by Vacquier (Vacquier, V. D. [1975]. The isolation of intact cortical granules from sea urchin eggs: Calcium ions trigger granule discharge. *Developmental Biology* **43**:62-74) and Schatten, G. and Mazia, D. [1976]. The surface events at fertilization: The movements of the spermatozoon through the sea urchin egg surface and the roles of the surface layers. *Journal of Supramolecular Structure* **5**:343-369). Calcium will then be added to the lawns to trigger the exocytosis of the **cortical granules**, simulating **fertilization**.

Obtain freshly spawned eggs as described in Exercise 1.1. To dejelly eggs, transfer them to seawater at pH 5 (add HCl to standard seawater to decrease its pH) for 2 minutes. Subsequently, wash them in two changes of seawater at pH 8. Attach eggs to glass microscope slides coated with poly-L-lysine by pipetting a drop of a concentrated slurry of eggs onto each slide letting them set for 5 minutes. Gently wash slides three times in *calcium-free* isolation medium. Direct a jet of isolation medium at the adhering eggs using a laboratory squirt bottle. If this is done properly, isolated discs of egg plasmalemma will remain adhering to the slides, and each disc will contain egg **cortex** on its upper surface, including thousands of cortical granules.

To observe the cortical reaction, drain off excess fluid from your slides and add a *supported* coverslip (as described in Exercise 1.1), but do not press down on the coverslip and do not establish contact with the fluid meniscus. While observing a plasmalemma disc under high, dry magnification (use phase contrast optics, Hoffman optics, or differential interference contrast optics), carefully add a drop of standard seawater (seawater containing calcium) to the edge of the coverslip. It should flow under the coverslip and cover the plasmalemma disc. Quickly refocus and observe the **cortical reaction**. This experiment provides direct proof of a role for calcium ions in the cortical reaction.

Describe what you observe in your preparations. For best results, prepare several slides of plasmalemma discs and repeat the experiments several times on different slides.

3. Exercise 1.3: Vegetalization, radialization, and exogastrulation

a. Materials

Materials from Exercise 1.1

10 mM, 20 mM, and 30 mM LiCl solutions

b. Procedures

In this exercise, early cleavage stages of sea urchin embryos will be treated with solutions of LiCl. Such treatment has three main effects. First, it causes **vegetalization** of the embryo (the development of the vegetal half of the embryo predominates over the development of the animal half, and embryos develop with expanded vegetal structures, such as gut). Second, it causes **radialization** of the embryo (the larva retains its **radial symmetry** rather than developing bilateral symmetry). Evidence of radialization can best be obtained by observing lithium-treated embryos through crossed polarizers as described in Exercise 1.1: more than two triradiate **spicules** develop after lithium treatment. Third, it causes **exogastrulation** the archenteron *evaginates* rather than *invaginates*).

Obtain living sea urchin embryos at early cleavage

stages (2-cell stage to 32-cell stage) as described in Exercise 1.1, and place some of them into each of the three lithium chloride solutions. Over the next day or two, periodically collect embryos and compare their development to that of untreated control embryos.

Draw examples of embryos at various stages of development. Attempt to find clear examples of embryos in which vegetalization occurred. How do these differ from control embryos? Additionally, try to find clear examples of embryos in which radialization occurred. Note the arrangement of spicules in these embryos. Finally, try to find clear examples of embryos in which exogastrulation occurred. How do these differ from control embryos, especially regarding development of the **archenteron**? Label your drawings with as many structures identified in the text as possible. Also list the number of minutes or hours postfertilization at which each stage was obtained.

C. EXERCISES 2.1-2.4: FROG EMBRYOS

1. Exercise 2.1: Spawning of eggs and observations of living embryos

a. Materials

Etheridge and Richter's _Xenopus laevis_: Rearing and Breeding the African Clawed Frog. 1978. Nasco, 901 Janesville Ave., Fort Atkinson, WI 53538. Phone: (414) 563-2446.

Mature male and female _Xenopus_ frogs

Covered holding tanks (see Etheridge and Richter's booklet)

Nasco frog brittle

Chorionic gonadotropin

Rubber gloves

Sterile 5-ml syringes

Sterile 30-gauge needles

Sterile distilled water

Paper towels

100% Steinberg solution (distilled water, sodium chloride, potassium chloride, calcium nitrate, magnesium sulfate, TRIZMA hydrochloride, TRIZMA base)

Dejellying solution (DL-Dithiothreitol)

HEPES buffer

Sodium hydroxide pellets

Petri dishes

Pasteur pipettes

Wide-mouth pipettes (diamond pencil, Pasteur pipettes)

Pipette bulbs

Safety glasses

Stereomicroscopes (dissecting microscopes)

Single-edge razor blades

Glass microscope slides (preferably depression slides)

b. Procedures

Observation of the development of living frog eggs is an exciting adventure for both student and professional embryologists. Methods for spawning eggs of _Xenopus laevis_, the African clawed frog, will be summarized in this exercise. Details on the rearing and breeding of _Xenopus_ can be obtained in Etheridge and Richter's superb, inexpensive booklet. The course instructor should obtain and read the booklet thoroughly before this exercise is attempted. It is recommended that several copies of the booklet be kept on hand for student use, especially during the staging of embryos according to the criteria of Nieuwkoop and Faber (1956; Normal Table of _Xenopus laevis_. Amsterdam: North Holland Publishing Co.), which are reproduced in the booklet.

Store _Xenopus_ frogs at room temperature in _covered_ holding tanks (see Etheridge and Richter's booklet). Fasten the cover or weight it to prevent frogs from escaping but be sure to allow air exchange to occur. Each tank should contain a minimum of 2 inches of water. Feed frogs Nasco frog brittle according to the schedule given in the booklet. Select a pair of mature frogs for breeding. The adult female is much larger than the male and will have **cloacal valves**; the mature male has "**nuptial pads**" on its forearms (see Figs. 2-4 in Etheridge and Richter's booklet).

Ovulation will be induced by injecting a solution of **chorionic gonadotropin** (Sigma Chemical Co., St. Louis, MO; stock no. CG-5). Gonadotropin is teratogenic, so _wear gloves and handle with care_. Each vial of gonadotropin contains 5,000 international units (IU). Use a sterile syringe to add 5 ml of sterile distilled water to each vial. Females will be injected with 1.5 ml of solution (1,500 units) and males with 1 ml (1,000 units). Injection will be made into the **dorsal lymph sac** with a 30-gauge needle attached to a 5-ml syringe. Wrap each frog in a paper towel to immobilize it (see Fig. 8 in Etheridge and Richter's booklet) and inject the dorsal lymph sac (see Figs. 9 and 10 in Etheridge and Richter's booklet). Place 1 or 2 pairs of injected frogs into a breeding tank (see Fig. 5 in Etheridge and Richter's booklet). Add 15 l distilled water, 1.5 l 100% Steinberg solution at pH 7.4 (to make a 5-l stock solution, take 5 l of distilled water and add 17 g sodium chloride, 0.25 g potassium chloride, 0.40 g calcium nitrate, 1.02 g magnesium sulfate, 3.3 g TRIZMA hydrochloride, and 0.48

g TRIZMA base; the last two components can be obtained from Sigma), and 1.3 g TRIZMA base to the tank. In our tanks this gives the optimal water level of 3-4 inches above the false bottom.

Store the breeding tanks, if possible, in an undisturbed cool room (20°C) in the dark. Frogs must not be disturbed during mating or fertilized eggs will not be obtained. If frogs are injected at 5 P.M. early cleavage stages will be obtained under these conditions by morning. Dejelly eggs to facilitate observation. To do this, first prepare a stock solution of 6.5 g DL-Dithiothreitol (DTT) (Sigma, no. D-0632) in 100 ml distilled water. Also prepare a buffer stock solution of 28.8 g HEPES (Sigma, H3375) in 200 ml distilled water. Adjust its pH to 8.9 with sodium hydroxide pellets. Mix 4 ml of the DTT stock solution with 20 ml of the buffer stock solution and add distilled water to make 200 ml of dejellying solution. Dejelly eggs by removing them from the breeding tank with a wide-mouth pipette. The blunt end of a Pasteur pipette works well. First, *put on safety glasses*. Then, wrap the tip of a Pasteur pipette in a paper towel; carefully break off the tip (facilitated by first scoring the glass with a diamond pencil), protecting your hands and eyes; and add a bulb to the pointed end. Place eggs in a petri dish containing dejellying solution for a *maximum* of 4 minutes (longer exposures result in abnormal development). Remove the dejellying solution and wash the eggs 3 times with 100% Steinberg solution. After washing, place the eggs in 70% Steinberg solution (dilute some of the 100% Steinberg stock solution with distilled water to obtain this concentration) and gradually dilute the Steinberg solution to 20% over the next hour. Remove nondeveloping eggs from the dish and reduce the density of the eggs to about 1 egg/ml solution.

Use the Nieuwkoop and Faber (1956) staging criteria (Figs. 11-13 in Etheridge and Richter's booklet) to stage your embryos. Examine embryos and tadpoles over several days (that is, at 1- to 2-day intervals) but especially during the first 24 hours after spawning (that is, at 1- to 2-hour intervals when possible). Fertilized eggs tend to rest on their vegetal poles with animal poles directed upward. Egg pigmentation is generally distributed more heavily in the *animal-ventral* quadrant, although this distribution may vary. Therefore, as you look at the embryo, you will see the animal hemisphere divided into *dorsal* (lightly pigmented) and *ventral* (heavily pigmented) quadrants. (Unfortunately, the gray crescent is poorly defined in *Xenopus*.) Sketch the stages you observe. Also record the times at which the various stages appear. Using single-edge razor blades, cut embryos at selected stages into thick transverse and sagittal sections. Place each section on a glass microscope slide (preferably a depression slide) in a drop of Steinberg solution and observe the cut surface with your

stereomicroscope.

2. Exercise 2.2: Mapping of morphogenetic movements

a. Materials

Materials from Exercise 2.1

Dejellied *Xenopus* blastulae

20% Steinberg solution

Modeling clay

Pencils

Nile blue sulfate agar staining slides (agar, sterile distilled water, sterile glass slides, paper tissues, aluminum foil, glass pans, medium-sized forceps, 70% ethanol, sterile paper towels, scalpel blades)

Watchmaker's forceps

b. Procedures

In this exercise **morphogenetic movements** will be mapped, using vital dyes, as the embryo passes from blastula to gastrula stages. Obtain dejellied and washed blastula-stage *Xenopus* embryos using the procedures described in Exercise 2.1. Place blastulae in a petri dish filled with 20% Steinberg solution and containing modeling clay on its bottom. Use a pencil tip to poke small wells in the clay and then pipette a single blastula into each well. Immobilizing eggs within the wells aids in subsequent staining. Blastulae will be vital stained through their **vitelline membrane** with chips of agar impregnated with Nile blue sulfate. First, make some agar staining slides. To do this, prepare a 1% solution of agar in sterile distilled water (heat the solution while stirring until the agar is dissolved). Have on hand some ordinary microscope *glass* slides, which have been washed with 70% ethanol, dried with tissues, and sterilized by autoclaving or heating. Pour a thin agar layer onto one side of each glass slide. Allow the slides to dry overnight (cover them with a canopy made from aluminum foil to prevent dust from accumulating on them). Soak the slides for one week in a *glass* pan (covered with aluminum foil) containing a 1% solution of Nile blue sulfate in sterile distilled water. Do not stack slides on top of one another or incomplete dye impregnation will occur. Finally, remove each slide from the pan with forceps, wash it with sterile distilled water, allow it to dry, and wrap it in a sterile (autoclaved) paper towel for storage. To stain embryos, unwrap a slide and place a drop of sterile distilled water on the coated side of the agar slide. Remove a *tiny* chip of agar from the moistened area using a scalpel blade (the chip should be as small as possible, preferably only slightly larger than the tip of the blade). Transfer the chip with a pair of watchmaker's forceps and lay it on the desired region of the surface of the blastula. Staining should occur within a minute or two. Remove the chip with a gentle stream of fluid generated with a pipette.

Sketch blastulae and the initial positions of the dye marks. Examine embryos periodically over the next several hours and resketch them at each interval, noting the changing positions of the marks. Pay particular attention to the shapes of marks placed in close vicinity to the forming **blastopore**. Cut selected stages of embryos into transverse and sagittal thick sections as described in Exercise 2.1.

3. Exercise 2.3: Exogastrulation

a. Materials

Materials from Exercises 2.1 and 2.2

Dejellied *Xenopus* blastulae

100% Steinberg solution

20% Steinberg solution

Watchmaker's forceps

Nile blue sulfate agar staining slides

b. Procedures

Morphogenetic movements during gastrulation result in changes in the original positions of cells composing the blastula and give rise to the germ layers: ectoderm on the outside of the embryo, forming the surface epithelium; endoderm on the inside, forming the archenteron; and mesoderm filling the space between ectoderm and endoderm. Cells are brought into new positions by this process, allowing them to undergo tissue interactions (inductions) and form new structures (such as the neural plate). Normal morphogenetic movements will be inhibited in this exercise, preventing cells from involuting over the lips of the blastopore and forming the archenteron (that is, **exogastrulation** will occur). Instead, cells will remain on the surface to form an epithelial vesicle. We will cause exogastrulation by allowing blastulae to develop in 100% Steinberg solution, a hypertonic salt solution.

Obtain dejellied and washed blastula-stage *Xenopus* embryos using the procedures described in Exercise 2.1. Place eggs in a petri dish containing 100% Steinberg solution and maintain them in this solution during the course of the experiment. Remove the vitelline membrane covering each blastula by using two pairs of fine-tipped watchmaker's forceps, one in each hand. Use one pair to poke and grasp the membrane and the other to grasp and tear it. This can be difficult to do without injuring the blastula, and it requires practice.

Sketch blastulae at the beginning of the experiment. Observe embryos periodically over the next several hours and sketch them at each interval. Compare mor-

phogenetic movements in embryos undergoing exogastrulation with those of control embryos (developing in 20% Steinberg solution with or without an intact vitelline membrane) by staining them with chips of agar impregnated with Nile blue sulfate (see Exercise 2.2). What differences can you detect? Also cut thick sections of selected embryos as described in Exercise 2.1.

4. Exercise 2.4: Neural crest ablation

a. Materials

Materials from Exercises 2.1-2.3

Dejellied *Xenopus* late gastrulae

100% Steinberg solution

20% Steinberg solution

Watchmaker's forceps

Cactus needles (magnifying glasses, cactus spines, medium-sized forceps, wooden sticks, nail polish)

70% ethanol

b. Procedures

A **neural fold** will be extirpated in this exercise, thereby eliminating a patch of **neural crest cells** on one side of the embryo. *Xenopus* embryos contain **pigment cells** in their **epidermis**. Neural crest cells are the source of this pigment (do not confuse crest-derived pigment *cells* with egg pigment *granules*; the latter are deposited in the egg by the ovary during oogenesis, whereas the former migrate from the roof of the neural tube during late neurulation and invade the epidermis). This fact will be confirmed by deleting neural crest cells and noting the absence of pigmentation in a portion of the embryo.

Obtain dejellied and washed late gastrulae using the procedures described in Exercise 2.1. Let embryos develop in 20% Steinberg solution until neurula stages are obtained. Then place them in 100% Steinberg solution to facilitate healing during microsurgery. Remove vitelline membranes by using 2 pairs (one in each hand) of fine-tipped watchmaker's forceps, as defined in Exercise 2.3. Use care to avoid damaging the embryo. Cactus needles will be used to extirpate a portion of one of the elevated neural folds. To make cactus needles, simply obtain a few cactus plants from a local florist (examine their needles carefully with a magnifying glass prior to purchase, and select several plants with needles of various sizes). *Wearing safety glasses*, pluck the desired needles from the plant with forceps and mount the needles onto wooden sticks with nail polish. Sterilize needles by dipping them into 70% ethanol followed by sterile water. Make two transverse cuts through a neural fold some distance apart (as you repeat the experiments on other embryos, vary the craniocaudal level and side of the embryo at which you extirpate and also the distance between the two transverse cuts). Connect these two cuts with a horizontal incision and remove the neural fold. Determine the depth at which you make the horizontal cut by examining Photo 2.6. Allow embryos to remain in 100% Steinberg solution for 30 minutes, and then transfer them to 70% and eventually to 20% Steinberg solution, as described in Exercise 2.1.

Sketch the appearance of neurulae at the time of surgery. Observe embryos periodically over the next few days and sketch their morphogenesis, especially during early tadpole stages. Pay particular attention to patterns of pigmentation on the operated side.

D. EXERCISES 3.1-3.13: CHICK EMBRYOS

1. Exercise 3.1: Preparation of a graphical reconstruction

a. Materials

Millimeter rulers

Colored pencils

b. Procedures

It is recommended that you prepare a graphical reconstruction of *your* transversely serially sectioned 33-hour embryo to help you visualize relationships of parts of an embryo to one another in three dimensions. A sample is given in Fig. 3.2. Examine your slides of serial transverse sections with the naked eye. Note that the sections are mounted in horizontal rows. Count and record the number of rows and the number of sections in each row. Add the number of sections in each row to determine the total number of sections.

On the sheet of graph paper provided on the next two pages (ruled with 1-cm squares), let each centimeter along the longest (horizontal) axis represent the thickness of *ten* sections. At the top of the paper, every 10 sections have been numbered off from left to right. Draw vertical dashed lines at the levels of the last section number in each row (determine millimeter markings with the aid of a millimeter ruler). Number the rows at the bottom of the paper. The paper has been subdivided into three areas by brackets at the left. The following labels have been placed at the right of the page within each subdivision. Under *ectodermal structures*, the first centimeter line from the bottom has been labeled *neural tube and groove*, and the third centimeter line has been labeled *subcephalic pocket*. Under *mesodermal structures*, the first centimeter line has been labeled *notochord*, and the second centimeter line has been labeled *head mesenchyme, somites, and segmental plates*. The space between the fifth and sixth centimeter lines has been labeled *heart and related vessels*. The ninth centimeter line has been labeled *right dorsal aorta*. Under *endodermal structures*, the second centimeter line has been labeled *foregut*. Instructions for plotting structures on the graph paper are given in

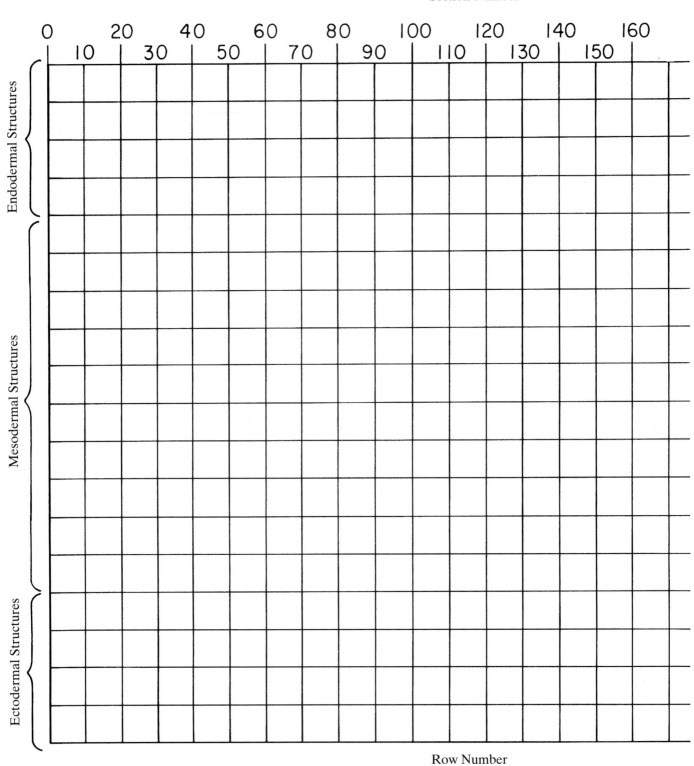

Graph paper for use in Exercise 3.1.

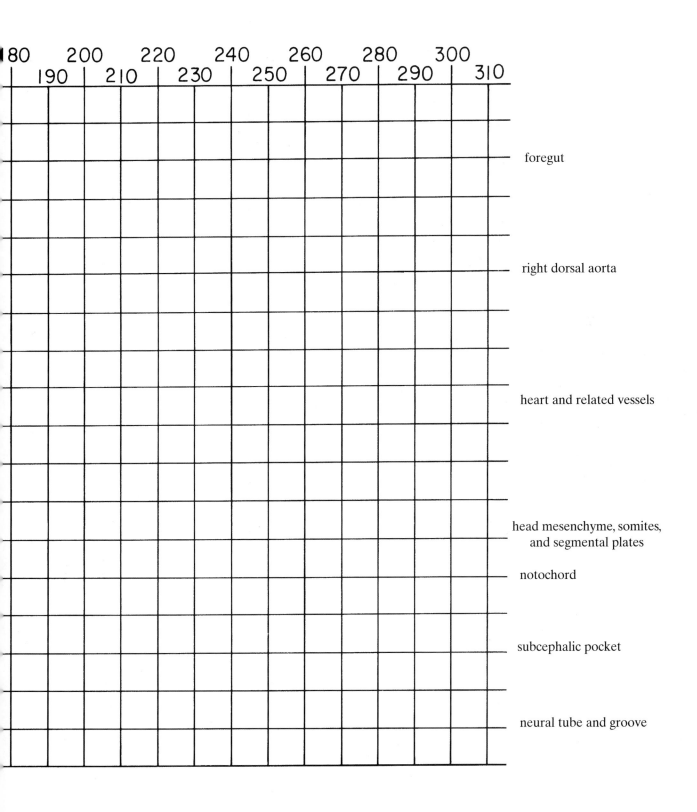

brackets in Chapter 3 (C. 33-hour chick embryos, 2. Serial transverse sections).

2. Exercise 3.2: Staging chick embryos

a. Materials

Photos 6.1-6.18

Optional: Complete copies of the Eyal-Giladi and Kochav and Hamburger and Hamilton stage series

b. Procedures

The avian egg is laid at a blastoderm stage during which cleavage is well advanced and the hypoblast is just beginning to form through the polyingression of small groups of cells from the future epiblast toward the subgerminal cavity. With incubation at about 38°C, further development of the blastoderm occurs. Photos 6.1-6.4 illustrate several stages of early development of the bird embryo following initial incubation. Such stages are defined using the criteria of Eyal-Giladi and Kochav (Eyal-Giladi, H. and S. Kochav [1976]. From cleavage to primitive streak formation: A complementary normal table and a new look at the first stages of the development of the chick. I. General morphology. *Developmental Biology* **49**:321-337), and stages are designated by using the Roman numerals I-XIV (stages I-IX occur within the hen prior to laying of the egg; only stages XI-XIV are illustrated). Older bird embryos are staged according to the criteria of Hamburger and Hamilton (Hamburger, V. and H. L. Hamilton [1951]. A series of normal stages in the development of the chick embryo. *Journal of Morphology* **88**:49-92)., and stages are designated by using the Arabic numerals 1-46. Hamburger and Hamilton's stage 1 has been subdivided in the Eyal-Giladi and Kochav stage series into stages X-XIV; because this substaging is more precise, Hamburger and Hamilton's stage 1 is no longer used. In this exercise, you should use Photos 6.1-6.4 and the accompanying description of the Eyal-Giladi and Kochav staging criteria to gain an understanding of the structure of the avian blastoderm during its first few hours of incubation after laying. Then, you should use Photos 6.5-6.18 and the accompanying description of the Hamburger and Hamilton staging criteria to gain an understanding of the structure of the avian embryo during the first 3 days of incubation. You will have an opportunity to test your staging skills on living embryos in Exercises 3.3-3.12.

The Eyal-Giladi and Kochav stage series defines three groups of stages: stages I-VI, cleavage; stages VII-X, formation of the area pellucida; and stages XI-XIV, formation of the hypoblast. A brief description of Eyal-Giladi and Kochav stages during early postincu-

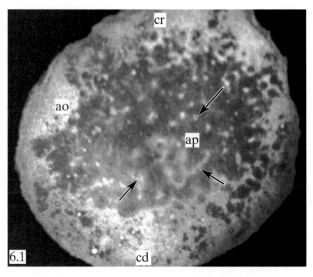

Stage XI. Early stage of hypoblast formation. Photo 6.1. Cells are polyingressing from the epiblast as well as migrating cranially from Koller's sickle (not visible in photographs). cr: cranial end of blastoderm; cd: caudal end of blastoderm; ao: area opaca; ap: area pellucida; arrows: polyingressing clusters of prospective hypoblast cells.

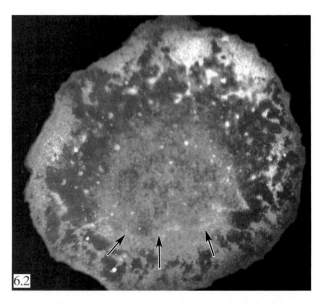

Stage XII. Intermediate stage of hypoblast formation. Photo 6.2. Cells continue their polyingression from the epiblast and their migration from Koller's sickle. Caudally, hypoblast cells derived from polyingression and the sickle are forming the sheet-like hypoblast (arrows).

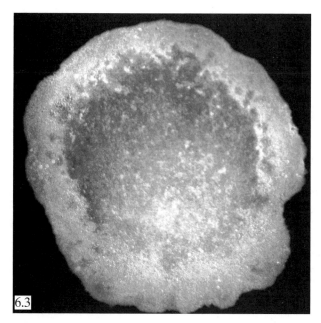

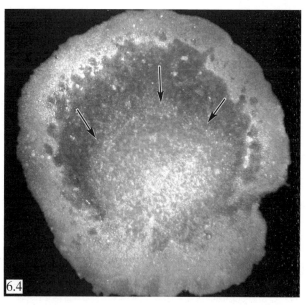

Stage XIII. Stage of fully spread hypoblast.
Photo 6.3. The sheetlike hypoblast now underlies most of the central portion of the area pellucida. Caudally, the hypoblast is well defined, but its craniolateral borders still are poorly demarcated.

Stage XIV. Stage of fully defined hypoblast.
Photo 6.4. The craniolateral borders of the hypoblast (arrows) are now becoming distinct.

The Hamburger and Hamilton stage series defines 46 stages from the freshly laid egg (stage 1) to the newly hatched chick (stage 46). A brief description of Hamburger and Hamilton stages covered in this guide follows. The number in the upper right-hand corner of each illustration indicates its Hamburger and Hamilton stage.

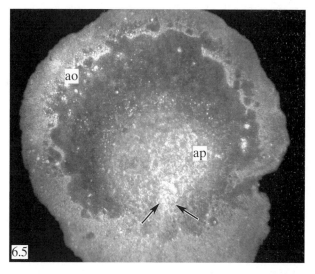

Stage 1. Prestreak stage. Photos 6.1-6.4.
See above illustrations and those on preceding page. Superseded by Eyal-Giladi and Kochav stages X-XIV. The primitive streak has not yet formed. The hypoblast is forming but is not readily visible from the dorsal surface. The egg is usually laid at this stage.

Stage 2. Initial streak stage. Photo 6.5. The primitive streak has just formed and consists of a short conical density, almost as wide as it is long. Arrows indicate the primitive streak; ao: area opaca; ap: area pellucida. 6-7 hours of incubation.

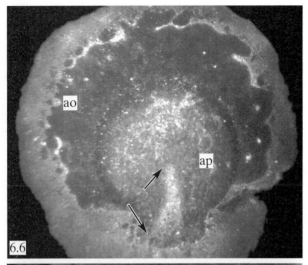

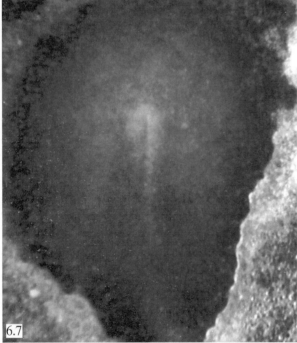

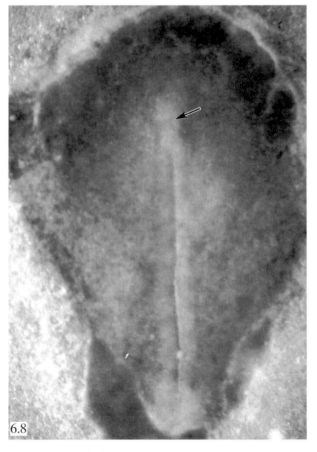

Stage 3. Intermediate streak stage. Photos 6.6, 6.7. The primitive streak now has greater length than width but has not yet fully elongated (that is, it does not yet extend craniad beyond about the center of the area pellucida). A primitive groove may (late Stage 3; Photo 6.7) or may not (early Stage 3; Photo 6.6) be present. Arrows in Photo 6.6 indicate the primitive streak; ao: area opaca; ap: area pellucida. 12-13 hours of incubation.

Stage 4. Definitive streak stage. Photo 6.8. Typical 18-hour embryo. The primitive streak has elongated fully and the primitive knot, pit, and ridges are distinct. A short tongue of cells extends slightly craniad from the primitive knot (arrow). 18-19 hours of incubation.

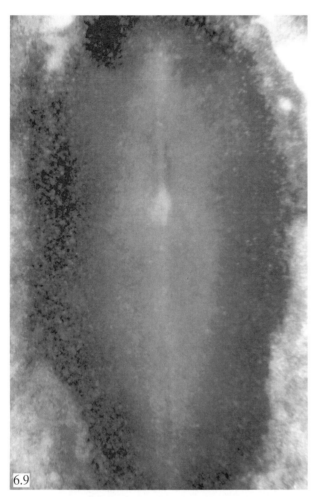

6.9

Stage 5. Head process stage. Photo 6.9.
Typical 18-hour embryo. A distinct, rodlike head
process (notochord) has formed and the primitive
knot is regressing. The head fold of the body has not
yet formed. 19-22 hours of incubation.

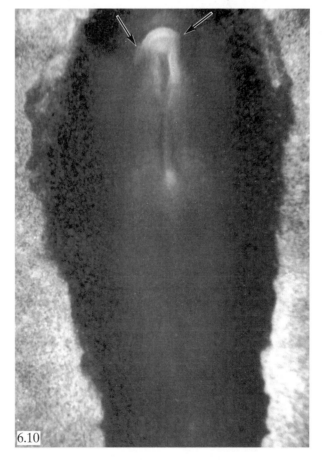

6.10

Stage 6. Head fold stage. Photo 6.10. Typical
18-hour embryo. The head fold of the body (arrows)
has formed cranial to the tip of the head process.
Somites have not yet formed. 23-25 hours of incubation.

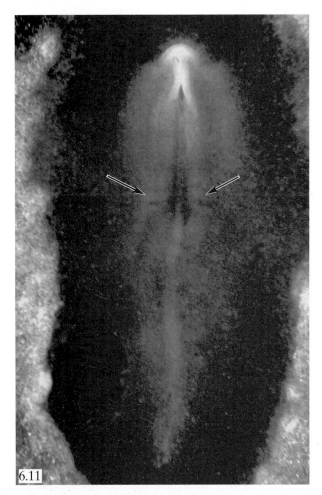

6.11

Stage 7. One-somite stage. Photo 6.11. The first pair of somites (arrows) has formed just cranial to the primitive knot. Stages 7-14 are designated principally by the number of somite pairs. A numbered stage is assigned to embryos when they acquire each third somite pair. Thus, embryos at Stage 8 have four somites, and embryos at Stage 9 have seven somites. Pluses and minuses are used to designate intermediate stages. Therefore, embryos at stage 7+ have two somite pairs, and those at 8- have three somite pairs. 23-26 hours of incubation.

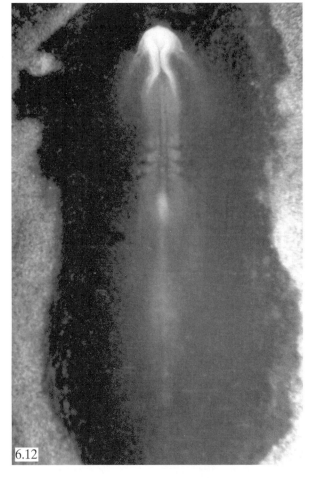

6.12

Stage 8. Four-somite stage. Photo 6.12. Typical 24-hour embryo. Neural folds have fused at the mesencephalon level. 26-29 hours of incubation.

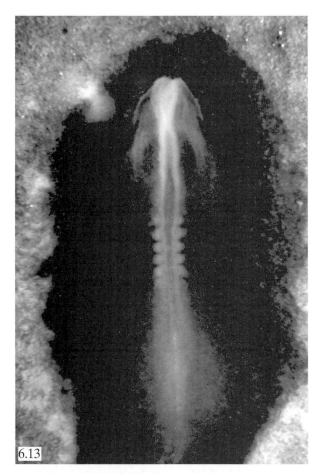

Stage 8+. Five-somite stage. Photo 6.13.
Typical 24-hour embryo. Neurulation is more advanced than at Stage 8.

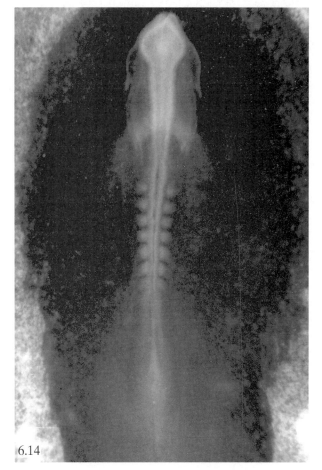

Stage 9. Seven-somite stage. Photo 6.14.
Optic vesicle formation is under way. 29-33 hours of incubation.

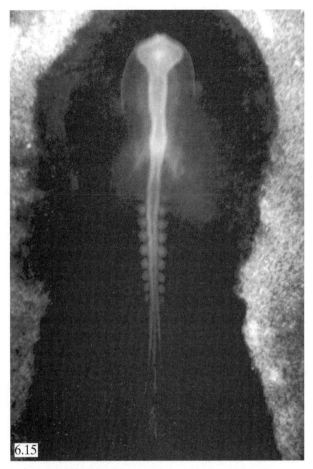

6.15

Stage 10. Ten-somite stage. Photo 6.15.
Typical 33-hour embryo. 33-38 hours of incubation.

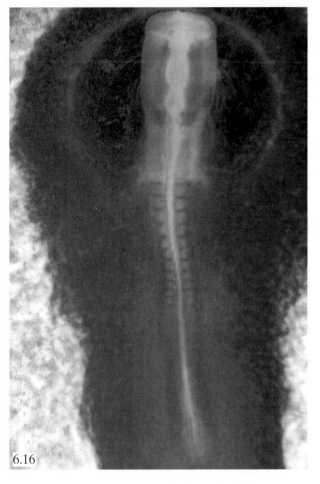

6.16

Stage 11. Thirteen-somite stage. Photo 6.16.
Typical 33-hour embryo. 40-45 hours of incubation.

Stage 12. Sixteen-somite stage. Not illustrated.
Typical 33-hour embryo. 45-49 hours of incubation.

Stage 13. Nineteen-somite stage. Not illustrated. 48-52 hours of incubation.

Stage 14. Twenty-two-somite stage. Not illustrated. The cranial flexure has partially formed so
that the longitudinal axes of the forebrain and hindbrain meet in the midbrain at about a right angle. 50-
53 hours of incubation.

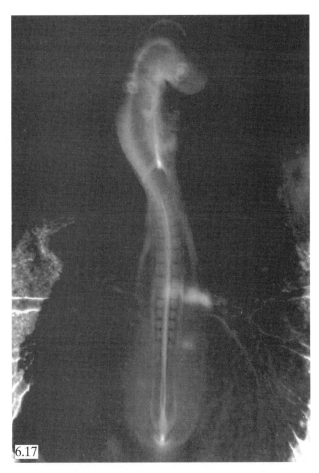

6.17

Stage 15. Twenty-four- to twenty-seven-somite stage. Photo 6.17. Typical 48-hour embryo. The longitudinal axes of the forebrain and hindbrain meet in the midbrain at an acute angle. The lateral body folds extend to the cranial border of the wing buds (that is, to about somite 15). 50-55 hours of incubation.

Stage 16. Twenty-six- to twenty-eight-somite stage. Not illustrated. Typical 48-hour embryo. The lateral body folds extend to a level midway between the wing and leg buds (that is, to about somite 20). 51-56 hours of incubation.

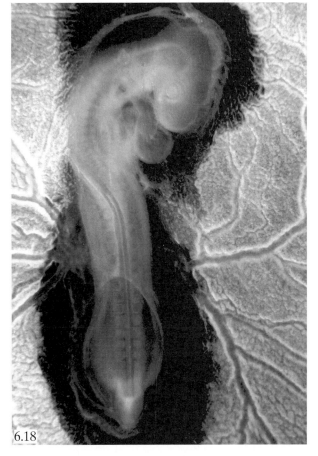

6.18

Stage 17. Twenty-nine- to thirty-two-somite stage. Photo 6.18. Typical 72-hour embryo. The cervical flexure is well formed so that the longitudinal axes of the brain and spinal cord meet at about a right angle. 52-64 hours of incubation.

bation follows. The number in the upper right-hand corner of each photograph indicates its Eyal-Giladi and Kochav stage.

3. Exercise 3.3: Observations of living embryos at 18-33 hours

a. Materials

> Materials from Exercise 3.2
>
> Fertile chicken eggs (incubated 18-25, 26-29, and 33-49 hours)
>
> 38°C incubators
>
> Fingers bowls
>
> Saline (NaCl, distilled water)
>
> Microscope illuminators or high-intensity lamps
>
> Stereomicroscopes (dissecting microscopes)
>
> 1% solution of neutral red in saline
>
> 1% solution of Nile blue sulfate in saline
>
> 0.05% solution of Nile blue sulfate in saline
>
> Carbon/melted agar mixture (animal charcoal powder, mortar and pestle, agar, saline, hot plate stirring rods)
>
> India ink
>
> 5-ml disposable syringes
>
> 18-gauge needles
>
> Single-edge razor blades
>
> Glass microscope slides (preferably depression slides)
>
> Petri dishes

b. Procedures

Obtain fertile chicken eggs incubated at 38°C for 18-25, 26-29, and 33-49 hours. These eggs should contain, respectively, typical 18-, 24-, and 33-hour embryos. (Typical 33-hour embryos are formed after a range of incubation times. The actual stage of the embryo after a fixed interval of incubation is dependent upon the season, the breed of the chickens that produced the egg, the humidity of the incubator, and the temperature of the incubator.) Carefully crack open the eggs against the rim of a finger bowl half filled with warm saline (123 mM NaCl in distilled water). The crack in the shell should be held beneath the surface of the saline solution, and the two halves of the shell should be slowly pulled apart, releasing the egg contents into the solution. The side of the yolk containing the blastoderm should float to the upper surface of the bowl where the embryo can be readily observed. If not, use your fingertips to gently rotate the egg until the blastoderm is uppermost. Illuminate the blastoderm by placing an illuminator above and to the side of the blastoderm, so

that light strikes its surface at about a 45° angle.

Two methods can be used to increase the contrast of the blastoderm relative to the yolk, aiding in the identification of embryonic structures. The simplest is to apply a drop of a 1% aqueous solution of vital stain (neural red or Nile blue sulfate) onto the vitelline membrane overlying the blastoderm. The second method requires injecting contrasting medium into the **subgerminal cavity**: a narrow cleft separating the area pellucida region of the blastoderm from the underlying yolk. Three solutions can be injected: a solution of 0.05% Nile blue sulfate in 123 mM saline, full-strength india ink, or a mixture of carbon and melted agar (grind animal charcoal powder with a mortar and pestle to a fine consistency; mix 3 g of the ground powder with 0.24 g agar and add 40 ml saline; heat the mixture almost to boiling while stirring; let the solution cool to 47° to 50°C and maintain it at this temperature to keep the agar melted; stir the mixture frequently). It is easier to inject a Nile blue sulfate solution or india ink than the carbon/agar mixture, but the former quickly diffuse away and multiple injections are required if blastoderms are viewed over substantial periods. To inject the subgerminal cavity, fill a 5-ml disposable syringe with one of the three solutions (work quickly if using the carbon/agar mixture, so that hardening does not occur). Attach an 18-gauge needle to the syringe, remove any air bubbles, insert the tip of the needle through the vitelline membranes slightly into the yolk just peripheral to the blastoderm, reorient the needle so that it lies within the yolk parallel to the surface of the blastoderm, and place the tip of the needle just beneath the center of the blastoderm. Slowly inject the solution and fill the subgerminal cavity.

Use the abridged Hamburger and Hamilton stage series in Exercise 3.2 to stage your embryos. Draw the stages that you identify.

Remove some blastoderms from the yolk and transfer them to petri dishes as follows. After cracking eggs into finger bowls containing warm 123 mM saline, manipulate the yolk and attached blastoderm with your fingertips until the blastoderm floats uppermost. Grasp the periphery of the blastoderm (that is, the area opaca) with a pair of fine forceps (watchmaker's or #5). Then use fine-pointed scissors to cut around the perimeter of the blastoderm. (Make sure that you hold on to the periphery of the blastoderm with your forceps; also make sure that the blastoderm is floating uppermost, centered on the yolk—eccentrically positioned blastoderms are more difficult to remove without tearing.) After the entire perimeter of the blastoderm is cut free from the yolk, gently grasp the blastoderm with forceps and *float* it away from the yolk. Place an ordinary spoon beneath the blastoderm while you hold the latter with forceps. Slowly lift the spoon and the cuddled blastoderm out of the finger bowl and transfer them to a second bowl containing fresh saline. Gently free the blastoderm from the spoon and remove the latter from the bowl. The blastoderm at this point is usually still covered dorsally by the vitelline membrane. (Although this latter structure seems to be a single membrane, it actually consists of both the inner and outer vitelline membranes.) To remove this membrane it is best to grasp the periphery of the blastoderm and gently waft the blastoderm back and forth in the saline until the vitelline membrane floats free. (The action involved is analogous to shaking out a rug, when you grasp one edge of the rug and the rug undulates in the air as you raise and lower your arms.)

After the blastoderm has been freed from its membranes, place a small dish (a dish approximately 50 mm in diameter works best) into the bowl of saline beneath the blastoderm and withdraw the dish slowly. This procedure results in removal of the blastoderm and some saline from the bowl. Use forceps to unwrinkle the blastoderm and to float it dorsal side up in the small dish. Remove excess saline from the dish with a pipette so that the blastoderm flattens onto the bottom of the dish.

Cut transverse and sagittal thick sections using single-edge razor blades. Place each section on a glass microscope slide (preferably a depression slide) in a drop of saline and observe the cut surface with your stereomicroscope.

4. Exercise 3.4: Blastoderm development in culture

a. Materials

Materials from Exercises 3.2 and 3.3

Fertile chicken eggs (incubated 18-25, 26-29, 33-49 hours)

Sterile paper towels

70% ethanol

Medium-sized scissors

Spoons

Blunt watchmaker's forceps

Fine-tipped watchmaker's forceps

Sterile pipettes

Sterile saline

Sterile finger bowls

Sterile petri dishes

Sterile beakers

Falcon, 35 x 10 mm, culture dishes

Thin albumen from fresh fertile eggs

Bacto-agar

Sterile saline (distilled water, NaCl)

Hot plate

Water bath

Plastic, covered food-storage container

Cotton

Safety glasses

Pasteur pipettes

Wide-mouth pipettes (diamond pencil, Pasteur pipettes)

Pipette bulbs

Razor blade knives (single-edge razor blades, two pairs of long-nose pliers, wooden handles, nail polish)

b. Procedures

Blastoderms will be cultured by modification of the method of Spratt (Spratt, N. T., Jr. [1947] Development *in vitro* of the early chick blastoderm explanted on yolk and albumen extract saline-agar substrate. *Journal of Experimental Zoology* **106**:345-365). All procedures must be performed using sterile conditions. First clean your working area with paper towels soaked in 70% ethanol. Clean the bench top, dissecting microscopes, and all instruments (scissors, spoon, and blunt and sharp watchmaker's forceps). Spread sterile (autoclaved) paper towels over the working area of the bench top. Have available sterile pipettes, saline (aqueous solution of 123 mM NaCl), and glassware (bowls, petri dishes, and beakers). Blastoderms will be cultured on an albumen/agar substrate in plastic culture dishes (Falcon, 35 X 10 mm).

Prepare the culture dishes by mixing together equal parts of *thin* albumen from fresh fertile eggs and a 0.6% solution of agar (Bacto-agar; Difco, Detroit, Michigan) in sterile 123-mM saline to produce a 0.3% agar concentration (the agar solution is first heated to boiling in a sterile beaker and then placed in a water bath at 47° to 50°C; the albumen also is placed in a sterile beaker in the same water bath and the temperatures of the two solutions are allowed to equilibrate before mixing). Maintain the mixture in a water bath at 47° to 50°C to keep the agar melted. Pour about 2.5 ml of the albumen/agar mixture into the bottom of each plastic dish. Cover each dish and refrigerate them in a humidified chamber until ready to use. (A plastic, covered food-storage container containing sterile paper towels soaked with sterile water makes an excellent, inexpensive chamber.)

Obtain typical 18-, 24-, and 33-hour embryos (18-25, 26-29, and 33-49 hours of incubation, respectively) as described in Exercise 3.3, *but carry out all procedures under sterile conditions.* Begin by taking a piece of cotton moistened with 70% ethanol and wiping the shell of each egg (the cotton should be relatively dry, not soaking wet; *embryos will die if shells are soaked with ethanol*). Then crack open the shells and remove blastoderms as described in Exercise 3.3, using sterile saline, glassware, and instruments. Transfer each blastoderm to a culture dish with a sterile wide-mouth pipette. A Pasteur pipette makes an excellent wide-mouth pipette. To make wide-mouth pipettes, first *put on safety glasses*. Then, wrap the tip of a Pasteur pipette in a paper towel; carefully break off the tip (facilitated by first scoring the glass with a diamond pencil), protecting your hands and eyes; and add a bulb to the pointed end. After transferring the blastoderm with a wide-mouth pipette, use sterile forceps to position the blastoderm on the albumen/agar substrate, either dorsal side up or or dorsal side down, and to remove wrinkles. Remove most of the area opaca from the blastoderm so that only a narrow rim remains attached to the area pellucida. To do this, fashion a small knife out of a single-edge razor blade: take a razor blade and two pairs of long-nose pliers; use one pair to hold the blade and the second to break off the smallest piece of edge possible. *Make sure you wear safety glasses to protect your eyes from flying splinters*. Attach the piece of edge to a wooden handle using nail polish. Sterilize the blade prior to use by dipping it into 70% ethanol and then washing it in sterile saline to remove the ethanol, which is toxic to the embryo. Make sure that the blade is wet with sterile saline before it is touched to the blastoderm, to prevent the latter from sticking to the blade. Remove all of the area opaca at the caudal end of the blastoderm (that is, cut into the area pellucida slightly); this facilitates craniocaudal elongation of the embryo. Use a fine-tipped pipette to remove excess saline from the dish, and cover each dish and return it to a humidified incubator at 38°C.

Sketch each blastoderm as it is set up in culture. Stage the embryo using the abridged Hamburger and Hamilton stage series in Exercise 3.2. Carefully pass one of the tips of a fine-tipped watchmaker's forceps beneath the head of a typical 33-hour embryo. How far caudally (with respect to brain levels) can you place your forceps? What is the name of the space containing the forceps tip? Repeat this procedure with younger embryos. In such embryos, can you place the forceps tip as far as in older embryos? Why or why not?

Examine the blastoderms at convenient intervals over the next 24 hours and sketch them at each interval. Note if craniocaudal elongation of the embryo is inhibited. Also look for abnormal development of the neural tube. A high incidence of **neural tube defects** occurs in this culture system, especially if insufficient area opaca is removed.

5. Exercise 3.5: Mapping of morphogenetic movements

a. Materials

Materials from Exercises 3.2-3.4

Fertile chicken eggs (incubated 18-25 hours)

Spratt cultures

Powdered animal charcoal

Mortar and pestle

Cactus needles (magnifying glasses, cactus spines, medium-sized forceps, wooden sticks, nail polish)

Nile blue sulfate agar staining slides (agar, sterile distilled water, sterile glass slides, paper tissues, aluminum foil, glass pans, medium-sized forceps, 70% ethanol, sterile paper towels, scalpel blades)

Sterile saline

Sterile scalpel blades

Sterile watchmaker's forceps

b. Procedures

Morphogenetic movements can be mapped in young blastoderms by applying marks to the surface of the blastoderm and then following their movements over time. Cells can be marked in a variety of ways. We will use two of the simplest: carbon particles and vital dyes. Collect typical 18-hour embryos (18-25 hours of incubation) using sterile conditions and set them up in modified Spratt cultures as described in Exercises 3.3 and 3.4. Blastoderms can be positioned either dorsal side up (to map morphogenetic movements in the epiblast) or ventral side up (to map morphogenetic movements in the endoderm).

Use powdered animal charcoal to mark blastoderms with carbon particles. Grind the powder to a fine consistency with a mortar and pestle. Use a fine needle to transfer a few of the particles to the surface of the blastoderm. Cactus needles work superbly for this purpose. To make cactus needles, simply obtain a few cactus plants from a local florist (examine their needles carefully with a magnifying glass prior to purchase, and select several plants with needles of various sizes). *Wearing safety glasses*, pluck the desired needles from the plant with forceps and mount the needles onto wooden sticks with nail polish. Sterilize needles by dipping them into a beaker containing 70% ethanol and a second beaker containing sterile saline; then allow them to dry. Stick the tip of a dried, sterile needle into the ground, powdered charcoal; a few particles will adhere to the tip. Gently touch the tip of the needle to the desired region on the surface of the blastoderm to deposit the particles.

Use agar impregnated with Nile blue sulfate to mark blastoderms with vital dyes. To do this, prepare a 1% solution of agar in sterile distilled water (heat the solution while stirring until the agar is dissolved). Have on hand some ordinary microscope *glass* slides, which have been washed with 70% ethanol, dried with tissues, and sterilized by autoclaving or heating. Pour a thin agar layer onto one side of each glass slide. Allow the slides to dry overnight (cover them with a canopy made from aluminum foil to prevent dust from accumulating on them). Soak the slides for one week in a *glass* pan (covered with aluminum foil) containing a 1% solution of Nile blue sulfate in sterile distilled water. Do not stack slides on top of one another or incomplete dye impregnation will occur. Finally, remove each slide from the pan with forceps, wash it with sterile distilled water, allow it to dry, and wrap it in a sterile (autoclaved) paper towel for storage. Stain blastoderms by unwrapping a slide and placing a drop of sterile saline on the coated side of the agar slide. Then, remove a small piece of agar with a sterile scalpel blade, pick up the piece with sterile forceps, and touch the *edge* of the agar to the blastoderm. Staining occurs within seconds.

Cover the cultures and reincubate them in humidified incubators at 38°C. Remove the cultures at desired intervals over the next 24 hours to observe the displacement of the carbon particles or stain. Sketch the initial position of the marks and their new positions at each interval.

6. Exercise 3.6: Separation of gastrulating and neurulating regions

a. Materials

Materials from Exercises 3.2-3.5

Fertile chicken eggs (incubated 18-25 hours)

Spratt cultures

Cactus needles

b. Procedures

Two major developmental processes are occurring simultaneously in young blastoderms: the cranial half of the blastoderm is undergoing neurulation and the caudal half is undergoing gastrulation. Are these two processes causally related? That is, is regression of the **primitive knot** during gastrulation required for **neural plate** elongation and folding during neurulation? Or, alternatively, is neural plate elongation responsible for primitive knot regression? The purpose of this exercise is to do an experiment to answer these questions, as well as to determine whether Hensen's node is required for gastrulation to occur.

Collect typical 18-hour embryos (18-25 hours of incubation) using sterile conditions and set them up in modified Spratt cultures as described in Exercise 3.4. Position blastoderms *dorsal* side up. Some blastoderms will be left intact (except for the removal of most of their area opaca), to serve as controls; others will be completely transected. Transections will be done with cactus needles (see Exercise 2.4 for instructions for making needles) at two craniocaudal levels: just cranial to the primitive knot, and just caudal to the primitive knot.

After transection, separate the two halves of the blastoderm with cactus needles so that the cut edges cannot heal together.

Cover the cultures and reincubate them in a humidified incubator at 38°C. Examine cultures after an additional 24 hours of development. Sketch the initial appearance of the transected blastoderms and their appearance 24 hours later. Determine whether the cranial half neurulates after transection. If so, is the primitive knot required for this process to occur? Determine whether the caudal half gastrulates after transection. If so, is the primitive knot required for this process to occur? The portion of the neural plate that will eventually form the caudal levels of the brain and the spinal cord lies alongside the cranial part of the primitive streak in 18-hour embryos (that is, the definitive primitive streak is partially flanked by the neural plate, although this is not visible in whole mounts). Does this region form a neural tube after transection? If so, is the primitive knot required for this process to occur?

7. Exercise 3.7: Formation of a double heart

a. Materials

Materials from Exercises 3.2-3.6

Fertile chicken eggs (26-29 hours of incubation)

Spratt cultures

Cactus needles

Stopwatches

b. Procedures

Recall that the heart develops from paired **cardiac primordia**, which are brought into apposition in the ventral midline by the action of the lateral body folds. This exercise will *demonstrate* the fact that paired rudiments contribute to the heart and that each rudiment is capable of developing into a miniature, beating heart.

Collect typical 24-hour embryos (26-29 hours of incubation) using sterile conditions and set them up in modified Spratt culture as described in Exercise 3.4. Position blastoderms *ventral* side up. Some blastoderms will be left intact (except for the removal of most of their area opaca) to serve as controls; the others will be subjected to surgery. Surgery will be performed with cactus needles (see Exercise 2.4 for instructions for making needles). Cut the **splanchnopleure** in the *midline* of the embryo beginning at the **cranial intestinal portal** and extending craniad. As you cut through the floor of the foregut, make sure you do not cut beyond its cranial end.

Cover the cultures and reincubate them in a humidified incubator at 38°C for an additional 48 hours. Examine cultures at about 24, 36, and 48 hours postsurgery. Sketch the initial appearance of the embryo and its appearance at each interval of observation. Do two beating hearts form? If so, do they beat in synchrony? Use a stopwatch to determine the heart rate (in beats per minute) of each beating heart. If paired beating hearts formed, do they have the same heart rate?

8. Exercise 3.8: Observations of living embryos at 48 hours

a. Materials

Materials from Exercises 3.3-3.7

Fertile chicken eggs (50-56 hours of incubation)

1% solution of neutral red

1% solution of Nile blue sulfate

Carbon/melted agar mixture

India ink

Single-edge razor blades

Glass microscope slides (preferably depression slides)

b. Procedures

Obtain fertile chicken eggs incubated at 38°C for approximately 50-56 hours. These eggs should contain typical 48-hour embryos. Carefully crack open the eggs against the rim of a finger bowl half filled with warm saline solution, as described in Exercise 3.3. Observe particularly the circulation of blood—especially its passage through the heart. If you watch the living heart, you can discern the relationships of its parts much better than if you examine only whole mounts of *fixed* embryos. Note the jerkiness of the circulation in the arteries and the smoother flow of blood in the veins. Observe as many as possible of the structures you previously identified in whole mounts. To make identification of most structures easier, apply a drop of a 1% aqueous solution of vital stain or inject eggs subblastodermically with Nile blue sulfate, india ink, or carbon/agar (see Exercise 3.2). Sketch the vitelline vessels and heart; add arrows to your sketch to indicate the directions of blood flow.

As described in Exercise 3.3, use single-edge razor blades to cut thick transverse and sagittal sections of your embryos. Place each section on a glass microscope slide (preferably a depression slide) covered with a drop of saline, and view the cut surface with your stereomicroscope.

9. Exercise 3.9: Chemical dissection of organ rudiments

a. Materials

Materials from Exercises 3.3-3.8

Fertile chicken eggs (33-49 and 50-56 hours of incubation)

EDTA/trypsin solution (ethylenediaminetetraacetic acid, trypsin, glycine, sodium hydroxide, HCl)

Glass microscope slides (preferably depression slides)

b. Procedures

We will use chemical dissection in this exercise to demonstrate the structure of axial and paraxial organ rudiments (that is, the **neural tube**, **notochord**, and **somites**).

Collect typical 33- to 48-hour embryos (33-49 and 50-56 hours of incubation, respectively) as described in Exercise 3.2, and place them in petri dishes (one per dish). Remove excess saline and flatten each blastoderm onto the surface of the dish. Cover each blastoderm with a large drop of EDTA/trypsin solution (0.4% ethylenediaminetetraacetic acid [Fisher Scientific Co., New Jersey] and 0.5% trypsin [Difco laboratories "1:250" trypsin) in an aqueous solution of 1 M glycine; to dissolve the EDTA, first add it to the glycine solution and then add 10 N NaOH dropwise to the glycine until the EDTA dissolves; finally, adjust the pH to 7 with 2 N HCL. After 10-15 minutes, place the tips of a partially closed pair of blunt watchmaker's forceps on either side of the midline of the blastoderm, just lateral to the somites, and allow the tips to gently spread apart. This tears the blastoderm and separates organ rudiments. Examine isolated somites and segments of the neural tube and notochord under high magnification. To do this, pipette individual rudiments in a drop of albumen/saline mixture (that is, 1 part thin egg albumen to 2 parts saline; this mixture stops the action of the trypsin by saturating the enzyme with excess protein) to ordinary glass microscope slides or depression slides (if available). Sketch the structures of the isolated rudiments. Compare the structures of cranial and caudal somites (in embryos at about Stage 15). Also note differences in the structures of each rudiment at early (Stage 10) and late (Stage 15) stages.

10. Exercise 3.10: Observations of living embryos at 72 hours

a. Materials

Materials from Exercises 3.2-3.9

Fertile chicken eggs (72 hours of incubation)

Single-edge razor blades

Glass microscope slides (preferably depression slides)

b. Procedures

Obtain fertile eggs incubated at 38°C for approximately 72 hours. These eggs should contain typical 72-hour embryos. Open and examine these eggs exactly as described in Exercises 3.3 and 3.8. Sketch the vitelline vessels and heart; and add arrows to your sketch to indicate the directions of blood flow. As described in Exercise 3.3, use single-edge razor blades to cut thick transverse and sagittal sections of your embryos. Place each section on a glass microscope slide (preferably a depression slide) covered with a drop of saline and view the cut surface with your stereomicroscope.

11. Exercise 3.11: Extraembryonic membrane development in culture

a. Materials

Materials from Exercises 3.2-3.10

Fertile chicken eggs (72-96 hours of incubation)

Sterile petri dishes (20 x 100 mm and 25 x 150 mm)

Sterile distilled water

b. Procedures

A culture system will be used in this exercise to examine **extraembryonic membrane** formation. The system we will use was developed by Auerbach and coworkers (Auerbach, R., L. Kubai, D. Knighton, and J. Folkman [1974]. A simple procedure for the long-term cultivation of chicken embryos. *Developmental Biology* **41**:391-394). Sterile conditions must be maintained for cultures to develop successfully. With this culture procedure, eggs can be cultured until embryos reach near-hatching stages. Obtain chicken eggs incubated for 72-96 hours. Begin by taking a piece of cotton moistened with 70% ethanol and wipe the shell of each egg. Carefully crack eggs into sterile petri dishes (20 x 100 mm). This procedure must be done without breaking the yolk; practice the procedure until this is possible. If the egg is cracked properly, the blastoderm will lie uppermost and the yolk and albumen will flatten over the bottom of the petri dish. Cover each petri dish and place it into the bottom of a larger sterile petri dish (25 x 150 mm). Add some sterile distilled water to the larger dish and cover it with its lid. Place the entire culture in a humidified incubator at 38°C. Observe cultures at daily intervals; discard those that fail to develop as soon as possible to prevent contamination of the incubator. Observe particularly the development of the **extraembryonic membranes**: the **amnion, chorion, yolk sac, allantois**, and **chorioallantoic membrane**. Also observe changes in the circulatory pattern over time. Finally, observe the embryo through its membranes. Watch for movement of the embryo caused by contraction of the **amnion**. When do such contractions begin? Also try to determine when the embryo begins moving actively. Its first movements occur jerkily, but

eventually coordinated movements develop. Try to determine when coordinated movements arise, and describe their appearance. Focus on a particular active movement, such as a leg kick. Determine the number of kicks over a 15-minute observation period. Does this value change with increase in developmental age?

12. Exercise 3.12: Somitogenesis in partial isolation from surrounding tissues

a. Materials

Materials from Exercises 3.2-3.11

Fertile chicken eggs (50-56 hours of incubation)

Cactus needles

Modified Spratt culture plates

Sterile saline

b. Procedures

In this exercise, we will examine whether segmental plates isolated from surrounding tissues can undergo normal somitogenesis. We will repeat an experiment originally performed by D. S. Packard, Jr., and Jacobson, A. G. [1976]. The influence of axial structures on chick somite formation. *Developmental Biology* **53**:36-48). Obtain typical 48-hour chick embryos (as described in Exercise 3.8) and place them dorsal side up in modified Spratt culture (as described in Exercise 3.4). Cactus needles will be used to prepare segmental plate isolates. Four types of experiments will be performed.

In the first type of experiment (performed either on a single embryo or on groups of embryos as available material allows), first identify the neural tube (spinal cord) in the trunk region, the flanking pairs of somites, the more caudal flanking segmental plates, and the tail bud. Recall that in the trunk region, the notochord lies just beneath (that is, ventral to) the midline of the neural tube. Use a cactus needle to make two transverse cuts through the *right* side of an embryo (or group of embryos): the cranial cut extending from the *caudal* end of the last somite (begin your cut just *lateral* to the last somite) laterally to the area pellucida/area opaca interface, and the caudal cut from the lateral edge of the segmental plate at the *caudal* end of the tail bud laterally to the area pellucida/area opaca interface. Finally, make a parasagittal cut interconnecting the *medial* sides of the cranial and caudal transverse cuts. Thus the segmental plate on the *right* side is now isolated from the more lateral mesoderm (that is, the lateral plate mesoderm). The *left* segmental plate is undisturbed in this experiment and will serve as a control.

In the second experiment on another embryo (or group

of embryos), make a transverse cut on the *right* side that is identical to the cranial one in the first experiment, but extend it more medially to the midline of the neural tube. Finally, make a second cut, a midsagittal cut, from the transverse cut down the length of the neural tube and through the midline of the tail bud, to the caudal midline at the area opaca/area pellucida interface. The segmental plate on the *right* side is now separated from the somites cranially and from one-half of the neural tube (and perhaps the notochord) medially. Possible influences of the left segmental plate on the right segmental plate are also eliminated by this experiment. Again, the *left* segmental plate is undisturbed and will serve as a control.

In the third experiment on another embryo (or group of embryos), repeat the transverse cut exactly as described in the *second* experiment. Also repeat the midsagittal cut described in the *second* experiment but stop it at the level of the *cranial* end of the tail bud. Finally, make a third cut, a transverse cut, from the caudal end of the midsagittal cut (just cranial to the tail bud) laterally to the area opaca/area pellucida interface. The segmental plate on the right side is now isolated from some medial structures as in the second experiment, but it is also isolated from the somites cranially and the tail bud caudally. The *left* segmental plate is again undisturbed and will serve as a control.

In the fourth experiment on another embryo (or group of embryos), make the two transverse cuts described in the *third* experiment but stop them medially at the interface between the neural tube and *right* segmental plate mesoderm. Make a third cut, a parasagittal cut, interconnecting the cranial and caudal transverse cuts and running parallel to and just adjacent to the *lateral* edge of the neural tube. Thus the segmental plate on the right side is now isolated from cranial and caudal structures, as in the third experiment, as well as from medial structures (that is, the neural tube and notochord). The *left* segmental plate is again undisturbed and will serve as a control.

After each experiment is completed, place cultures into the incubator for an additional 24 hours at 38°C. Then examine each embryo, count the number of somites formed from the *right* segmental plate (that is, the segmental plate on the operated side), and compare this number to that formed from the *left* segmental plate (that is, the segmental plate on the control side). Note any differences in somite number and shape; draw examples of operated and control sides. Also compare the results obtained among the four experimental groups. Are midline structures required for somitogenesis? Lateral structures? Cranial structures? Caudal struc-

tures? If possible, select some of your operated embryos for paraffin serial sectioning (either as transverse or sagittal sections) to study somitogenesis in more detail (see Advanced Hands-On Studies. 1. Preparing Embryos for Light Microscopy. b. Serial sections). Can any differences be detected among the different experimental groups at the histological level?

13. Exercise 3.13: Selective labeling in embryos of cells, organ rudiments, and expression domains of positional identity genes

a. Materials

Recent journals in developmental biology and embryology such as *Anatomy and Embryology, Development, Developmental Biology, Developmental Dynamics, Genes and Development, International Journal of Developmental Biology,* or *Mechanisms of Development*

b. Procedures

In this exercise you will do a literature search at the library or on-line and will then write a brief research report on your findings. In your search, you will find a recently described molecular marker that is expressed in a temporally and/or spatially distinct pattern during embryogenesis. Examples of such markers are included in Chapter 6, F. Advanced Hands-On Studies, Photos 6.19-6.38. Two restrictions are to be placed on your search. First, the molecular markers must be expressed in embryos. Although this exercise is placed with those on chick embryos, the actual embryos chosen can be plant or animal embryos of any type and age. Second, the marker must be described in a recent journal, such as those listed above ("recent" will be defined as a journal issue published within the last three months).

Your research report should have the following sections: Abstract, Introduction, Materials and Methods, Results, Discussion, and Literature Cited. The Abstract should provide summary information such as the name of the molecular marker, the species in which it is examined, the ages of the embryos examined, a very brief account (two or three sentences) of its temporal and spatial pattern of expression, and a concluding statement of the importance of the findings. The Introduction should provide background information about the development of the organism studied, the molecular marker examined, and the particular developmental problem the marker deals with. The Materials and Methods should describe the species studied, the number and ages of embryos studied, and the methods used for identifying the molecular marker. The Results should describe the temporal and spatial pattern of expression of the molecular marker. The Discussion should consider the importance of the marker and the new findings in relation to the particular developmental problem under study and, perhaps, other similar markers. The Literature Cited should provide a full citation of the paper or papers you have actually read.

The entire report should be between 5 and 10 typewritten, double-spaced pages. Include drawings showing the distribution of the marker, and reference these in the Results. Begin by outlining the main topics to be covered in your report.

E. EXERCISES 4.1, 4.2: MOUSE EMBRYOS

1. Exercise 4.1: Breeding of mice

a. Materials

Male and female mice of breeding age (6 weeks to 2 years of age); 4 females/male; the total number needed is based on the number of stages to be studied and the class size

An approved animal facility, with appropriate light cycle, cages, food, and water bottles

b. Procedures

Sexually mature male and female mice will be paired in this exercise in the ratio of 1 male to 4 females/cage. Female mice will be examined daily each morning to check for a vaginal plug, that is, coagulated semen which suggests that fertilization has occurred. For Exercises 4.1 and 4.2, it is mandatory that apropriate animal-use (IACUC) reviews have taken place, and that an approved protocol has been obtained. Also, animals need to be housed in approved animal facilities in rooms with appropriate light cycles.

To check for **vaginal plugs**, each female mouse is picked up by the tail and allowed to grip the top of the cage with her front paws. As you hold the tail, twist your hand so as to place the mouse in a lordotic position (that is, with the dorsum of the back curved downward), elevating the **vagina** so that it can be examined with forceps. Look for a crusty obstruction (that is, the vaginal plug), and if present, remove the impregnated female and place her in a separate cage (be sure that adequate food is present in the cage and that a full water bottle is added), where she will remain in isolation until embryos are collected. To prevent miscarriage, allow each pregnant mouse to remain undisturbed (except for feeding, watering, and cage cleaning) until she is sacrificed to obtain embryos. Label the cage with the date on which the vaginal plug was observed.

2. Exercise 4.2: Mouse embryo observation and staging

a. Materials

Copies of Theiler and Downs and Davies stage series (Theiler, K. [1972]. The House Mouse. Development and Normal Stages from Fertilization to 4 Weeks of Age. New York, NY: Springer-Verlag; Downs, K. M., and Davies, T. [1993]. Staging of gastrulating mouse embryos by morphological landmarks in the dissecting microscope. *Development* **118**:1255-1266)

Pregnant dams at GD 7, 7.5, 8, 8.5, 9, 9.5, and 10 days of gestation

70% ethanol

Petri dishes

Phosphate-buffered saline (to make this solution, dissolve 8 g of NaCl, 0.2 g KCl, 1.44 g Na_2HPO_4, and 0.24 g KH_2PO_4 in 1 l distilled water; adjust the pH to 7.4, and store the solution until use in the refrigerator)

Single edge razor blades

Watchmaker's forceps

Hemostats or blunt forceps with gripping teeth on their tips

Coarse and fine scissors with pointed tips

b. Procedures

To collect mouse embryos, pregnant dams (that is, female mice) are first killed by cervical dislocation at desired days of gestation (have your instructor complete this part of the exercise to ensure that mice are killed as humanely as possible; mice can also be killed by placing them in a container in which air has been displaced by CO_2, but this method cannot be used if embryos will be subsequently cultured, as described in Chapter 6, F. Advanced Hands-On Studies, 9. Mouse whole-embryo culture). Collect mouse embryos at half-day intervals from gestation day (GD) 7-10 (with the day on which a vaginal plug was observed equaling GD 0) using the following procedures. First, wet the abdomen of the killed dam with 70% ethanol. Then using a hemostat or a blunt forceps with gripping teeth on its tips to, grab the midline skin of the abdomen and pull it away from the underlying viscera. Using coarse scissors with pointed tips, cut the skin along the midline from the pelvis to the rib cage. Inspect to abdominal contents and identify the paired **ovaries** and **uterine horns**. Note where the two horns join one another at the **uterine cervix**. If the dam is pregnant, several swellings (implantation sites contain the conceptuses) should be visible along each uterine horn. Grab a uterine horn with forceps and using fine scissors with pointed tips, cut the horn away from the ovary and the body wall; transect the horn at the cervix. Transfer the horn to a Petri dish containing phosphate-buffered saline (PBS) and then repeat the process, removing the other uterine horn.

Wash the uterine horns in fresh saline to remove as much blood as possible (excessive blood will obscure the subsequent dissection), and transect the two horns between each of the swellings. Then while observing each conceptus with a dissecting microscope, use two pairs of watchmaker's forceps (one in each hand) to remove carefully the uterine tissue covering each conceptus. Again using watchmaker's forceps, remove each conceptus and transfer it to a fresh dish of PBS.

Dissect embryos while observing them with a dissecting microscope by carefully tearing away the extraembryonic membranes using two pairs of watchmaker's forceps. Examine embryos at GD 7, 7.5, 8, 8.5, 9, 9.5, and 10, and stage embryos using the criteria of Theiler and Downs and Davies. Then identify as many of the structures described in Chapter 4 as possible. Next, use single-edge razor blades to cut free-hand slices through embryos. Examine the cut surface of the slice in a drop of PBS with your microscope. Try to identify the internal anatomy of your embryos at the level of each slice. Draw representative features.

F. ADVANCED HANDS-ON STUDIES

1. Preparing embryos for light microscopy

Although there are many different methods for preparing embryos for light microscopy, we describe here only those methods that routinely yield good results. Specific procedures for processing chick embryos are described. Similar (and in many cases identical) procedures are also adequate for amphibian, mammalian, and sea urchin embryos. *Use extreme care when processing tissues for microscopy because many of the solvents are toxic. To be safe, use a fume hood whenever possible, wear protective (safety) eyeglasses, wear disposable surgical gloves, and do not attempt any of the procedures without the direct supervision of an experienced person.*

a. Whole mounts

Incubate fertile eggs at 38°C and about 50 to 60% relative humidity (keeping incubator water pans filled throughout the entire period of incubation gives adequate humidity) until desired stages are obtained (that is, 18-25 hours—primitive streak, head process, and early neural fold stages; 26-29 hours—typical 24-hour embryos; 33-49 hours—typical 33-hour embryos; 50-56 hours—typical 48-hour embryos; about 72 hours—typical 72-hour embryos). Two procedures can be used to collect embryos; the first usually works best for embryos 24 hours or older, whereas the second works best for younger embryos. It is best to start with older embryos

(which are easier to collect than are younger embryos) and then to progress to younger stages as your skills improve.

To do the first procedure, carefully crack eggs into finger bowls containing warm 123 mM saline (0.72% aqueous solution of NaCl). Take care not to break the yolk (its fragility increases with increased length of incubation) because breaking the yolk clouds the saline and usually results in lost embryos. To avoid this problem, crack the egg against the rim of the finger bowl, submerse the egg in the saline contained in the bowl, and gently separate the two halves of the shell. Then manipulate the yolk and attached blastoderm with your fingertips until the blastoderm floats uppermost. Grasp the periphery of the blastoderm (that is, the area opaca) with a pair of fine forceps (watchmaker's or #5). Then use fine-pointed scissors to cut around the perimeter of the blastoderm. (Make sure that you hold on to the periphery of the blastoderm with your forceps; also make sure that the blastoderm is floating uppermost, centered on the yolk—eccentrically positioned blastoderms are more difficult to remove without tearing.) After the entire perimeter of the blastoderm is cut free from the yolk, gently grasp the blastoderm with forceps and *float* it away from the yolk. Place an ordinary spoon beneath the blastoderm while you hold the latter with forceps. Slowly lift the spoon and the cuddled blastoderm out of the finger bowl and transfer them to a second bowl containing fresh saline. Gently free the blastoderm from the spoon and remove the latter from the bowl. The blastoderm at this point is usually still covered dorsally by the vitelline membrane. (Although this latter structure seems to be a single membrane, it actually consists of both the inner and outer vitelline membranes.) To remove this membrane it is best to grasp the periphery of the blastoderm and gently waft the blastoderm back and forth in the saline until the vitelline membrane floats free. (The action involved is analogous to shaking out a rug, when you grasp one edge of the rug and the rug undulates in the air as you raise and lower your arms.)

To do the second procedure, use a dull pair of watchmaker's forceps to excise a cap of shell at the blunt end of the egg (hold the blunt end of the shell up, insert one tip of the forceps *slightly* into the shell about one-third the distance from the top of the egg [keep the tip parallel, not perpendicular, to the shell to avoid penetrating too deeply], and then hold the forceps stationary and rotate the egg around until the cap can be removed; this action is much the same as that used by an electric can opener). Also use forceps to remove chalazae (the white strands at the poles of the egg) and as much of the thick albumen as possible. Decant most of the thin albumen into a bowl and discard it. Gently pour the yolk, its attached blastoderm, and enveloping vitelline membranes into a bowl containing saline, and cut along the equator of the yolk (with the blastoderm floating uppermost) with a pair of fine scissors. Peel the vitelline membranes and attached blastoderm off the yolk and transfer them (with a spoon) to a petri dish filled with saline. Separate the blastoderm from the vitelline membranes by gently squirting saline between the blastoderm and membranes with a pipette.

After the blastoderm has been freed from its membranes (using either the first or second procedure) place a small dish (a dish approximately 50 mm in diameter works best) into the bowl of saline beneath the blastoderm and withdraw the dish slowly. This procedure results in removal of the blastoderm and some saline from the bowl. Use forceps to unwrinkle the blastoderm and to float it dorsal side up in the small dish. Remove excess saline from the dish with a pipette so that the blastoderm flattens onto the bottom of the dish; slowly replace the saline with fixative. (Begin by placing one drop directly onto the surface of the blastoderm and then slowly adding several drops peripheral to the blastoderm until the latter is completely covered.) Many fixatives can be used. We prefer a mixture (1:2:7) of glacial acetic acid, 37% formaldehyde solution, and absolute ethanol because it consistently yields good results. Fix embryos for 2 hours to overnight in *covered* dishes. (Add enough fixative so that the embryo does not dry out as evaporation occurs.) Then trim the blastoderm (with a fine scissors or razor or scalpel blade) to a convenient size, removing most of the tissue peripheral to the area pellucida.

Next begin dehydrating the embryo with ethanol by removing the fixative and adding 70% ethanol. (To avoid damaging your embryo, it is best to transfer liquids to the dish containing the embryo rather than transferring the embryo to different dishes; remove most of the previous liquid with a pipette before adding the succeeding liquid, but be sure the embryo does not dry out between changes.) Embryos should be placed in two to three changes of 70% ethanol for 5 to 10 minutes each. Then stain the embryo with a borax carmine solution (to make the staining solution, add 3 g carmine and 4 g sodium borate to 100 ml distilled water; boil for 10 to 15 minutes; cool; add 100 ml 70% ethanol; add distilled water, if necessary, to bring the volume up to 200 ml; let the solution set for 2 days; then filter it twice). After staining the embryo overnight, partially destain it for about 6 hours with six changes of acid-alcohol (100 ml 70% ethanol containing 5 drops of concentrated hydrochloric acid); dehydrate it (95% ethanol, one change, 10 minutes; 100% ethanol, two changes, 10 minutes each); and finally clear it (xylene, toluene, or Histosol, two changes, 10 minutes each). Then transfer each embryo with forceps to a glass slide; immediately surround it with a large drop of mounting media (Permount, Pro-Texx, or the equivalent), and cover it with a circular coverslip (18 mm in diameter). For embryos older than about 33 hours, coverslips usually need to be supported with small pieces of a broken coverslip placed around the periphery of the blastoderm; otherwise, air bubbles will form between the slide and coverslip (such supports may also be needed for supporting 24- to 33-hour embryos, depending on the thickness of your mounting media, to prevent the coverslip from crushing the embryo).

b. Serial sections

Treat embryos as described for making whole mounts, with the following exceptions. After embryos are fixed and partially dehydrated with 70% ethanol, stain them for 30 seconds (to aid in locating and orienting them during subsequent processing) with an eosin solution (0.5 g in 100 ml 70% ethanol). Then dehydrate and clear embryos as described. Infiltrate embryos with melted paraffin (Paraplast X-tra, 52-54°C, seems to work best) in the following sequence (embryos being infiltrated with paraffin should be kept in an oven at about 58°C— *do not allow the temperature to become higher because the embryo will be destroyed and the flash point of the paraffin might be reached, resulting in a fire*): three changes, 100% paraffin, 20 minutes each after paraffin has completely melted. Then orient embryos with a warm dissecting needle (for transverse or sagittal sections) in fresh 100% melted paraffin in molds (for example, Peel-A-Way molds) and leave them at room temperature until the blocks harden. Trim the blocks with razor blades and mount them in the chuck of a rotary microtome for serial sectioning at 8-10 μm. The microtome slices (sections) the entire embryo into thin sections and simultaneously fuses the section edges together (end to end) in the order in which they were cut, forming a delicate ribbon. Pick up the ribbons with forceps or fine paintbrushes and float them onto glass slides coated with albumen fixative and flooded with distilled water (albumen fixative consists of a 50:50 mixture of egg albumen and glycerine; use your index finger to spread a *very light* film of fixative on each slide; store the albumen fixative in the refrigerator to retard bacterial growth). Allow slides to dry overnight on hot plates at 40°C.

Process and stain sections in the following way (place groups of slides in slide racks and transfer slides to staining dishes containing the solutions listed below). First remove paraffin from sections with two changes of xylene, toluene, or Histosol (10 minutes each) and gradually hydrate sections (100% ethanol, two changes, 2 minutes each; 95% ethanol, 2 minutes; 70% ethanol, 2 minutes; water, 5 minutes). Then stain sections with Harris' hematoxylin (Harleco, Gibbstown, New Jersey; 30 seconds to 3 minutes), wash in running tap water (3 minutes), rinse in 2% aqueous sodium bicarbonate (2 minutes), wash in running tap water (3 minutes), and gradually dehydrate and clear (70% ethanol, 5 minutes; 95% ethanol, 5 minutes; 100% ethanol, two changes, 5 minutes each; xylene, toluene, or Histosol, two changes, 5 minutes each). Next, mount coverslips with Permount,

Pro-Texx, or the equivalent.

c. Frozen (cryostat) sections

A quicker method for obtaining sections that is advantageous for use in combination with other techniques, such as immunocytochemistry or in situ hybridization (see Chapter 6, F. Advanced Hands-On Studies, 3 and 4) is the cutting of frozen (cryostat) sections. To do this, remove blastoderms from the yolk at desired stages and flatten them onto the bottoms of plastic or glass dishes, exactly as described in Chapter 6, F. Advanced Hands-On Studies, 1. Preparing embryos for light microscopy. Fix the blastoderms for 1-2 hours with 4% formaldehyde in 0.2 M phosphate buffer. (Add 27.6 g NaH_2PO_4:H_2O to 1 l distilled water; mix thoroughly; label the solution as Solution 1 and store it in the refrigerator. Add 28.4 g $NaHPO_4$ to 1 l distilled water; mix thoroughly; label the solution as Solution 2 and store it in the refrigerator. To make the buffer, mix 280 ml Solution 1 with 720 ml Solution 2. Adjust the pH to 7.4.) Trim the blastoderms to isolate embryos from extraembryonic areas, and wash them three times for 5 minutes each in phosphate-buffered saline (PBS). (To make this solution, dissolve 8 g of NaCl, 0.2 g KCl, 1.44 g Na_2HPO_4, and 0.24 g KH_2PO_4 in 1 l distilled water; adjust the pH to 7.4, and store the solution until use in the refrigerator.) Next, transfer embryos to the following solutions: 5% sucrose in PBS for 2 hours to overnight, 15% sucrose for 2 hours to overnight, and 15% sucrose and 7.5% gelatin in PBS at 37°C (to keep the gelatin melted) for 2 hours to overnight. Orient embryos in embedding molds containing fresh sucrose/gelatin solution and allow the solution to cool, forming a block containing each embryo. Trim the block as desired and then freeze it in liquid nitrogen (*wear protective clothing, including gloves and safety glasses, and use extreme care when handling liquid nitrogen; when liquid nitrogen contacts the skin, it causes a severe burn*). Mount the block to a cooled cryostat chuck using a drop of Tissue-Tek O.C.T. (Catalog no. 25608-903; VWR, Salt Lake City, UT) and then section tissue at about 20 μm. Collect sections in the order in which they were cut on glass slides, place slides into slide ranks contained in a bath of PBS at 37°C (to remove the gelatin), and mount coverslips to slides using Fluoromount-G (Catalog no. 100-01; Southern Biotechnology Associates, Inc., Birmingham, AL). Sections are then ready for viewing with brightfield microscopy (if previously labeled with immunocytochemistry or in situ hybridization using non-fluorescent labels), fluorescence microscopy (if previously labeled using fluorescent labels; caution, *the UV light from the fluorescence microscope can damage your eyes; make sure you use the fluorescence microscope only with proper supervision of an experienced person*), or other types of microscopy (such as phase microscopy) if embryos are unlabeled.

2. Preparing embryos for scanning electron microscopy

There are many different ways to prepare embryos for scanning electron microscopy. We describe here methods for producing the various types of images of chick embryos shown in the plates. *Use extreme care when processing tissues for microscopy because many of the solvents are toxic. To be safe, use a fume hood whenever possible, wear protective (safety) eyeglasses, wear disposable surgical gloves, and do not attempt any of the procedures without the direct supervision of an experienced person.*

a. Intact specimens

Remove blastoderms from the yolk at desired stages and flatten them onto the bottoms of plastic or glass dishes, exactly as described in Chapter 6, F. Advanced Hands-On Studies, 1. Preparing embryos for light microscopy. Fix the blastoderms for 2 hours with a solution of 2.5% glutaraldehyde in 0.1 M phosphate buffer. (Add 13.8 g NaH_2PO_4:H_2O to 1 l distilled water; mix thoroughly; label the solution as Solution 1 and store it in the refrigerator. Add 14.2 g $NaHPO_4$ to 1 l distilled water; mix thoroughly; label the solution as Solution 2 and store it in the refrigerator. To make the buffer, mix 280 ml Solution 1 with 720 ml Solution 2.) Trim the blastoderms to isolate embryos from extraembryonic areas, and gradually dehydrate them (35% ethanol, 5 minutes; 50% ethanol, 5 minutes; 70% ethanol, 5 minutes; 80% ethanol, 5 minutes; 95% ethanol, two changes, 10 minutes each; 100% ethanol, two changes, 10 minutes each).

After dehydration, embryos can be dried in one of two ways. The approach we strongly recommend is to transfer embryos from 100% ethanol to two changes (10 minutes each) of hexamethyldisilazane (HMDS; Catalog no. 16700; Electron Microscopy Sciences, Fort Washington, PA). *This should be done in a hood using care not to inhale the vapors.* HMDS is allowed to evaporate over night, leaving a dried sample. The second method should be used only if your instructor directs you to do so. In this method, embryos are dried using a critical-point-drying apparatus (a pressure chamber into which samples infiltrated with 100% ethanol are placed for drying). *This procedure can be dangerous because high pressures are generated during critical-point-drying. Do not attempt to use a critical-point-dryer without the direct supervision of an experienced person.* After placing samples in the pressure chamber, seal the chamber and then cool it with liquid carbon dioxide. Slowly leak liquid carbon dioxide into the chamber while the 100% ethanol is drained out. The liquid carbon dioxide quickly replaces the ethanol that formerly saturated the sample. Heat the sealed pressure chamber. As the temperature of the chamber increases, so does its pressure. The "critical point" eventually is reached (that is, about 31⁰C and 1,100 lb./in.² pressure; *do not exceed such pressures or the chamber may explode*). Liquid carbon dioxide is converted to gaseous carbon dioxide at the critical point; carbon dioxide is then slowly bled off, eventually reducing the pressure of the chamber to atmospheric pressure. At the end of this process a dried sample is available for examination in the vacuum of a scanning electron microscope. (A vacuum must be used so that streams of electrons can flow to, and then be scanned across, the sample. Samples must be dry because wet samples give off water vapor, which destroys the vacuum. HMDS or critical-point-drying is used instead of simple air drying. The latter process grossly distorts the surfaces of cells as evaporation occurs at the air-water interface.)

Affix the dried samples to stubs (usually made of aluminum) with adhesive carbon-conductive tabs (Catalog no. 16084-1; Ted Pella, Inc., Redding, CA) and coat the surfaces of the samples with a thin layer of metal (for example, gold) to make them conductive. This is done in a vacuum evaporator or a sputter coater. *Both these instruments generate high voltages, so use extreme care.* If the metal coating is insufficient, samples "charge" or glow in the scanning electron microscope as they are bombarded with electrons. Charging obscures the details of viewed samples.

b. Slices

Slices of embryos are prepared in much the same way as intact specimens, except that embryos are cut with scissors or razor or scalpel blades after fixation. These cuts can be made transversely or sagittally, and the resulting slices can be dehydrated, dried, mounted onto stubs, and coated with metal, as already described.

The internal morphology of embryos can be readily viewed in slices. One advantage of this technique is that the surfaces of cells are displayed (that is, the fracture plane generated by slicing passes *between* rather than through cells; this is not true of other techniques used to expose internal areas of embryos). One disadvantage of this technique is that slicing sometimes distorts embryos by displacing cells to ectopic areas, separating adjacent organ rudiments from one another, or tearing organ rudiments.

c. Cryofractures

Cryofractures are also prepared in much the same way as are intact specimens. However, after embryos are dehydrated, and before they are critical-point-dried, place them in small cylinders fashioned from Parafilm. Crimp shut the two ends of each cylinder with a hemostat so that each cylinder contains an embryo surrounded by 100% ethanol. Next, plunge these cylinders into liquid nitrogen (*wear protective clothing, including gloves and safety glasses, and use extreme care when handling liquid nitrogen; when liquid nitrogen contacts the skin, it causes a severe burn*), where they quickly freeze, and position a razor blade (gripped with a hemostat and cooled in liquid nitrogen) on each frozen cylinder at a desired level (that is, if a transverse cryofracture through the heart is desired, position the razor blade at a level estimated to overlie the heart). Then hit the razor blade briskly with a hammer, creating two cryofractures (that is, two fractured surfaces are produced with each blow of the hammer). Place both pieces in fresh 100% ethanol (at room temperature), dry them in HMDS, mount them to stubs, and coat them with metal.

Cryofractures, like slices, readily show the internal morphology of embryos. Two advantages of this technique are that cells are rarely displaced to other areas by fracturing, and organ rudiments usually do not tear (although they sometimes chip or crack). One disadvantage of this technique is that the fracture plane passes *through* cells rather than between them. Thus the surfaces of cells are not revealed in the plane of the fracture, and the fractured surface characteristically has a flat appearance. A second disadvantage is that it is difficult to obtain fractures exactly through desired areas, so many attempts must be made.

d. Sectioned blocks

The internal morphology of embryos can be revealed in yet a third way: embryos can be embedded in paraffin and the blocks sectioned until desired levels are reached. Paraffin can then be removed from the remainder of the sectioned block, creating a cut surface suitable (after appropriate processing) for viewing with scanning electron microscopy.

The specific procedure for producing sectioned blocks is to fix and dehydrate embryos exactly as described for intact embryos. Clear embryos in xylene, toluene, or Histosol and embed them in paraffin. Next, cut sections as for light microscopy until the desired areas are reached. Place sectioned blocks in several changes of xylene, toluene, or Histosol to remove the paraffin, and finally into two changes of 100% ethanol (10 minutes each). Dry the dehydrated, partially sectioned tissue blocks in HMDS, mount them to stubs, and coat them with metal as already described.

One advantage of using sectioned blocks to examine the internal aspects of embryos is that precise areas within embryos can be readily exposed, unlike the situation with cryofractures. Thus, sectioned blocks are particularly useful for demonstrating sagittal views of embryos. Sectioned blocks have a disadvantage shared by cryofractures: the section plane passes *through* cells rather than between them. Therefore, the cut surface of a sectioned block appears to be flat.

3. Preparing embryos for immunocytochemistry

Although there are many different methods for preparing embryos for immunocytochemistry, we describe here only one method for whole-mount labeling that routinely yields good results. Specific procedures for processing chick embryos are described. Similar (and in many cases identical) procedures are also adequate for amphibian, mammalian, and sea urchin embryos. *Use extreme care when processing tissues for immunocytochemistry because some of the solutions are toxic. To be safe, use a fume hood whenever possible, wear protective (safety) glasses, wear disposable surgical gloves, and do not attempt any of the procedures without the direct supervision of an experienced person.*

For whole-mount labeling of chick embryos, use the following procedures which require three consecutive days to complete. On day 1, collect embryos as described in Chapter 6, F. Advanced Hands-On Studies, 1. Preparing embryos for light microscopy. Fix them for 1-2 hours with 4% formaldehyde in 0.2 M phosphate buffer. (Add 27.6 g $NaH_2PO_4:H_2O$ to 1 l distilled water; mix thoroughly; label the solution as Solution 1 and store it in the refrigerator. Add 28.4 g $NaHPO_4$ to 1 l distilled water; mix thoroughly; label the solution as Solution 2 and store it in the refrigerator. To make the buffer, mix 280 ml Solution 1 with 720 ml Solution 2. Adjust the pH to 7.4.) Transfer embryos to either 48-well plates or 96-well plates (Corning, Corning, NY) depending upon the stage and, consequently, the size of the embryo.

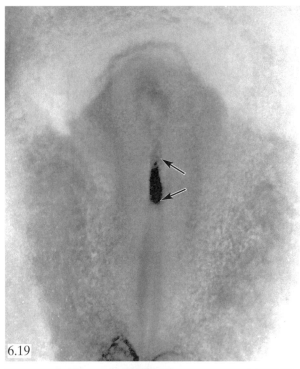

6.19

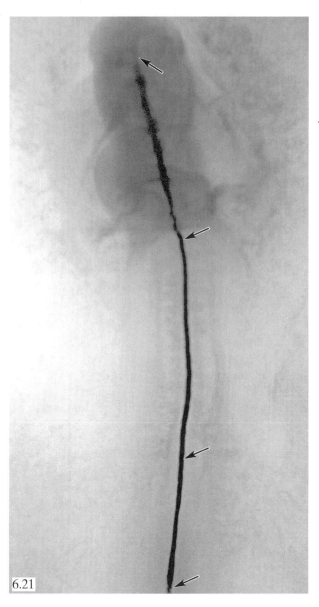

6.21

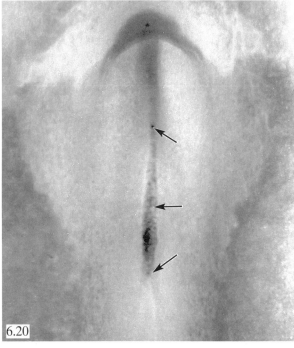

6.20

Photos 6.19-6.21. Whole mounts of 18-, 24-, and 48-hour quail embryos labeled immunocytochemically with an antibody to the notochord (not-1); arrows indicate the extent of the notochord.

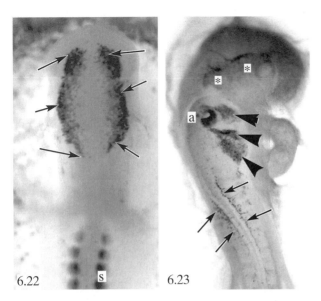

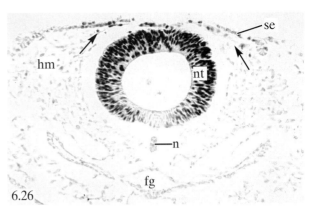

6.22　s　6.23

Photos 6.22, 6.23. Whole mounts of 33- and 48-hour quail embryos labeled immunocytochemically with an antibody to neural crest cells and their derivatives (HNK-1). Arrows in Photo 6.22 indicate the extent of the neural crest cells as they leave the prosencephalon, mesencephalon, and cranial myelencephalon of the neural tube; in quail embryos, there is also some labeling of the somite(s) with this antibody. Arrows in Photo 6.23 indicate the extent of the neural crest cells as they leave the myelencephalon and spinal cord; arrowheads in Photo 6.23 indicate neural crest cells contributing to cranial nerve ganglia cranial to and caudal to the auditory vesicle (a); asterisks in Photo 6.23 indicate neural crest cells contributing to the future craniofacial mesenchyme.

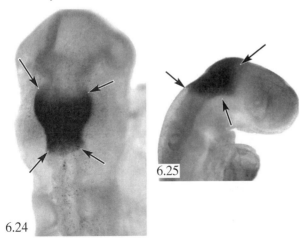

6.25

6.24

Photos 6.24, 6.25. Whole mounts of 33- and 48-hour chick embryos labeled immunocytochemically with an antibody to a homeobox gene product (Engrailed-2). Arrows indicate the extent of the domain of gene expression within the mesencephalon and cranial metencephalon.

6.26

Photo 6.26. Transverse section of a chick embryo labeled immunocytochemically as a whole mount with an antibody to a homeobox gene product (Engrailed-2). The notochord (n), foregut (fg), and head mesenchyme (hm) are unlabeled. The neural tube (except for its floor plate, which overlies the notochord), the mid-dorsal skin ectoderm (se), and the neural crest (arrows) are labeled.

Wash embryos first for two times for 5 minutes each in phosphate-buffered saline (PBS). (To make this solution, dissolve 8 g of NaCl, 0.2 g KCl, 1.44 g Na_2HPO_4, and 0.24 g KH_2PO_4 in 1 l distilled water; adjust the pH to 7.4, and store the solution until use in the refrigerator.) Continue washing embryos for two times for 5 minutes each in PBS containing Triton X-100 and bovine serum albumin (combined solution called PBT). (Add 0.2 g bovine serum albumin and 100 μl Triton X-100 to 100 ml PBS; store the solution until use in the refrigerator; Triton and bovine serum albumin can be purchased from Sigma Chemical Co., St. Louis, MO.) Complete the washing by placing embryos for 30 minutes in fresh PBT on an orbital shaker. After washing, soak embryos in the refrigerator for 30 minutes in PBT containing normal goat serum (PBT + N). (Add 20 μl normal goat serum to 4 ml PBT; keep this solution in the refrigerator and only use it fresh; N can be purchased from Sigma Chemical Co.) PBT is used to punch holes in cell membranes, thereby allowing large proteins to penetrate the tissue during subsequent processing. N is used to block nonspecific binding of the antibody to tissue. Finally, remove the PBT + N and replace it with your primary antibody and soak embryos in this solution in the refrigerator overnight. In this step, your antibody should bind specifically to its appropriate antigen. For each batch of antibody, the appropriate concentration must be determined empirically. Labeling patterns obtained with three different antibodies are illustrated in Photos 6.19-6.26: anti notochord (not-1; available from Developmental Studies Hybridoma Bank, Iowa City, IA; approximate dilution, 1:10 in PBT + N), anti neural crest (HNK-1; available from American Type Culture Collection, Rockville, MD;

use undiluted); and anti Engrailed-2 protein (4D9; available from American Type Culture Collection; approximate dilution, 1:1 in PBT + N).

On day 2, wash embryos as follows. First, two times for 5 minutes each in PBT. Second, four times for 30 minutes each in PBT on an orbital shaker. After washing, soak embryos for 30 minutes at room temperature in PBT + N on an orbital shaker. Finally, remove the PBT + N and replace it with your secondary antibody (for the three antibodies whose labeling patterns are illustrated in Photos 6.19-6.26, use goat anti-mouse IgG conjugated to peroxidase (1:200 dilution in PBT + N; goat anti-mouse IgG can be purchased from Jackson ImmunoResearch Lab., Inc., West Grove, PA); soak embryos in this solution in the refrigerator overnight. In this step, the second antibody should bind specifically to the primary antibody (note that the secondary antibody must be chosen based on the isotype of the primary antibody, for example, an IgG, and on the species in which the antibody was made, for example, a mouse).

On day 3, wash embryos as follows: first, three times at 5 minutes each in PBT; second, four times at 30 minutes each in PBT on an orbital shaker. During the final 30-minute wash, make up the following two solutions and place them in 4-ml test tubes. Solution A: 2 ml PBT, 1 ml DAB/PBT stock solution (place one DAB tablet [diaminobenzidine; product no. D-5905, Sigma Chemical Co.] in 10 ml PBT), 75 µl 1% CoCl, and 60 µl 1% NiNHSO$_4$. *Caution: DAB is a carcinogen; avoid direct contact with it. Wear safety glasses and disposable surgical gloves at all times. Inactivate any spilled DAB or DAB on your instruments using liquid bleach [for example, Clorox], and dispose of all contaminated liquid and dry materials in hazardous waste containers).* DAB is a chromogen, which will yield an insoluble reaction product when exposed to the enzyme peroxidase and the substrate hydrogen peroxide (see below); CoCl and NiNHSO$_4$ are used to intensify the DAB reaction. Solution B: (2.5 ml distilled water plus 25 µl 30% hydrogen peroxide (Sigma Chemical Co.). After the last 30-minute wash, remove the PBT and soak embryos in Solution A for 10 minutes. Label a third test tube "final." Place 1 ml of Solution A into this tube and then add 12 µl of Solution B. Pour this solution into a watch glass and incubate embryos one at a time in the watch glass, observing them with a stereomicroscope until an appropriate level of labeling is achieved. Use care not to leave embryos in the labeling solution too long or heavy background labeling will occur, obliterating your specific labeling. When satisfactory labeling is achieved, transfer embryos to PBS and store them in the refrigerator. Labeled whole mounts may be subsequently processed for paraffin serial sectioning, if so desired, using procedures described in Chapter 6, F. Advanced Hands-On Studies, 1. Preparing embryos for light microscopy.

4. Preparing embryos for *in situ* hybridization

Several excellent techniques exist for preparing embryos for *in situ* hybridization. Whole-mount procedures provide superb results on young embryos. To begin, we will discuss the general steps involved in *in situ* hybridization and what is accomplished at each step. Two excellent references provide detailed procedures to follow to actually conduct *in situ* hybridization (Wilkinson, D.G. [1993] *In Situ* Hybridization. A Practical Approach. IRL Press, Oxford; Nieto, M.A., Patel, K., and Wilkinson, D.G. [1996] *In situ* hybridization analysis of chick embryos in whole mount and tissue sections. In "Methods in Cell Biology, Vol. 51, Methods in Avian Embryology." M. Bronner-Fraser, Ed., pp. 219-235, Academic Press, Inc., New York).

Four main steps are followed in *in situ* hybridization: synthesis of a labeled RNA probe complementary to the RNA one is trying to detect; fixation and permeabilization of tissue; hybridization of the probe to the target RNA and washing to remove the unhybridized probe; and immunocytochemical detection of the probe.

For whole-mount *in situ* hybridization, the RNA probe is typically labeled with the hapten digoxigenin (DIG). This allows subsequent detection of the hybridized probe within the tissues of the embryo using immunocytochemistry (see Chapter 6, Advanced Hands-On Studies, 3. Preparing embryos for immunocytochemistry) by using an alkaline phosphatase-conjugated anti-DIG antibody and the chromogenic substrate mixture of BCIP (5-bromo-4-chloro-3-indolylphosphate) and NBT (4-nitro blue tetrazolium chloride). Labeled areas turn a dark-blue/purple color.

Begin by collecting chick embryos for *in situ* hybridization using the procedures described in Chapter 6, Hands-On Studies, D. Exercise 3.3 (the *in situ* hybridization procedures we describe here also work with other vertebrate embryos with minor modification). Because mRNA is detected during *in situ* hybridization and mRNA is readily degraded by the enzyme RNase, it is important that precautions be taken to minimize the possibility of RNase contamination. RNase is present on our skin, so wear disposable gloves during the collection of embryos and the handling of all reagents and glassware. Glassware should be thoroughly cleaned and sterilized before use. We will not describe here the first step of the *in situ* hybridization procedure, the making of probes (the above references provide detailed procedures). Rather, we suggest that your obtain riboprobes, if available, from a neighboring lab to ensure that a strongly expressed (that is, a "good") probe is available for the stage of chick embryos that you have collected. Whenever possible, place embryos on a rotating shaker during the following procedures to facilitate the exchange of reagents. Fix embryos for 2 hours (at room temperature) to over night (at 4°C) with 4%

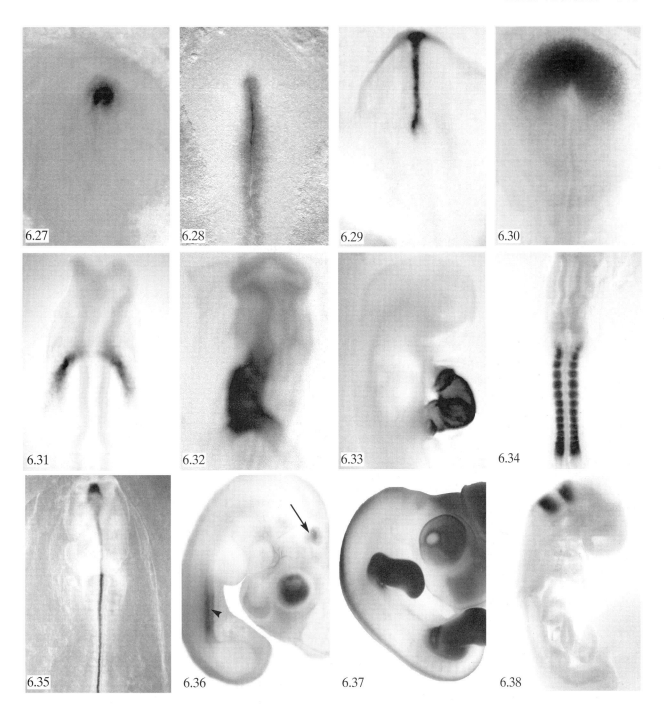

Photos 6.27-6.38. Whole mounts of embryos labeled by in situ hybridization. Photos 6.27-6.30 show 18- to 21-hour chick embryos with the primitive knot labeled with a transcription factor (*cNot*; Photo 6.27), the primitive streak labeled with a transcription factor (*brachyury*; Photo 6.28), the notochord and prechordal plate labeled with a secreted factor (*Sonic hedgehog*; Photo 6.29), and the neural plate labeled with a transcription factor (*Otx-2*; Photo 6.30). Photos 6.31-6.33 show 24- to 48-hour chick embryos with the heart labeled with a transcription factor (*vMHC*). Photo 6.32 is a ventral view. Photo 6.34 shows a 33-hour chick embryo with the somites labeled with a transcription factor (*paraxis*). Photo 6.35 shows a 33-hour chick embryo with the notochord and prechordal plate labeled with a secreted factor (*Sonic hedgehog*). Photo 6.36 shows a 4-day chick embryo with the mesonephroi (arrowhead) and auditory vesicles (arrow) labeled with a transcription factor (*Pax-2*; the eyes contain pigment and are not labeled with the probe). Photo 6.37 shows a 4-day chick embryo with the limb buds and dorsal neural tube labeled with a transcription factor (*Lmx-1*; the eyes contain pigment and are not labeled with the probe). Photo 6.38 shows a 9-day mouse embryo with the third and fifth rhombomeres of the hindbrain labeled with a transcription factor (*Krox-20*).

paraformaldehyde in phosphate-buffered saline (PBS; to make this solution, dissolve 8 g of NaCl, 0.2 g KCl, 1.44 g Na_2HPO_4, and 0.24 g KH_2PO_4 in 1 l distilled water; adjust the pH to 7.4, and store the solution until use in the refrigerator). Cut open enclosed spaces, such as the heart and brain ventricles, to prevent the trapping of reagents and non-specific labeling. Rinse embryos twice with autoclaved PBS, following by two more rinses with PBT (autoclaved PBS plus 0.1% Tween 20–the latter is a detergent that digests holes in membranes, allowing reagents to penetrate tissues). Dehydrate embryos through a graded series of PBT and methanol (75% PBT/25% methanol, followed by 50%/50%, 25% PBT/75% methanol, 100% methanol, and a second wash of 100% methanol. Embryos can be stored at -20°C in 100% methanol for up to several weeks before processing for *in situ* hybridization (cooling tissues in methanol also favors the subsequent penetration of reagents, so embryos should be kept at -20°C in 100% methanol for at least 1 hour before continuing). After cooling, rehydrate embryos through a graded series of PBT/methanol (75% methanol/25% PBT, 50%/50%, 25% methanol/75% PBT, 100% PBT, and a second wash of 100% PBT).

To begin *in situ* hybridization, embryos are first soaked in a prehybridization buffer (50% formamide [25 ml], 5X SSC [use 12.5 ml of a 20X stock solution], 2% blocking powder [1 g], 0.1% Triton X-100 [use 250 μl of a 20% stock solution], 0.1% CHAPS [use 500 μl of a 10% stock solution], 50 μg yeast RNA [use 250 μl of a 10 mg/ml stock solution], 5 mM EDTA [use 500 μl of a 0.5M stock solution], 50 μg/ml heparin [use 50 μl of a 50 mg/ml stock solution], and DEPC-treated water [11 ml]; these and subsequent reagents can be purchased from any of several common vendors; *use caution when handing reagent as some are toxic*). To do this, remove most of the PBT and add enough prehybrization buffer to allow embryos to sink. After embryos have sunk below the surface of the buffer, remove most of the buffer and add fresh prehybridization buffer; incubate embryos at 55-65°C for 2 hours to over night.

Next, 1-5 μl of riboprobe (the concentration needs to be determined empirically) is added to the buffer to allow hybridization to occur. Embryos are incubated with the probe and buffer at 55-65°C for over night to 3 days (the length of time needs to be determined empirically). Make sure during that during incubation, solutions do not evaporate and that embryos do not dry out.

After hybridization, wash embryos at 55-65°C with prewarmed 2X SSC and 0.1% CHAPS for three times, 20 minutes each (use 10 ml 20X SSC, 1 ml 10% CHAPS, and 89 ml distilled water). Repeat the washes at 55-65°C using prewarmed 0.2X SSC and 0.1% CHAPS, again for three times, 20 minutes each (use 1 ml 20X SSC, 1 ml 10% CHAPS, and 89 ml distilled water), and then rinse embryos with KTBT (50 mM Tris-HCl at pH

7.5 [use 10 ml of a 1M stock solution], 150 mM NaCl [use 6 ml of a 5M stock solution], 10 mM KCl [use 1 ml of a 2M stock solution], 1% Triton X-100 [use 10 ml of a 20% stock solution], and distilled water [173 ml]) for 2 times, 20 minutes each at room temperature. Preblock embryos in 20% sheep serum in KTBT at 4°C for 2 hours to over night, and then wash embryos with KTBT at room temperature, 5 times, for 1 hour each (embryos can be stored over night at 4°C in KTBT, if desired). Incubate embryos with a 1:2000 dilution of anti-digoxigenin antibody in 20% sheep serum in KTBT for over night.

To complete the *in situ* hybridization procedure, wash embryos 2 times for 15 minutes each in NTMT (100 mM NaCl [use 2 ml of a 5M stock solution], 100 mM Tris at pH 9.5 [use 10 ml of a 1M stock solution], 50 mM $MgCl_2$ [use 2.5 ml of a 2M stock solution], 0.1% Triton X-100 [use 500 μl of a 20% stock solution], 1 mM levamisol [0.48 g], and distilled water [84.5 ml]). Then incubate embryos in NTMT containing 4.5 μl NBT and 3.5 μl BCIP/ml in the dark, to allow the color reaction to occur (for a strong probe, the color reaction can be run at 4°C to prevent over staining; for a weak probe, the color reaction can be run at room temperature to enhance staining). After the color reaction reaches the desired intensity, stop the reaction by washing embryos in several rinses in KTBT or PBS. Shift the color of the reaction product to purple by rinsing embryos for a few minutes in 100% methanol. Finally, rinse embryos in PBS and postfix them in 4% paraformaldehyde in PBS for 2 hours to over night. Then examine embryos as whole mounts and, if desired, as frozen (see Chapter 6, F. Advanced Hands-On Studies, Preparing embryos for light microscopy, c. Frozen [cryostat] sections) or paraffin (see Chapter 6, F. Advanced Hands-On Studies, 1. Preparing embryos for light microscopy, b. Serial sections) sections.

5. Chick New whole-embryo culture

New culture is the most widely used culture for studying chick development during gastrula and neurula stages. It was developed by Denis New (New, D.A.T. [1955] A new technique for the cultivation of the chick embryo *in vitro*. *Journal of Embryology and Experimental Morphology* 3:326-331). Unincubated blastoderms can be placed in New culture and incubated for periods of up to 2 days. Blastoderms at later stages (e.g., 18- to 33-hours) can also be cultured by this method. Unfortunately, the vitelline circulation does not develop in New or Spratt culture, so embryos die prior to reaching the 48-hour stage. Development in New culture is excellent, and if done properly, neural tube development occurs normally. For New culture, prepare albumen/agar culture plates as described in Exercise 3.4 for Spratt culture (Chapter 6, Hands-On Studies, D). As in Spratt culture, blastoderms are cultured on an albumen/agar substrate in 35 mm dishes under sterile conditions. However in contrast to Spratt culture, the blastoderm is left attached to its vitelline membranes to maintain its normal tension during culture.

To collect blastoderms, clean the blunt end of the end with 70% ethanol in a manner similar to that as described in Exercise 3.4. Then use blunt watchmaker's forceps to carefully remove the shell at the blunt end of the egg. Use a motion similar to that of a can opener, piercing the shell at an acute angle to prevent damaging the underlying contents, and progressing around the circumference of the shell to remove the blunt-end cap in one piece. Carefully decant the thin albumen into a waste container, and use a medium-sized pair of forceps with rounded ends to remove the coat of thick albumen covering the yolk. It is essential that you remove this covering as completely as possible to avoid problems in subsequent steps. Next, pour the remaining contents of the egg (the yolk and blastoderm enclosed within the vitelline membranes) into a finger bowl containing saline. Rotate the yolk using your fingers until the blastoderm floats uppermost. Add sufficient saline so that the yolk can be submerged.

The next step requires practice. Using medium-sized scissors with pointed tips, quickly cut through the vitelline membranes along the equator of the yolk, making sure that the blastoderm remains uppermost (to keep your cut exactly at the equator will require that you adjust the position of the yolk with your fingers or forceps so that the blastoderm remains uppermost). It is necessary to cut along the equator so that the surface area of vitelline membranes removed with the blastoderm is sufficient to stretch the membranes over a ring for culture. After you have cut along the entire equator, use blunt watchmaker's forceps to peel the cut edge of the vitelline membranes off the yolk toward the blastoderm. Rotate the egg as you do this so that the entire perimeter is freed. Then carefully pull the vitelline membranes and attached blastoderm off the yolk. To do this, make sure that saline covers the top of the yolk and that the vitelline membranes are keep beneath the surface of the saline as you are pulling. Also, bend the vitelline membranes back upon themselves as you pull over the surface of the yolk in the area of the blastoderm. With practice, you will be able to remove the vitelline membranes with the blastoderm attached.

Blastoderms and the attached vitelline membranes are next transferred to fresh Petri dishes containing saline by using a spoon as described in Exercise 3.4. Position the vitelline membrane in the dish (make sure you transfer enough saline so that this is easy to do) with the blastoderm uppermost. Then place a ring of the appropriate diameter (these can usually be made inexpensively by University glass shops by cutting rings with a lathe from plastic or glass tubing of approximately 20-mm diameter) onto the vitelline membranes and center it with respect to the blastoderm. Using blunt watchmaker's forceps, pull the cut edges of the vitelline membranes over the external diameter of the ring, making sure that the blastoderm is centered and wrinkles are eliminated. Remove all saline from the well created by the ring, as well as from the bottom of the culture dish outside of the ring, using a pipette. Then carefully pick up the ring and attached vitelline membranes/blastoderm using forceps and transfer it to a cultue dish; remove any excells saline after transfer and make sure that the vitelline membranes remain attached to the ring and wrinkles are absent. Cover the dishes with their lids and place them into a humidified incubator; examine cultures periodically over the next 1-2 days, depending on the stage of the embryos at the time of culture.

6. Dye injections to follow cell movements

Fluorescent dyes provide excellent tools for labeling cells and following their movements over time. Such dyes are used extensively for constructing prospective fate maps. Recently, several fluorescent dyes have become available for cell labeling. We will describe a procedure in which mixtures of fluorescent dyes are injected into embryos. We use a mixture of dyes, including in the mixture a dye that is intensely fluorescent (for example, DiI or DiO) so that labeled cells can be readily detected at the time of injection and over the course of the next 24 hours. We also include in the mixture a dye that contains either fluorescein or rhodamine molecules. These molecules can be detected subsequently using immunocytochemistry with a primary antibody (that is, either antifluorescein or antirhodamine) and a peroxidase-conjugated secondary antibody (see F. Advanced Hands-On Studies. 3. Preparing embryos for immunocytochemistry), increasing the sensitivity of the technique severalfold and allowing the label to be detected in paraffin serial sections. Use of immunocytochemistry also makes the labeling permanent so that samples can still be studied months or even years later.

Incubate chick eggs until embryos reach the 18-hour (that is, primitive-streak) stage (see Chapter 6, Hands-On Studies, D. Exercise 3.3). Place embryos into Spratt or New whole-embryo culture (see, respectively, Exercise 3.4 in Chapter 6, Hands-On Studies, D and Chapter 6, F. Advanced Hands-On Studies. 5. Chick New whole-embryo culture).

Prepare two mixtures of dyes. One mixture, consisting of the dyes DiI and CRSE, will fluoresce red using a fluorescence microscope and a rhodamine filter set. The other mixture, consisting of the dyes DiO and CFSE, will fluoresce green using a fluorescence microscope and a fluorescein filter set. To prepare the first mixture, make a stock solution (called solution 1.1) by dissolving 2.5 mg DiI (1, 1'-dioctadecyl-3, 3, 3', 3'-tetramethylindocarbocyanine perchlorate; Molecular Probes, Inc., Eugene, OR; catalog no. D-282) in 50 µl DMSO (dimethyl sulfoxide; Baker analyzed reagent, VWR Scientific Products, Salt Lake City, UT). Then add 1 ml of 100% ethanol and mix with a vortex mixer. Make a second stock solution (called solution 1.2) by dissolving 2.5 mg CRSE (5-carboxytetramethylrhodamine, succinimidyl ester; Molecular Probes, Inc.; catalog no. C-2211 or C-1171) in 1 ml DMSO. Both stock solutions can be stored in a freezer until the time of use (we store the stock solutions in eppendorf tubes covered in aluminum foil to prevent light degradation of the dye). To prepare the second misture, make a stock solution (called solution 2.1) by dissolving 2.5 mg DiO (3,3'-dioctadecyloxacarbocyanine perchlorate; Molecular Probes, Inc.; catalog no. D-275) in 500 µl DMSO. Then add 1 ml of 100% ethanol and mix with a vortex mixer. Make a second stock solution (called solution 2.2) by dissolving 2.5 mg CFSE (5-[and -6] carboxyfluorescein diacetate, succinimidyl ester; Molecular Probes, Inc.; catalog no. C-2211 or C-1171) in 1 ml DMSO.

Just before use, mix 50 µl aliquots of solutions 1.1 and 1.2 to make the red-fluorescing dye mixture and 50 µl aliquotes of solutions 2.1 and 2.1 to make the green-fluorescing dye mixture.

Embryos in whole-embryo culture will now be labeled by microinjecting dye using microinjection pipettes. For careful studies, pull fine pipettes (capillaries with Omega dot fiber; Fredrick Haer & Co., Brunswick, ME) using a pipette puller, and load the tip of the pipette using a Microfil 28 AWG needle (World Precision Instruments, Inc., Sarasota, FL). Pipettes are then mounted onto holders connected to an air-pressure microinjector (e.g., Picospritzer; General Valve Corp., Fairfield, NJ) mounted on a micromanipulator. If such equipment is not available, micropipettes can be hand-pulled from 100 lambda pipettes (Becton, Dickinson and Co., Parsippany, NJ; catalog no. 4625; each tube of pipettes comes with a rubber tube and mouth piece for manual microinjection) using a microburner. To construct the microburner, connect the syringe-mounting side of an 18-gauge syringe needle to a rubber hose, carefully pierce through the edge of a large rubber or cork bottle stopper (*avoiding piercing your hand*) with the point of the needle at about a 45° angle pointing up, forming a holder for the needle; connect the free end of the rubber tube to a gas source. With the tap of the gas valve just barely turned on, light the gas escaping from the tip of the needle. *Caution: Do not turn the valve more than the minimum required to leak gas from the tip of the needle or the excess pressure may whip the end of the rubber tube, possibly causing injury. Also, be careful as the tip of the needle is sharp and can easily impail you. Finally, carefully monitor the flame of the microburner—it can be difficult to see so you can burn yourself; also the flame can go out, causing toxic and explosive gas to leak into the room. As always, make sure that you are properly supervised by an experience person to prevent injury.* Pipettes can then be pulled in the flame of

the microburner by holding the two ends of the pipette, one hand per end, heating the center of the pipette in the flame while rotating the pipette to heat the glass evenly, and then when the glass is suffficiently softened, quickly removing the pipette from the flame and pulling your two two hands apart to form a slender filament of glass in the center of the pipette. The glass is then allowed to cool for several minutes (*remember, hot glass looks identical to cool glass, but causes nasty burns*). *Wearing safety glasses,* use a diamond pencil to score the glass filament and then snap the filament, forming two micropipettes. Load micropipettes with dye by sucking the dye into their tips by using the mouth piece and rubber tubing supplied with the 100 lambda pipettes. Insert the tip of the pipette into an embryo were desired, and inject the dye by gently blowing into the mouthpiece.

Examine embryos immediately after injection using a fluorescence microscope (*see cautions for using a fluorescence microscope listed in F. Advanced Hands-On Studies. 1C. Frozen [cryostat] sections*) and again at selected intervals of culture for up to 24 hours postinjection.

Finally, the fluorescent label can be converted into a permanent label by using a primary antibody to rhodamine (solution 1) or fluorescein (solution 2). If you have not yet done immunocytochemistry, see Chapter 6, F. Advanced Hands-On Studies. 3. Preparing embryos for immunocytochemistry, and especially note the cautions listed; here we provide information only on the source, concentration, and length of treatment for the primary and secondary antibodies.

For dye solution 1, use anti-rhodamine as the primary antibody (= a rabbit IgG polyclonal antibody used at a 1:100 dilution; Molecular Probes; catalog no. A-6397) and horseradish peroxidase-conjugated goat anti-rabbit IgG as the secondary antibody (used at a 1:200 dilution; Boehringer Mannheim, Indianapolis, IN; catalog no. 605-220). For dye solution 2, use anti-fluorescein as the primary antibody (= a rabbit IgG polyclonal antibody used at a 1:200 dilution; Molecular Probes; catalog no. A-889) and horseradish peroxidase-conjugated goat anti-rabbit IgG as the secondary antibody (Boehringer Mannheim; catalog no. 605-220).

After immunocytochemistry, labeled cells can be readily identied in whole-mount embryos. Additionally, embryos can be embedded in paraffin and sectioned to study the distribution of the label in detail (see F. Advanced Hands-On Studies. 1. b. Serial sections).

7. BrdU labeling to assess cell proliferation

Rates of cell division in different tissues or at different stages of development can be determined by pulsing embryos with a labeled nucleoside or an analog of a nucleoside. To assess cell proliferation, we generally use the thymidine analog BrdU (5-Bromo-2'-

Deoxyuridine); the radioactively labeled nucleoside, tritiated thymidine can also be used. BrdU (and tritiated thymidine) is incorporated into DNA during the S (synthetic) phase of the cell cycle. Thus only dividing cells are labeled, and an antibody to BrdU can be used to detect such labeled cells. Comparisons can then be made, for example, between the percentage of dividing (that is, BrdU-labeled) cells within the neural plate of the 18-hour chick and those within the neural tube of the 33-hour chick. A decrease in the number of labeled cells in the neural tube of the 33-hour chick as compared to the neural plate of the 18-hour chick (assuming both embryos were pulsed with BrdU for identical periods of time; for example, for 1 hour), would suggest that the rate of proliferation decreased in the neural ectoderm over time.

To assess cell proliferation, collect chick embryos at 18- to 33-hours of incubation (see Chapter 6, Hands-On Studies, D. Exercise 3.3). Place embryos into Spratt or New whole-embryo culture (see, respectively, Exercise 3.4 in Chapter 6, Hands-On Studies, D and Chapter 6, F. Advanced Hands-On Studies. 5. Chick New whole-embryo culture). Pulse embryos by placing a drop of BrdU solution (that is, 200 µl of 0.1 mM BrdU in saline or distilled water; Sigma Chemical Co., St. Louis, MO; catalog no. B9285) directly onto the surface of the embryo. *Use appropriate care when handling BrdU; it is a mutagen.* Place embryos into humidified incubators for exactly 1 hour. Then fix embryos as described for immunocytochemistry (see Chapter 6, F. Advanced Hands-On Studies. 3. Preparing embryos for immunocytochemistry) and process them for paraffin histological sectioning (see F. Advanced Hands-On Studies. 1. b. Serial sections). Allow sections to become firmly attached to glass slides (by incubating them sufficiently on hot plates) before proceeding.

Process sections to remove the wax and rehydrate the tissue. Then hydrolyze sections by placing slides in a solution of 2N HCl for 15 minutes at 37^{0}C. Immediately after hydrolysis is completed, wash slides in two changes of 0.1 M sodium borate, 10 minutes each.

To detect BrdU-labeled cells, cover sections with the primary antibody (= a mouse IgG monoclonal antibody; 1:1000 anti-BrdU; Sigma Chemical Co.; catalog no. B2531) and place them in a humidified chamber (a Tupperware container lined with paper towels soaked in sterile water plus a secured lid) at room temperature for over night. Wash sections with buffer (for all solutions, see Chapter 6, F. Advanced Hands-On Studies. 3. Preparing embryos for immunocytochemistry) and then cover them with the secondary antibody (= a goat anti-mouse antibody in PBT + N; 1:200; Jackson ImmunoResearch Laboratories, Inc., West Grove, PA; catalog no. 115-035-003). Again, incubate sections over night at room temperature in a humidified chamber.

For subsequent processing, follow the third-day procedure described under Chapter 6, F. Advanced Hands-On Studies. 3. Preparing embryos for immunocytochemistry.

To assess the rate of cell proliferation, examine sections and count the number of labeled nuclei (brown colored) and unlabeled nuclei (not colored) in an organ rudiment such as the neural plate. Add these numbers together, and then divide the number of labeled nuclei by the total number of nuclei to determine the percentage of labeled cells. In general, a higher percentage of labeled cells in an organ rudiment compared to a lower number in another rudiment (or at a different stage of development of the same rudiment) indicates that the rate of cell proliferation is greater in the rudiment with the higher number.

8. TUNEL labeling to assess cell death (apoptosis)

Cell death or apoptosis is a normal developmental process in which half or more of the cells of developing populations are eliminated. To assess cell death, we will use the TUNEL method. Because the TUNEL method involves a procedure similar to *in situ* hybridization, see Chapter 6, F. Advanced Hands-On Studies. 4. Preparing embryos for in *situ* hybridization.

The acronym TUNEL stands for terminal transferase dUTP nick end labeling, and it refers to the fact that during apoptosis DNA undergoes fragmentation to form 3'-hydroxyl ends; these ends are labeled enzymatically (by terminal deoxynucleotidyl transferase) with modified nucleotides (that is, nucleotides labeled with the small protein digoxigenin).

To conduct TUNEL labeling, we use the ApopTag kit (Intergen Co., Purchase, New York; catalog no. 7110). Briefly, the 3'-hydroxyl ends of DNA in cells undergoing apoptosis are tagged with digoxigenin-labeled nucleotides. A alkaline phosphatase-conjugated anti-dioxigenin antibody is used to label the incorporated digoxigenin, and substrate and chromogen is added to produce an insoluble reaction product that can be readily detected (that is, by its color).

To perform TUNEL labeling, fix embryos in 1% paraformaldehyde in phosphate-buffered saline (PBS; to make this solution, dissolve 8 g of NaCl, 0.2 g KCl, 1.44 g Na_2HPO_4, and 0.24 g KH_2PO_4 in 1 l distilled water; adjust the pH to 7.4, and store the solution until use in the refrigerator) for 20 minutes at 4°C. Wash embryos in PBS for 10 minutes at 4°C, and sequentially dehydrate embryos to 100% ethanol (that is, 50%, 70%, 95%, and 100% ethanol, 2 changes of each concentration for 5 minutes each change for chick embryos up to two days of incubation; use longer times for older embryos). Dehydration makes embryos permeable to the enzyme used in a subsequent step. If embryos cannot be processed immediately, store them in 100% ethanol at -20°C until needed. Rehydrate embryos by reversing the sequence used for dehydration, ending in PBS. Next, incubate embryos in equilibration buffer (catalog no. S7110-1) at room temperature for 5 minutes, then remove the buffer and replace it with working strength TdT enzyme solution (77 µl reaction buffer, catalog no. S7110-2, plus 33 µl TdT enzyme, catalog no. S7110-3) for 2 hours at 37°C. Stop the enzyme reaction by incubating embryos in stop/wash buffer (catalog no. S7110-4) for 40 minutes at 37°C.

For the remainder of the protocol follow similar procedure used for *in situ* hybridization (see Chapter 6, F. Advanced Hands-On Studies, 4. Preparing embryos for in *situ* hybridization). First, inactivate the endogenous alkaline phosphatase by washing embryos twice for 5 minutes each in Tris-buffered saline (TBST; 0.14 M NaCl, 10 mM KCl, 25 mM Tris-HCl, and 0.1% Tween 20; add the Tris-HCl first and then adjust the pH to 7.0 before adding the other components; always prepare fresh before use) and 1 mM levamisol. Then wash embryos 3 times for 5 minutes each in Tris-magnesium buffer (NTMT; 0.1 M NaCl, 0.1 M Tris-HCl at pH 9.5, 50 mM $MgCl_2$, 0.1% Tween 20; always prepare fresh before use) and 2 mM levamisol. Second, pre-block embryos in TBST containing 10% sheep serum. Third, treat embryos for 40 minutes in alkaline phosphatase-conjugated anti-digoxigenin antibody (Boehringer Mannheim, Indianapolis, IN) diluted 1:2000. Then wash embryos in TBST, 3 times for 10 minutes each, followed by washes in NTMT, 3 times for 10 minutes each.

Finally, incubate embryos with BCIP and NBT exactly as described for *in situ* hybridization (see Chapter 6, F. Advanced Hands-On Studies, 4. Preparing embryos for in *situ* hybridization). A strong color reaction usually occurs after 10-20 minutes of incubation.

9. Mouse whole-embryo culture

Mouse embryos during stages of gastrulation, neurulation, and early morphogenesis (7-9 days) can be removed from the uterus and some of their covering layers, and then cultured for periods of up to 1-2 days

Obtain embryos at 7- to 9.5-days of gestation as described in Exercise 4.2 (see Chapter 6, Hands-On Studies, E) except that all instruments should be sterilized in 70% ethanol and sterile Petri dishes should be used. Also, use sterile Tyrode's solution (Sigma Chemical Co., St Louis, MO) in place of PBS.

The conceptus will now be dissected in a manner dependent upon its age. In embryos at 7-8.5 days of gestation, the embryonic part of the egg cylinder is covered with an acellular, transparent membrane called Reichert's membrane. Use two pairs of watchmaker's forceps to remove this layer. In older embryos, Reichert's membrane is no longer present, but a well development yolk sac has formed. Using watchmaker's forceps open the yolk sac at the distal side of the embryonic part of the egg cylinder, leaving it attached prox-

imally. At all stages, try to preserve the ectoplacental cone intact.

Embryos are culture in a medium composed of 1:1 Tyrode's solution and heat-inactivated rat serum (obtained from blood immediately centrifuged upon collection; available from Harlan Bioproducts for Science, Indianapolis, IN; to heat-inactivate serum, place it in a water bath at 56°C for 40 minutes) plus 50 U penicillin and 50 µg streptomycin per milliliter (Gibco BRL, Grand Island, NY; catalog no. 600-5140). Additionally, the medium must be gassed with 5% CO_2 in air, and embryos must be rotated throughout incubation to allow proper exchange to occur. To do this, add 1 ml of gassed media to a 50-ml tube (Falcon 2098 Blue Max conical tube; Baxter diagnostic Inc., McGraw Park IL) and add several embryos (4-8, depending on the age). Using the tip of a Pasteur pipette attached to a rubber hose attached to a tank of 5% CO_2/95% air, gently blow gas onto the surface of the medium for 2-3 minutes, replacing the air in the tube with 5% CO_2/95%; quickly cap the tubes and incubate them at 37°C on a rotator (30 revolutions per minute) for 12 to 48 hours.

Cultured embryos can be used for a number of different purposes. Embryos can be injected with dye and fate mapped (for example, see Chapter 6, F. Advanced Hands-On Studies, 6. Dye injections to follow cell movements). Additionally, embryos can be treated with various drugs, chemical inhibitors, blocking antibodies, or anti-sense mRNA. Whole-embryo culture alleviates the main disadvantage to working with mammalian embryos: lack of accessibility owing to the fact that mammalian embryos develop within the uterus of their mother. Thus the use of whole-embryo culture greatly increases the value of the mouse model.

Use the space below for notes.

References

Although this manual is thoroughly illustrated by photomicrographs, scanning electron micrographs, and line drawings, certain textbooks and atlases provide additional helpful illustrations. Because textbooks vary in terminology, the student should follow the terminology used in this manual to avoid confusion. Several of the listed textbooks are now, unfortunately, out of print. Nevertheless, they are still referenced here because of their timeless quality and because on the author's travels to various institutions, they frequently appear on faculty members' bookshelves, in teaching laboratories, and in university and departmental libraries. Faculty and students are urged to seek out these valuable resources.

Arey, L. B. 1974. *Developmental Anatomy*. Rev. 7th Ed. Philadelphia: Saunders.

Balinsky, B. I. 1981. *An Introduction to Embryology*. 5th Ed. Philadelphia: Saunders.

Bard, J. 1994. *Embryos: Color Atlas of Development*. London, England: Mosby-Year Book Europe Limited.

Bellairs, R., and M. Osmond 1998. *The Atlas of Chick Development*. San Diego, CA: Academic Press.

Gilbert, S. F. 1997. *Developmental Biology*. 5th Ed. Sunderland, MA: Sinauer Assoc. Inc.

Gilbert, S. F., and A. M. Raunio 1997. *Embryology. Constructing the Organism.*. Sunderland, MA: Sinauer Associates, Inc.

Hamilton, H. L. 1952. *Lillie's Development of the Chick*. New York: Holt, Rinehart and Winston.

Hausen, P., and M. Riebesel 1991. *The Early Development of Xenopus laevis. An Atlas of the Histology*. New York, NY: Springer-Verlag.

Huettner, A. F. 1949. *Comparative Embryology of the Vertebrates*. Rev. Ed. New York: Macmillan.

Kalthoff, K. 1996. *Analysis of Biological Development*. New York, NY: McGraw-Hill, Inc.

Kaufman, M. H. 1992. *The Atlas of Mouse Development*. San Diego, CA: Academic Press.

Kaufman, M. H., and J. B. L. Bard 1999. *The Anatomical Basis of Mouse Development*. San Diego, CA: Academic Press.

Mathews, W. W., and G. C. Schoenwolf 1998. *Atlas of Descriptive Embryology*. 5th Ed. Upper Saddle River, NJ: Prentice-Hall, Inc.

Patten, B. M. 1971. *Early Embryology of the Chick*. 5th Ed. New York: McGraw-Hill.

Patten, B. M. 1948. *Embryology of the Pig*. 3rd Ed. New York: McGraw-Hill.

Rugh, R. 1968. *The Mouse. Its Reproduction and Development.* Minneapolis, MN: Burgess Publishing Co

Rugh, R. 1951. *The Frog. Its Reproduction and Development.* New York, NY: McGraw-Hill Book Co.

Theiler, K. 1972. *The House Mouse. Development and Normal Stages from Fertilization to 4 Weeks of Age.* New York, NY: Springer-Verlag.

Wolpert, L., R. Beddington, J. Brockes, T. Jessell, P. Lawrence, and E. Meyerowitz 1998. *Principles of Development.* New York, NY: Current Biology, Ltd.

Books or videos of methods for experimental embryology are also useful for working with living embryos. The following methods books and videos are highly recommended.

Bronner-Fraser, M. 1996. *Methods in Cell Biology. Vol. 51. Methods in Avian Embryology*. San Diego, CA: Academic Press.

Hamburger, V. 1960. *A Manual of Experimental Embryology*. Rev. Ed. Chicago: University of Chicago Press.

Hogan, B., R. Beddington, F. Costantini, and E. Lacy 1994. *Manipulating the Mouse Embryo. A Laboratory Manual*. 2nd. Ed. Plainview, NY: Cold Spring Harbor Laboratory Press.

Stern, C.D., and P. W. H. Holland 1993. *Essential Developmental Biology. A Practical Approach.* Oxford, England: Oxford University Press.

Grainger, R. M., and H. L. Sive 1999. *Manipulating the Early Embryo of Xenopus Laevis. A Video Guide. Tapes 1-3.* VHS, color, sound. Available from Cold Spring Harbor Laboratory Press, Plainview, New York. Phone: (800) 843-4388.

Pedersen, R. A., V. Papaioannou, A. Joyner, and J. Rossant 1996. *Targeted Mutagenesis in Mice. A Video Guide.* VHS, color, sound. Available from Cold Spring Harbor Laboratory Press (address and phone are given above).

Pedersen, R. A., and J. Rossant 1988. *Transgenic Techniques in Mice. A Video Guide.* VHS, color, sound. Available from Cold Spring Harbor Laboratory Press (address and phone are given above).

Sive, H. L., R. M. Grainger, and R. M. Harland 2000. *Early Development of Xenopus laevis. A Laboratory Manual.* Plainview, NY: Cold Spring Harbor Laboratory Press.

Audiovisual aids are also very useful for studying developmental anatomy, especially in helping students to visualize developmental events three-dimensionally and changes in morphology with time. The following audiovisual aids are highly recommended.

A Dozen Eggs. Time-Lapse Microscopy of Normal Development. Edited by R. Fink. VHS, 45 min, color, sound. Available from: Sinauer Associates, Inc., Publishers, 108 North Main Street, Sunderland, MA 01375-0407. Phone: (413) 665-3722.

Amphibian Embryo (Frog, Toad and Salamander). 16-mm film, 16 min., color, sound. Available from: Encyclopedia Britannica Education Corp., 310 South Michigan Ave., Chicago, IL 60604. Phone: (800) 621-3900.

Development of the Cardiovascular System of the Chick: The Heart. 16-mm film, 20 min., color, sound. Available from: Indiana University, Instructional Support Services, Film Library, Bloomington, IN 47405-5901. Phone: (812) 855-3396.

Development of the Cardiovascular System of the Chick: The Blood Vessels. 16-mm film, 23 min., color, sound. Available from: Indiana University, Instructional Support Services, Film Library (address and phone are given above).

Development of the Chick: Extra-embryonic Membranes. 16-mm film, 20 min., color, sound. Available from: Indiana University, Instructional Support Services, Film Library (address and phone are given above).

Development of the Normal Heart. 16-mm film, 15 min., color, sound. Available from Classroom Support Services, DG-10, University of Washington, Seattle, WA 98195. Phone: (206) 543-9909.

Embryo. CD Color Atlas for Developmental Biology. By G. C. Schoenwolf. CD-ROM, Macintosh and Windows compatible. ISBN 0-13-594011-7. Available from Prentice-Hall, Upper Saddle River, NJ 07458; http://www.prenhall.com. Phone: (201) 236-7000.

Recently, several web sites have appeared that deal with certain aspects of developmental biology. The following web sites are particularly useful for the material covered in this guide.

Home page, Society for Developmental Biology: http://sdb.bio.purdue.edu/.

K. Sulik's Embryo Images: http://www.med.unc.edu/embryo_images/unit-welcome/welcome_htms/contents.htm.

Intact, fixed embryos, whole-mount slides, and sections of most of the animal models covered in this guide can be purchased from biological suppliers. The following suppliers are recommended.

Carolina Biological Supply Company. 2700 York Road, Burlington, North Carolina. Phone: (800) 334-5551.

Nebraska Scientific. 3823 Leavenworth Street, Omaha, Nebraska 68105-1180. Phone: (800) 228-7117

Ward's Biology. Post Office Box 92912, Rochester, New York 14692-9012. Phone: (800) 962-2660.

History

The eighth edition of this guide is proudly dedicated to the memories of two of its former authors and two of my mentors, Robert Milton Sweeney and Ray Leighton Watterson. My purpose in the following few paragraphs is to provide brief biographical sketches of these two distinguished scientists and teachers, whose lives had a significant impact on mine. In addition, my comments provide an historical account of the evolution of this guide. Some of my comments on Ray Watterson are adapted from an article I coauthored shortly after his death (Stocum, D. L., and G. C. Schoenwolf [1985]. Ray Leighton Watterson, 1915-1984. *Anatomical Record* **213**:266-269).

Ray Watterson was born in Greene, Iowa, on April 15, 1915. He grew up in a small town in Iowa, living a life in close communion with nature. His initial interest in biology stemmed from his boyhood, but his love for the embryo began when he first viewed a collection of human embryos and fetuses at the 1933 World's Fair in Chicago. From that moment on (at age 18!) he wanted above all to become an embryologist. He more than achieved this goal. At the Proceedings of the Fourth International Conference on Limb Development and Regeneration, held in Asilomar, California, in July 1992, the distinguished scientist, John W. Saunders, Jr., said in his address, "Among members of his generation, he (that is, Ray Watterson) was almost certainly the most broadly knowledgeable about the entire field of embryology."

Ray attended grammar school and high school in Iowa. He received a B.A. degree from Coe College in 1936 and a Ph.D. from the University of Rochester in 1941. His Ph.D. advisor was the renowned Benjamin Harrison Willier (1890-1972), the mentor of a number of distinguished embryologists including (but not limited to) Alfred J. Coulombre, James D. Ebert, Casimer T. Grabowski, Howard L. Hamilton, Irwin R. Konigsberg, Clement L. Markert, Mary E. Rawles, Dorothea Rudnick, John W. Saunders, Jr., Nelson T. Spratt, Jr., and J. Philip Trinkaus. (For further details about the life of Willier, see Watterson, R. L. [1985]. Benjamin Harrison Willier, 1890-1972. Biographical Memoirs, Vol. 55, National Academy of Sciences. The manuscript was in press at the time of Ray's death in 1984.). Ray married Evelyn Lily Goddard shortly *after* completing his Ph.D. (Willier's conviction was that marriage and graduate study were incompatible, and he apparently let his students know this in no uncertain terms). Ray and Evelyn had four children in quick succession: Richard Dean (Dick), James Robert (Jim), Donald Kent (Donnie), and Jean Marie.

Ray obtained his first faculty position at Dartmouth College. Thereafter, he held faculty positions at the University of California at Berkeley, the University of Chicago, and Northwestern University, where he progressed to the level of Chairman of the Biology Department (Ray probably would have said that he *digressed* rather than progressed, because he found administrative work to be less than enjoyable). Several of his summers were spent teaching embryology at Marine labs, especially the Marine Biological Laboratory at Woods Hole. In 1961 he joined the faculty at the University of Illinois at Urbana-Champaign, first as Professor, Department of Zoology, and then as Professor in the newly created Department of Genetics and Development. From 1981 to the time of his death on September 1, 1984, he also held the appointment of Professor in the College of Medicine at Urbana-Champaign. During his final years at Illinois, Ray increasingly focused on his original, burning scientific interest—the study of human embryology. His lectures in this area were cherished by the thousands of premed undergraduate students, graduate students, and first-year medical students who were privileged to take his classes.

Ray loved to teach. He remained actively involved in teaching throughout his career and was thoroughly committed to his students. He once wrote, "Being useful to young people through teaching is my life's blood." As part of his teaching, Ray published his laboratory manual. The first draft was reputedly written while Ray was hospitalized for depression following the sudden death of his six-year-old son, Donnie, from an adverse penicillin reaction. Even during this period of intense personal suffering, Ray thought of his students. He knew that without his participation in the laboratory, the students and teaching assistants would be guideless. Consequently, he wrote a text and had it mimeographed for distribution to the students. Unfortunately, the death of a child was a theme to be repeated in Ray's life: his two other sons preceded him in death.

Throughout his career, Ray always trained graduate students. Nineteen students earned their Ph.D. degrees under his close supervision; several more earned M.S. degrees. Two of his doctoral students played a sustaining role in the evolution of this guide. Robert Milton Sweeney and the current author (Gary C. Schoenwolf).

Robert Sweeney was born in Blue Island, Illinois, on April 27, 1941. He attended grammar school and high school in Chicago, Illinois. He received a B.S. degree from St. Joseph College in 1963 and an M.S. degree and a Ph.D. from the University of Illinois at Urbana-Champaign under the supervision of Ray Watterson (his Ph.D. was granted in 1968). During his tenure as a graduate student, he met and married Jean Wisman (Ray, unlike Willier, did not believe that marriage and graduate study were incompatible). They had three children: David, Jennifer, and Jeff.

In 1969 Bob was appointed Assistant Professor at Elmhurst College, Elmhurst, Illinois. In 1973 his contract at Elmhurst was terminated abruptly, an event clearly not based on Bob's teaching performance. He left academics, holding administrative positions at Turtox, Inc., and Lab-Line Instruments, both in the Chicago area. His life was tragically ended on June 1, 1978, when he died of cancer at age 37. His oldest child at the time of his death was seven. His wife remarried in 1981 to Don Ceithaml, a chemistry teacher and widower who had a son and a daughter; in 1982, an additional daughter was added to the family.

Although Bob's life was cut short, it was full. He thrived on interactions with students. At Elmhurst, he was named Instructor of the Year, an honor bestowed by both faculty and students. He enjoyed gardening and opera. He was a dedicated father and husband. He was a friend to many.

My involvement with this guide began in 1969, my junior year at Elmhurst College. Dr. Robert M. Sweeney appeared in the Biology Department in the fall as a bright, young, energetic new recruit. His teaching assignments included Developmental Biology, Comparative Vertebrate Anatomy, Advanced Developmental Biology, and General Zoology. I registered for developmental biology that fall and was immediately impressed with Bob (of course, I then called him Dr. Sweeney). Like most biology majors at Elmhurst, I was a premed student who was moving through the biology curriculum, taking the most medicine-related courses. Bob's love of development, a discipline he taught with a vigorous, questioning, and charismatic style, was infectious; simply stated, I was enthralled by his presentation of the embryo. I completed his Developmental Biology course (with an A, I'm proud to say), moved on to his Comparative Vertebrate Anatomy course (with a somewhat less stellar outcome), and finally took his Advanced Developmental Biology course, which included laboratories in experimental embryology (happily, the A returned). In the fall of 1969 when I took Bob's Developmental Biology course, the second edition of this guide was still in press, with Bob as coauthor. Consequently, we used the *first* edition (then in its ninth printing), which was illustrated with only three simple line drawings (Watterson, R. L. [1955]. Laboratory Studies of Chick and Pig Embryos [Minneapolis, MN: Burgess Publishing Company]). Bob's major contribution to the manual was to convert it from a largely provincial booklet, intended for

use only by Ray's own students, to a full-fledged textbook. He did this in close collaboration with Ray by making three major changes: supplementing the studies of chick and pig embryology with frog embryology, including 38 plates of photomicrographs, and adding 6 new line drawings. In addition at this juncture, the manual was transformed from essentially a photo-offset version of mimeographed handouts to a typeset publication. Both authors were extremely proud of the new edition (Watterson, R. L. and R. M. Sweeney 1970. Laboratory Studies of Chick, Pig & Frog Embryos, 2nd Edition. Burgess Publishing Company, Minneapolis). In the fall of 1970 Bob again taught his Developmental Biology course. I was greatly honored when Bob asked me to serve as his teaching assistant. I was also very proud to teach the laboratory from his new manual; that is, I was proud to be assisting an author of such a fine teaching tool. In the spring of 1971 I also assisted Bob in his General Zoology course.

I had many long talks with Bob during our two-year overlap at Elmhurst College. One of the several positive characteristics that set him apart from other faculty was his accessibility—he had an open-door policy. I was thrilled that he was willing to spend time talking with me, a mere undergraduate student. Our talks dealt with a variety of subjects, essentially anything I was interested in discussing. These talks eventually led to the topic of graduate school, and I ultimately decided to pursue a doctorate in developmental biology. I chose three large midwestern schools as possibilities for graduate study, based on Bob's advice, and I was interviewed by embryology faculty at all three. When I was interviewed by Ray (who I then, and throughout my tenure as a graduate student, called Dr. Watterson), I felt that I had found what I was looking for—a mentor who genuinely cared about his students and who was an absolute scholar and gentleman (to use a trite, yet most appropriate phrase). There was one problem, however: the Vietnam war was raging. With graduation from Elmhurst College I lost my student deferment, and my draft number from the lottery was "9" (draft numbers ranged from 1 to 365 and were based on birth date; the lower the number, the more likely one was to be drafted). Nevertheless, I decided to move from the Chicago suburbs to Urbana-Champaign to start graduate school. Incidentally, I also married just before moving south. Despite the pessimistic scenario (I mean the draft, not the marriage, which is approaching its 30th year), I was able to matriculate and earn an M.S. degree in 1973 and a Ph.D. in 1976, both under the tutelage of Ray Watterson. I was spared from the draft by President Nixon, who proclaimed a moratorium on the draft shortly after I received an "invitation" letter from my local draft board.

At the University of Illinois, I served as Ray's head TA, teaching one laboratory section of Vertebrate Embryology, Zoology 333, and coordinating the other sections. The second edition of this manual was used as our guide. During the summer of 1972, while I was busy collecting data for my M.S. thesis, Bob and his family were living in Urbana-Champaign, working with Ray on the third edition (Watterson, R. L., and R. M. Sweeney [1973]. Laboratory Studies of Chick, Pig, and Frog Embryos, 3rd ed. [Minneapolis, MN: Burgess Publishing Company]). The third edition included many changes, several of which I suggested (perhaps, in retrospect, my "suggestions" for changes were made in a much too cocky and self-assured fashion; nevertheless, I was acknowledged for my assistance in the preface, a generous gesture that I was and am still very proud of and thankful for). Numerous heated discussions between the authors and the publisher resulted in improvements in the reproductions of the 38 plates of photomicrographs. In addition, 10 new line drawings were added; the text was updated, expanded, and clarified; and an abridged index was added.

I remained at the University of Illinois for one year after finishing my Ph.D. and was appointed Visiting Lecturer. Ray decided to take a sabbatical during this time, entrusting to me his two beloved development courses (that is, Vertebrate Embryology and Human Embryology). I used his teaching materials, adopted his style (with some measure of Bob's style thrown in), and consequently was successful. Ray later told me that my success in teaching *his* courses caused him a "twinge of jealousy," but I never detected it. Throughout that year, we worked side by side as equal partners revising the guide for a fourth edition (that is, Ray treated me as a peer rather than a student, although the latter may have been more accurate). The plan was that Ray and I would revise the text, and Bob would provide input as his commitments allowed. By this time, Bob had left academics and he felt that without continued exposure to the material his expertise might wane. Yet he enthusiastically endorsed my authorship and the new edition, fully realizing that his royalties would be reduced. Horribly, Bob died as the revision was under way. In a letter to me dated six days after Bob's death, Ray wrote, "I feel as though I have lost another son."

In the summer of 1977, I moved to Albuquerque to begin a postdoctoral fellowship at the University of New Mexico School of Medicine. Ray and I completed the manuscript for the fourth edition a few months thereafter (drafts were shuttled back and forth by mail) and the new edition appeared about one year later (Watterson, R. L., G. C. Schoenwolf, and R. M. Sweeney [1979]. Laboratory Studies of Chick, Pig, and Frog Embryos, 4th ed, Minneapolis, MN: Burgess Publishing Company]). Major changes in this edition included a thorough revision of the text, as well as considerable condensation. These changes were made to reduce the time required for students to read the text and to comprehend the material. Most photomicrographs were relabeled in part, almost all text figures were modified,

and the total number of text figures was increased to 21. Finally, drawings were placed on plate legends to aid students in understanding the orientations and levels of sections depicted by the photomicrographs.

Three editions remain to be described here: the 5th (Watterson, R. L. and G. C. Schoenwolf [1984]. Laboratory Studies of Chick, Pig, and Frog Embryos. Guide and Atlas of Vertebrate Embryology, 5th ed. [Minneapolis, MN: Burgess Publishing Company[); later published by Macmillan Publishing Company, New York), sixth (Schoenwolf, G. C., and R. L. Watterson [1989]. Laboratory Studies of Chick, Pig, and Frog Embryos. Guide and Atlas of Vertebrate Embryology, 6th ed. [New York: Macmillan Publishing Company]), and seventh (Schoenwolf, G. C. [1995]. Laboratory Studies of Vertebrate and Invertebrate Embryos. Guide and Atlas of Descriptive and Experimental Development, 7th ed. [New Jersey: Prentice Publishing Company). In 1979 I moved to Salt Lake City to join the faculty at the University of Utah School of Medicine. Ray's failing health precluded much participation by him in the fifth edition. For this edition, I added 13 new plates of scanning electron micrographs, with accompanying descriptions. I felt the enhanced depth of field provided by scanning electron microscopy would aid students in comprehending changes in embryonic morphology in three spatial dimensions over time. Additional changes included two new supplementary text figures, a list of suggested audiovisual aids, and methods for preparing embryos for histological and ultrastructural studies. Ray was extremely supportive of the changes I made in the fifth edition, and I know from correspondence that he was very proud of the new edition. My joy in this was dampened with sadness; Ray died less than one year after its publication.

The sixth edition was the first edition I did without the benefit of Ray's input. A number of changes were made. These changes were made not with the previous self-assured cockiness I possessed as a graduate student, but rather with the nervousness of a child who has to step out on his or her own without the guidance of a caring and ever watchful parent. Changes in the sixth edition included the addition of 14 exercises on experimental embryology of the frog and chick, an expansion of the list of audiovisual aids, the inclusion of a new section comparing the early development of amphibian and avian embryos, and the addition of an abridged Hamburger and Hamilton stage series. Also, the number of text figures was increased to 24.

The seventh edition also included a number of changes. A new chapter was added on development of the sea urchin embryo, necessitating a change in the guide's title. Five exercises, 2 line drawing, and 9 plates of photographs were added, and the appendix was expanded (now part of Hands-On Studies). Lists of terms appearing in the text in boldfaced type were added after each major section to aid students in reviewing for tests, and the entire text was updated. Finally, this section on history was added.

You have before you the eighth edition of the guide. It is my hope that it will help you to become excited by embryonic development and the field of developmental biology. If so, we may one day be colleagues, working for the common goal of understanding the ways of the embryo. I wish you every success and joy in your studies.

Jean and Robert Sweeney (taken at the reception following my wedding in August of 1971).

Ray Watterson (taken in his office at the University of Illinois in about 1975).

Credits

With each new edition, there are many to acknowledge who have provided much help. In particular, I thank former and present members of my laboratory, especially Diana Darnell, Michael Stark, and Shipeng Yuan, who provided technical information and illustrations. I also thank Raj Ladher, a visiting postdoc from England, and Aaron Lawson, a visiting scholar from Ghana, both of whom provided technical information.

Special appreciation is due to three individuals who graciously provided original illustrations that allowed us to reproduce several of their outstanding micrographs: Dr. John B. Morrill (Photos 1.1-1.42; Dr. Morrill also played an instrumental role in shaping the text and exercises of Chapter 1), Dr. Patrick Tam (Photos 4.14-4.23, 4.68, and 4.69), and Dr. Robert E. Waterman (Photos 2.19-2.27, 2.31, 2.32, 3.76, 3.106, 4.24, 4.38-4.4.41, 4.47-4.51, and 4.65-4.67).

The following illustrations have been modified from figures that have appeared in research publications. **Chapter 1: Fig. 1.1** (Hörstadius, S. [1939]. The mechanics of sea urchin development, studied by operative methods. *Biological Reviews* **14**:132-179), **Fig. 1.2** (Schoenwolf, G. C., and I. S. Alvarez [1992]. Role of cell rearrangement in axial morphogenesis. In Pedersen, R. A., Ed., *Current Topics in Developmental Biology*, Vol. 27, Orlando, FL: Academic Press, Inc., pp. 129-173), **Photos 1.30, 1.34-1.38** (Morrill, J. B., and L. L. Santos [1985]. A scanning electron microscopical overview of cellular and extracellular patterns during blastulation and gastrulation in the sea urchin, *Lytechnius variegatus*. In Sawyer, R. H. and R. M. Showman, Eds., *The Cellular and Molecular Biology of Invertebrate Development*, Columbia, SC: University of South Carolina Press, pp. 3-33), and **Photos 1.43, 1.44** (Amemiya, S., K. Akasaka, and H. Terayama [1982]. Scanning electron microscopical observations on the early morphogenetic processes in developing sea urchin embryos. *Cell Differentiation* **11**:291-293). **Chapter 3: Photos 3.17, 3.18, 3.65** (Schoenwolf, G. C. [1979]. Observations on closure of the neuropores in the chick embryo. *American Journal of Anatomy* **155**:445-466), **Photo 3.30** (Schoenwolf, G. C. [1978]. An SEM study of posterior spinal cord development in the chick embryo. *Scanning Electron Microscopy* **1978/II**:739-745), **Photos 3.43, 3.44, 3.72, 3.73** (Smith, J. L., and G. C. Schoenwolf [1997]. Neurulation: coming to closure. *Trends in Neuroscience* **20**:510-517), **Photos 3.47, 3.107-3.109** (Schoenwolf, G. C. [1983]. The chick epiblast: a model for examining epithelial morphogenesis. *Scanning Electron Microscopy* **1983/III**:1371-1385), **Photo 3.71** (Schoenwolf, G. C., and J. L. Smith [1990]. Mechanisms of neurulation: traditional viewpoint and recent advances. *Development* **109**:243-270), and **Photo 3.74** (Schoenwolf, G. C. [1991]. Neurepithelial cell behavior during avian neurulation. In Gerhart, J.. Ed., *Cell-Cell Interactions in Early Development*, New York, NY: Wiley-Liss, pp. 63-78). **Chapter 4: Photos 4.14-4.18** (Tam, P. P. L., E. A. Williams, and W. Y. Chan [1993]. Gastrulation in the mouse embryo: ultrastructural and molecular aspects of germ layer morphogenesis. *Microscopy Research and Technique* **26**:301-328), **Photos 4.19-4.22, 4.68, 4.69** (Jacobson, A. G., and P. P. L. Tam [1982]. Cephalic neurulation in the mouse embryo analyzed by SEM and morphometry. *The Anatomical Record* **203**:375-396), and **Photos 4.24, 4.41, 4.49** (Waterman, R. E. [1976]. Topographical changes along the neural fold associated with neurulation in the hamster and mouse. *American Journal of Anatomy* **146**:151-172). **Chapter 6: Photo 6.17** (Schoenwolf, G. C., N. B. Chandler, and J. Smith [1985] Analysis of the origins and early fates of neural crest cells in caudal regions of avian embryos. *Developmental Biology* **110**:467-479), and **Photos 6.19-6.21** (Sausedo, R. A., and G. C. Schoenwolf [1993] Cell behaviors underlying notochord formation and extension in avian embryos: Quantitative and immunocytochemical studies. *Anatomical Record* **327**:58-70).

Finally, we acknowledge the following scientists who kindly provided constructs for the *in situ* hybridizations illustrated in Photos 6.27-6.38 Michael Kessel (*cNot*; Photo 6.27), Ray Runyan (*brachyury;* Photo 6.28), Cliff Tabin (*Sonic hedgehog*; Photos 6.29, 6.35), Laure Bally-Cuif (*Otx-2*; Photo 6.30), Katherine Yutzey (*vMHC*; Photos 6.31-6.33), Eric Olson (*paraxis*; Photo 6.34), Peter Gruss (*Pax-2*; Photo6.36), Randy Johnson (*Lmx-1*; Photo 6.37), and Angela Nieto (*Krox-20*; Photo 6.38).

Use the space below for notes.

Glossary

A

abducens (VII) cranial nerves

Somatic motor nerves whose axons grow out on each side from the floor of the rhombencephalon and innervate the lateral rectus eye muscles.

aboral surface of sea urchin

The side opposite the mouth.

acoustic ganglia

The most dorsal ganglia of the auditory (VIII) cranial nerves; they arise from the acousticofacialis ganglia.

acousticofacialis ganglia

The source of sensory neurons for the future facial (VII) and auditory (VIII) cranial nerves; they give rise to future acoustic and geniculate ganglia.

acrosomal enzymes

The enzymes contained in the acrosomal granule of the sperm.

acrosomal filament

A filamentous rod covered with acrosomal membrane that extends from the head of the sperm.

acrosomal granule

A single, membrane-bound granule in the acrosome of the sperm.

acrosomal process

A structure consisting of the acrosomal filament and the acrosomal membrane.

acrosomal reaction

A reaction occurring when the sperm enters the jelly layer surrounding the egg; two events are involved: the acrosomal membrane fuses with sperm plasmalemma and exocytosis occurs; and actin undergoes rapid polymerization to form the acrosomal filament.

acrosome

A structure contained in the head of the sperm; it contains the acrosomal granule.

actin

A cytoskeletal protein located beneath the acrosomal granule that undergoes rapid polymerization to form the acrosomal filament.

adenohypophysis

See *pituitary gland, anterior.*

adhesive glands

Also called *ventral suckers*; in 4-mm frog embryos, they are prominent ectodermal thickenings that are located to either side of the stomodeum.

alar plates

The paired regions of the intermediate zone in the dorsal half of the spinal cord.

albumen

Also known as *egg white*; a mixture of proteins, most of which protect the embryo from infection; it is added to the chick ovum in the magnum of the oviduct.

allantoic arteries

See *umbilical arteries.*

allantoic veins

See *umbilical veins.*

allantois

One of the extraembryonic membranes that originates from the splanchnopleure; it is formed by expansion of the endodermal allantois rudiment and surrounding splanchnic mesoderm; in 72-hour chick embryos, it is a saclike structure somewhat encircled by the tail; in mouse embryos, it extends from the caudal end of the embryo into the extraembryonic part of the egg cylinder; in 10-mm pig embryos, it is an endoderm-lined cavity that emerges from the umbilical cord to join the urogenital sinus ventrally.

allantois rudiment

An endoderm-lined cavity that later gives rise to the allantois.

amnion

An extraembryonic membrane formed in chick embryos by the elevation and fusion of the amniotic folds; it all vertebrates containing an amnion, it originates from somatopleure (somatic mesoderm and ectoderm); contraction of the amnion causes the early movement of the embryo.

amniotic cavity

An ectoderm-lined cavity that encloses the embryo.

amniotic fluid

A protective fluid that fills the amniotic cavity and enables the embryo to float freely.

amniotic folds, caudal or posterior; cranial or anterior; lateral

In chick embryos, folds that arch over the body of the embryo and fuse to form two extraembryonic membranes: the chorion and the amnion.

amniotic folds, boundary of the

In 48-hour chick embryos, the distinct curvature between the covered and uncovered parts of the cranial half of the embryo; in 72-hour chick embryos, an oval or circular boundary over the caudal trunk region.

anal arms

Two projections of the body in the pluteus larval stage of the sea urchin embryo.

animal hemisphere

The half of the egg that contains the animal pole; in frog embryos, it contains a heavily pigmented cortex; it consists of a layer four or five cells thick; it is composed of blastomeres that contain very little yolk.

animal plate

A structure formed during secondary invagination of the sea urchin gastrula, when the ectoderm at the animal pole thickens.

animal pole

The uppermost part of the pigmented region of the outer portion of the egg (cortex); it corresponds to the cranial end of the future embryo; in the amphibian egg, the blastocoel is displaced toward the animal pole; the first meiotic division occurs in the animal pole of the primary oocyte.

animal-vegetal axis

The axis defined by the animal and vegetal poles.

anus

The structure formed when the cloacal membrane ruptures.

aorta, arch of the

The blood vessel that carries blood to all regions of the body caudal to the heart, other than the lungs.

aorta, ascending

See *aortic trunk*.

aorta, descending

The blood vessel formed as a result of fusion of the paired dorsal aortae.

aortae, dorsal

A pair of blood vessels that lie dorsolateral to the pharynx.

aortae, ventral

A pair of blood vessels that lie ventral to the pharynx.

aortic arches, first-sixth

Paired blood vessels that interconnect the dorsal and ventral aortae and lie lateral to the pharynx within the correspondingly numbered branchial arches.

aortic sac

The blood vessel formed by the fusion of the paired ventral aortae.

aortic trunk

Also known as *ascending aorta*; the blood vessel at the right (apparent left in transverse sections) of the bulbar septum.

apical ectodermal ridges

The distal thickenings of the limb bud ectoderm; these ridges are necessary for normal outgrowth of the limb buds.

apical tuft

Long cilia present at the animal pole of the sea urchin embryo.

arch of the aorta

See *aorta, arch of the*.

archenteron

Also known as the *primitive gut*; in frog embryos, it is a narrow cavity located above the mass of endodermal cells forming its floor.

area opaca

The darker, peripheral region of the chick blastoderm; by the 48-hour stage, it is further divided into an inner area vasculosa and outer area vitellina.

area pellucida

The lighter, central region of the chick blastoderm.

area vasculosa

The inner, mottled region of the chick area opaca that contains developing blood islands.

area vitellina

The outer, nonmottled region of the chick area opaca that overgrows the yolk.

arterial circle of Willis

See *circle of Willis*.

arytenoid cartilages

The cartilages fromed in the larynx by the arytenoid swellings flanking the glottis.

arytenoid swellings

The prominent swellings on either side of the glottis that later form the arytenoid cartilages.

atria

Broad, saccular regions of the heart that are continuous caudally with the sinus venosus region of the heart.

atrioventricular canals

The canals that lie on either side of the endocardial cushion of the heart; the cavities of the atria and ventricles are continuous via these canals.

auditory (VII) cranial nerves

The sensory nerves that are associated with the acoustic ganglia originating from the auditory placodes; they innervate the inner ears.

auditory placodes or pits

Also known as *otic placodes or pits*; the rudiments of the inner ears that are formed by the thickened skin ectoderm alongside part of the myelencephalon; their formation is induced by the myelencephalon and adjacent head mesenchyme; the acoustic part of the acousticofacialis ganglia originates from these.

auditory vesicles

Also called *otocysts*; the paired vesicles that originate by invagination of the auditory placodes; they later differentiate into the inner ears.

auricles

See *pinnae*.

autonomic nervous system

The part of the peripheral nervous system that supplies the viscera.

axons

The sensory nerve fibers that grow out from ganglia; they transmit nerve impulses away from their cell bodies.

axons, preganglionic

The nerve fibers that grow out from motor nuclei developing within the floor of myelencephalon.

azygos vein

The vein formed by the proximal portion of the right postcardinal vein.

B

basal lamina

In sea urchin embryos, the basal side of the ectoderm where filopodia grope and search for the target region.

basal plates

The paired regions of the intermediate zone in the ventral half of the spinal cord.

basilar artery

A single artery lying in the midline, beneath the rhombencephalon; it is one of three vessels that brings blood into the circle of Willis.

bicuspid valve

Also called *mitral valve*; a valve that develops within the left atrioventricular canal, the entrance to the left ventricle.

bilateral symmetry

The characteristic of having right and left sides.

blastocoel

An internal cavity that forms at the blastula stage; in frog gastrulae, it allows cells to involute over the blastopore lips, where they move into the interior to form the archenteron and mesoderm; it is later squeezed out of existence.

blastocyst

In mouse embryos, the stage following the morula in which the conceptus consists of a trophoblast, inner cells mass, and blastocoel.

blastoderm

The structure in chick embryos that is formed when the blastodisc initiates cleavage (that is, subdivision by mitosis into cells called blastomeres).

blastodisc

The structure in chick embryos consisting of a circular disc approximately 35mm in diameter and containing the germinal vesicle; the blastodisc is a cytoplasmic cap that forms at one pole of the egg, which later cleaves to form the multicellular blastoderm.

blastomeres

A group of cells formed through mitosis during cleavage; collectively, they form the blastula and surround the blastocoel.

blastomeres, an^1

In sea urchin embryos, the upper tier of eight cells (nearest the animal pole) formed after division of eight mesomeres in the animal hemisphere; they further divide to form two tiers of an$_1$ cells (lower and upper), for a total of 16 cells.

blastomeres, an^2

In sea urchin embryos, the lower tier of eight cells (second tier of animal hemisphere cells) formed after mesomere division in the animal hemisphere; they further divide to form two tiers of an$_2$ cells (lower and upper), for a total of 16 cells.

blastomeres, central

In chick embryos, the blastomeres that appear at the 32-cell stage to be completely bounded in surface view from the marginal blastomeres; they are separated from the yolk by the subgerminal cavity.

blastomeres, marginal

In chick embryos, the blastomeres that appear at the 32-cell stage to be incompletely separated from one another in surface view; marginal blastomeres are in contact with the yolk.

blastomeres, veg^1

In sea urchin embryos, the cells in the upper tier (those nearest the animal pole) formed after macromeres latitudinally divide during the sixth cleavage to form two tiers of cells, for a total of 16 cells; like an$_1$ and an$_2$ blastomeres, they contribute to the ectoderm of the embryo.

blastomeres, veg^2

In sea urchin embryos, the cells in the lower tier formed after macromeres latitudinally divide during the sixth cleavage to form two tiers of cells, for a total of 16 cells; they form the endoderm of the archenteron and become secondary mesenchyme cells, thus contributing to the endoderm and mesoderm of the embryo.

blastoporal lip, dorsal

A liplike structure that forms just above the blastopore during the gastrula stage; it constitutes the organizer of the frog embryo.

blastoporal lip, ventral

In frog embryos, the blastoporal lip that forms after the dorsal and lateral blastoporal lips form during gastrula stages.

blastoporal lips, lateral

In frog embryos, the blastoporal lips that form after the dorsal lip forms.

blastopore

In frog embryos, a depression that forms below the gray crescent as cells initiate involution; it represents the future caudal end; throughout gastrulation, it is occupied by the yolk plug; in sea urchin and frog embryos, it is the first opening; it forms the anus.

blastula

The stage of the developing embryo at which blastomeres become arranged around the blastocoel; the egg reaches this stage near the end of cleavage.

blastula, mesenchyme

In sea urchins, the stage during which primary mesenchyme cells are ingressing into the blastocoel; this occurs during the late blastula stage.

blastulation

The process during which the blastula and blastocoel form; it results form a series of rapid mitotic divisions.

blood cells, primitive

The cells derived from the central cells of each blood island.

blood islands

The structures contained within the area vasculosa and that are responsible for its mottled appearance; they consist of irregularly shaped accumulations of mesodermal cells; they are most numerous lateral and caudal to the body of the embryo, and they fuse together to form blood vessels.

body axes

The axes of the body, consisting of the rostrocaudal, dorsoventral, mediolateral, and right-left axes.

body folding

The process whereby the somatopleure and splanchnopleure undergo folding to transform the essentially two-dimensional blastoderm into the three-dimensional body plan.

body folds

See *body folds, lateral; head fold of the body; tail fold of the body.*

body folds, lateral

Paired folds, each of which consists of two components: somatopleure and splanchnopleure; the lateral body folds are direct continuations of the head fold of the body; they delimit the body from the extraembryonic regions laterally.

body of embryo

In the 33-hour chick embryo, a dark region that runs lengthwise within the area pellucida; the darker and wider part of the body is the head (cranial) end of the embryo, and the lighter part is the tail (caudal) end.

body plan

The structure of the body of the embryo; the tube-within-a-tube body plan is characteristic of vertebrate embryos; it consists of an outer ectodermal tube, forming the skin, and an inner endodermal tube, forming the gut.

bottle cells

Also known as *flask cells;* cells that become bottle shaped by narrowing one end and widening the other end.

Bowman's capsules

See *glomerular capsules.*

brachial plexus

A nerve plexus formed as a result of the complex interconnection of the ventral rami of adjacent spinal nerves; it forms at the level of the foreleg/wing buds.

brain

See also specific divisions of; the cranial end of the neural tube.

branchial arches, first

See also mandibular processes; maxillary processes; the masses of tissue cranial to the first branchial groove; each is partially split into two processes by the stomodeum, called the maxillary and mandibular processes; they are innervated by the trigeminal (V) cranial nerves.

branchial arches, second

The masses of tissue located between the first and second branchial grooves; the acousticofacialis ganglia lie above them; they are innervated by the facial (VII) cranial nerves.

branchial arches, third

The masses of tissue located between the second and third branchial grooves; they are innervated by glossopharyngeal (IX) cranial nerves.

branchial arches, fourth

The masses of tissue located between the third and fourth branchial groves; they are innervated by the vagus (X) cranial nerves.

branchial clefts, first and second

The openings to the outside of the first and second pharyngeal pouches that arise from the rupture of the first and second closing plates, respectively.

branchial grooves, first

Grooves at the same level as the first pharyngeal pouches; the first branchial groove on each side forms the adult external auditory meatus.

branchial grooves, second

Grooves at the same level as the second pharyngeal pouches; they are located between the second and third branchial arches.

bulbar ridges

The two ingrowths of the wall of the conotruncus (bulbus cordis); they fuse to form the bulbar septum.

bulbar septum

The fused bulbar ridges; the bulbar septum separates the conotruncus (bulbus cordis) into two vessels.

bulbus cordis

See *conotruncus*.

C

C cells

Cells that arise from the fifth pharyngeal pouches (ultimobranchial bodies) and become incorporated into the thyroid gland as the parafollicular cells; they produce the hormone calcitonin.

calcitonin

The hormone produced by c (parafollicular) cells in the thyroid gland.

calcium ions

The ions that enter the egg during fertilization; they are required for the cortical reaction.

calyces, major

The first major buds of each ureter.

calyces, minor

The structures formed after the major calyces, when the epithelial walls of the ureters bud repeatedly.

cardiac jelly

The extracellular material that widely separates and fills the space between the endocardium and myocardium of the heart.

cardiac primordia

In 24-hour chick embryos, the thickenings of the splanchnic mesoderm on either side of the midline; at a later stage, cells emerge medially from this thickening on each side and organize into an endocardial tube; the heart tube develops from paired cardiac primordia, which are brought into apposition in the ventral midline by the action of the lateral body folds.

cardinal veins, caudal or posterior

See *postcardinal veins*.

cardinal veins, common

The blood vessels continuous with the sinus venosus; each is contained within a bridge of mesoderm connecting the lateral body wall to the sinus venosus.

cardinal veins, caudal or posterior

See *postcardinal veins*.

cardinal veins, cranial or anterior

See *precardinal veins*.

cardiogenesis

The process during which the heart is formed.

carotid arteries, common

Arteries formed from the third aortic arches.

carotid arteries, external

A pair of small arteries that form as ventral extensions from the third aortic arches.

carotid arteries, internal

The arteries formed in part from the third aortic arches; the remainder of each artery is formed from an extension of each dorsal aorta, cranial to each connection between the third aortic arch and dorsal aorta.

caudal arteries

In 72-hour chick embryos, the arteries at the level where the dorsal aortae become markedly reduced in diameter within the tail region (caudal to the leg buds).

caudal end of embryo

See *tail*.

cavitation

The process through which the solid morula forms a cavity (the blastocoel) and becomes the blastocyst.

celiac artery

A single midline artery continuous with the descending aorta ventrally; it forms at the level of the stomach.

cell adhesion, changes in

One of at least three factors involved in invagination during gastrulation.

cell bodies

Clusters of cells that form the motor nucleus within the intermediate zone of the neural tube.

cell marker

A substance used to label single cells or groups of cells; a prospective fate map can be constructed by using cell markers.

cell rearrangement

In sea urchin embryos, a rapid event that occurs during secondary invagination in which cells change position.

cell shape, changes in

One of at least three factors involved in invagination during gastrulation.

central nervous system

The adult derivative of the neural tube; it consists of the brain and spinal cord.

centrioles of the sperm

The structures in the midpiece of the sperm that after fertilization organize the mitotic spindles within the zygote for cleavage.

centrum of vertebra

A region of each vertebra that is formed by a mass of sclerotome cells encasing the notochord.

cerebellum

An adult brain component formed by the metencephalon of the neural tube.

cerebral aqueduct

The cavity of the mesencephalon; it is continuous with the fourth ventricle.

cerebral arteries, anterior

The arteries continuous with the internal carotid arteries; they can be identified in sections, where they lie lateral to the diencephalon.

cerebral hemispheres

The lateral oval-shaped expansions of the telencephalon; they contain cavities called the lateral ventricles, which are broadly continuous with the cavity of the middle of the telencephalon.

cerebral peduncles

An adult brain component formed by the mesencephalon of the neural tube.

cerebrospinal fluid

The fluid secreted by the choroid plexus; it fills the cavities of the brain and spinal cord.

cerebrum

An adult brain component formed by the telencephalon of the neural tube.

cervical sinuses

The small and deep sinuses confined to the third and fourth branchial grooves and caudal to the third branchial arches.

chain ganglia

the ganglia formed from the accumulations of neural crest cells near the visceral ramus.

cheeks

The portions of the face formed by the maxillary processes,, along with the lateral portions of the upper jaw.

chemoreceptor cells

In sea urchin embryos, the specialized cells that respond to a stimulant and trigger the release of gametes prior to fertilization.

chimeras

Embryos formed by using a technique in which quail cells are transplanted into chick embryos, resulting in an embryo containing a mixture of cells derived from two species; the transplanted quail cells and their descendants can be distinguished from chick cells using special techniques.

chorioallantoic membrane

In chick embryos, an extraembryonic membrane that is formed when the fully-developed allantois fuses with the chorion; fusion occurs wherever the splanchnic mesoderm of the allantois contacts the somatic mesoderm of the chorion.

chorion

In chick embryos, an outer extraembryonic membrane formed simultaneously with the amnion by the elevation and fusion of amniotic folds; in chick, mouse, and pig embryos, the chorion is an extraembryonic membrane formed from somatopleure.

chorionic cavity

In mouse embryos, the cavity contained within the chorion of the gastrulating/neurulating embryo, within the extraembryonic part of the egg cylinder; in chick and pig embryos, the chorionic cavity is called the extraembryonic coelom.

chorionic gonadotropin

A hormone used to induce ovulation.

choroid

A portion of the eye; it is formed from the head mesenchyme (both neural crest and mesoderm) surrounding each pigmented retina.

choroid plexus

A structure formed by the invagination of the thin roof plate of the myelencephalon and the pia mater into the cavity of the rhombencephalon (the fourth ventricle); it is the source of cerebrospinal fluid.

cilia

In sea urchin embryos, the structures that develop on the outer surface of the blastula; they are particularly long at the animal plate where they are called the apical tuft; their contraction is responsible for the motility of the embryo and early larva.

ciliary bands

In sea urchin embryos, groups of cilia that develop on the arms and along the circumference of the body proper in the pluteus larva; these bands beat synchronously to propel plankton toward the mouth opening, and to propel the larva through the water.

ciliary tufts

In frog embryos, long groups of cilia that cover the skin ectoderm and establish a current around embryos as they beat, circulating fluids; they also function in primitive locomotory movements before swimming begins.

circle of Willis

A complex of blood vessels that delivers blood to the ventral surface of the brain; the anterior cerebral and posterior communicating arteries collectively constitute the arterial circle of Willis.

cleavage

A series of rapid mitotic divisions that results in blastulation; cleavage is initiated following fertilization, and it results in the formation of blastomeres.

cleavage, discoidal

In chick embryos, a type of cleavage that is restricted to the circular disc of the cytoplasm.

cleavage furrows

Spaces that separate the blastomeres during cleavage; in holoblastic cleavage, they pass through the entire egg, whereas in meroblastic cleavage, they do not extend all the way to the periphery of the cytoplasm; cleavage furrows pass through the vegetal hemisphere much more slowly than through the animal hemisphere.

cleavage, partial or meroblastic

In chick embryos, a type of cleavage that is restucted to the circular disc of the cytoplasm; with such cleavage, the cleavage furrows neither extend all the way to the periphery of the cytoplasm nor cut entirely through its thickness.

cleavage, total or holoblastic

A type of cleavage in which the cleavage furrows pass through the entire egg; for example in sea urchin embryos, the entire egg undergoes cleavage, not just the animal pole, as in some organisms.

cloaca

In 72-hour chick embryos, a portion of the hindgut continuous ventrally with the allantois; in 10-mm pig embryos, it is a large cavity that is subdividing into a dorsal rectum and ventral urogenital sinus.

cloacal membrane

A membrane formed by the fusion of the endoderm of the hindgut and the ectoderm of the proctodeum; it ultimately ruptures to form the anus.

cloacal septum

A mesodermal ingrowth that separates the cloaca into two chambers: the dorsal (lower) rectum and ventral (upper) urogenital sinus.

cloacal valves

The structures located near the cloaca that are used to sex female frogs.

closing plates, first-fourth

The double-layered membranes formed by the endoderm of the pharyngeal pouches and the ectoderm of the branchial grooves; when they rupture, they give rise to the branchial clefts; they are numbered with the correspondingly numbered pharyngeal pouches and branchial grooves; the tympanic membranes are formed from the first closing plates.

clutch

A group of ovulated ova; each clutch consists of a series of ova, with each being ovulated daily over a period of two or more days; ovulation will then cease for at least a day, after which it will resume in a new clutch of ova.

cochleas

The ventral portions of the auditory vesicles, which form the receptor organs for sound (the spiral organs of Corti).

cochlear ganglia

See spiral ganglia.

coelom

See also the specific divisions of; a space that forms within the lateral plate mesoderm, between the somatic and splanchnic mesodermal layers.

coelom, extraembryonic

Prior to body folding, it is the space contained within the extraembryonic lateral plate mesoderm, and it is continuous medially with the intraembryonic coelom; it becomes separated from the intraembryonic coelom due to the action of the somatopleural component of the lateral body folds; with the initiation of formation of the amnion, it extends into the somatic mesoderm-lined cavity within each amniotic fold; after fusion of the amniotic folds, it is the space contained within the chorion, into which the allantois expands to form the chorioallantoic membrane; in 10-mm pig embryos, the intestinal loop herniates into the extraembryonic coelom of the umbilical cord, forming the temporary umbilical hernia.

coelom, intraembryonic

A space within the embryo that forms the adult pericardial, peritoneal, and pleural cavities.

coelomic peritoneum

In sea urchin embryos, the layer covering the five gonads.

coeloms of sea urchin embryos

The two spaces formed from secondary mesenchyme cells associated with the esophagus region of the archenteron of sea urchin embryos.

collateral ganglia

The sympathetic ganglia of the autonomic nervous system formed from accumulations of neural crest cells.

collecting tubules

The portions of the metanephric kidneys, along with the major and minor calyces, that are formed when the epithelial walls of the renal pelvis bud repeatedly.

colliculi, inferior and superior

Components of the corpora quadrigemina formed by the mesencephalon region of the neural tube.

colon (ascending, transverse)

Regions of the gut formed mostly from the caudal limb of the intestinal loop.

commissures

The transverse bundles of nerve fibers that form within the lamina terminalis to interconnect the cerebral hemispheres.

common bile duct

The duct that joins the duodenum and is continuous with the hepatic and cystic ducts; the ventral pancreatic rudiment buds off the common bile duct.

communicating arteries, posterior

Blood vessels, along with the anterior cerebral arteries, that collectively constitute the arterial circle of Willis.

compaction

The process that occurs during the morula stage as it begins its transformation into the blastocyst; during compaction, blastomeres become tighly apposed to one another.

conotruncus (bulbus cordis)

The rostral (outflow) end of the heart, which connects to the aortic sac.

copula

One of two midline tongue rudiments; it is derived from the second branchial arches.

cornea

A portion of the eye; the cornea is derived from both skin ectoderm and head mesenchyme (neural crest cells).

corneal epithelium

A portion of the cornea of the eye; the transparent epithelium formed by the induction of the ectoderm by the lens.

coronary sinus

A venous sinus formed by the left common cardinal vein as it enters the right atrium.

corpora quadrigemina

Brain structures formed by the mesencephalon level of the neural tube; the inferior and superior colliculi collectively constitute the corpora quadrigemina.

corpora striata

Brain structures formed by the telencephalon level of the neural tube.

cortex of egg

The outer portion of the egg; in sea urchin embryos, it contains thousands of cortical granules; in frog embryos, it is heavily pigmented in the animal pole.

cortical granules

The vesicles just beneath the egg plasmalemma; each vesicle is 1/2 to 1 micrometer in diameter; exocytosis of these granules is triggered by adding calcium.

cortical reaction

The reaction initiated by the fusion of the sperm with the egg; it requires the entrance of calcium ions into the egg; it changes the egg's surface properties, providing a block to polyspermy.

cranial end of embryo

See *head*.

cranial ganglia

See names of specific ones.

cranial nerves

See names of specific cranial nerves.

cranial nerves, roots of

See names of specific cranial nerves.

cystic duct

The duct connecting to the gallbladder and the common bile duct

cytoplasm of egg

The yolk is evenly distributed in the cytoplasm of the homolecithal sea urchin egg.

D

delaminate

The process whereby cells separate from a layer and move inward; for example, cells delaminate from the chick epiblast to contribute to the hypoblast.

dendrites

Portions of nerve cells (neurons) that generally conduct nerve impulses toward the cell body.

depolarization

A rapid process to block polyspermy in which the egg's plasmalemma is depolarized (that is, the electrical potential across the plasmalemma is changed to a less negative state).

dermatomes

A subdivision of the somites; dermatomes consist of plates of darkly stained cells lying just beneath the skin ectoderm; they are the source of the dermis in the trunk and tail.

dermis

A layer of the skin formed in part from the dermatomes of the somites.

dermomyotomes

An early subdivision of the somites (for example, in mouse embryos), prior to the formation of distinct dermatomes and myotomes; they form both the dermatomes and myotomes of the somites.

deuterostomes

The organisms in which the second opening into the embryo forms the mouth, and the first opening forms the anus; sea urchins are deuterostomes.

diaphragm

A structure that develops from the septum transversum; it is a muscular partition that assists in breathing.

diencephalon

A level of the neural tube formed by the prosencephalon; it is the region of the brain from which the optic vesicles evaginate.

differential growth

Localized growth in a portion of an embryo that causes morphogenesis to occur.

diffuse-type placenta

An epitheliochorialis type of placenta with a placental membrane (barrier) composed of many layers; the placenta of pig embryos.

DiI

A fluorescent dye used to label groups of cells; using a fluorescence microscope and the appropriate wavelength of light, the movement of cells can be observed over time; this technique can be used to construct prospective fate maps.

dioecious

A type of organism in which males and females are separate individuals.

DNA

The genetic material; it replicates early in the primary oocyte stage; in quail/chick chimeras, it can be stained using the Feulgen procedure.

dorsal lymph sac

The area where male and female frogs can be injected with chorionic gonadotropin to induce ovulation and the release of sperm.

dorsal surface of sea urchin

The side opposite the ventral (or oral) surface, or the side opposite the mouth, in the prism larva stage.

duct of Santorini

See *pancreatic duct, dorsal.*

duct of Wirsung

See *pancreatic duct, ventral.*

ductus arteriosus

The distal end of the left sixth aortic arch (the portion that connects to the left dorsal aorta); it persists until birth as an important blood channel acting as a shunt; it is converted after birth into a fibrous ligament called the ligamentum arteriosum.

ductus caroticus

The portion of each dorsal aorta lying between its connection to the 4th and 3rd aortic arches; it eventually degenerates.

ductus venosus

A single vessel continuous with the sinus venosus; in 72-hour chick embryos, it was formed by the fusion of the vitelline veins; after formation and growth of the liver, it consists of enlarged hepatic sinusoids; it carries blood from the left umbilical vein, through the liver, and into the inferior vena cava.

duodenohepatic ligament

See *hepatoduodenal ligament.*

duodenum

The level of the foregut that gives rise to the liver rudiments; in 10-mm pig embryos, the common bile duct connects to the duodenum.

E

ear cavities, middle

See *tympanic cavities.*

ear ossicles

Three bones in the middle ear; they consist of the malleus and incus, which are derivatives of the first branchial arches, and the stapes, which is derived from the second branchial arches; they are eventually surrounded by the tympanic cavity.

ears, external

The portions of the ears formed by the fusion of elevations from the first and second branchial arches on either side of the first branchial grooves.

ears, inner

The portions of the ear formed from the auditory (otic) placodes, which invaginate to form auditory pits and vesicles.

ectoderm

One of three germ layers; in frog and chick embryos, it is the outermost germ layer; in mouse embryos, it is the innermost germ layer, lining the amniotic cavity of the egg cylinder.

ectoderm, neural or ectodermal cells, neural

The ectoderm composing the neural plate, neural groove, and neural tube.

ectoderm, prospective

The layer of the embryo that forms the ectoderm after gastrulation.

ectoderm, skin or surface

The non-neural portion of the ectoderm; it contributes to the skin.

ectodermal layer, inner or deep

In frog embryos, one of two layers that constitute the ectoderm; it is far less pigmented than the outer ectodermal layer.

ectodermal layer, outer or superficial

In frog embryos, one of two layers that constitute the ectoderm; it is heavily pigmented.

ectoplacental cone

In mouse embryos, a portion of the extraembryonic part of the egg cyclinder that contributes to the placenta.

efferent ductules

Also known as *vasa efferentia;* a component of the male reproductive system, they are derived from mesonephric tubules that failed to degenerate.

egg-binding protein

A species-specific protein on the sperm's acrosomal membrane that mediates the adhesion of the egg and sperm during initial contact at fertilization.

egg capsule

In frog embryos, a multilayered, gelatinous capsule secreted outside the vitelline membrane by the cells lining the oviduct.

egg cylinder

The name for the elongated mouse conceptus during stages of gastrulation and neurulation.

egg white

See *albumen.*

eggs

See *ova.*

eggs, homolecithal

Eggs containing little yolk, and that which is present is evenly distributed throughout the cytoplasm; the sea urchin egg is a homolecithal egg.

ejaculatory ducts

A portion of the male reproductive system formed by the mesonephric ducts.

end bud

See *tail bud.*

endocardial cushion

A midline mass of lightly stained tissue in the heart; it separates the right and left atrioventricular canals.

endocardial tubes

The two primitive heart tubes; they are organized when cells emerge medially from each cardiac primordium; these tubes fuse ventral to the foregut in craniocaudal sequence, due to the action of the splanchnopleural component of the lateral body folds.

endocardium

The inner layer of the heart; it is derived from the splanchnic mesoderm.

endocrine gland

A ductless gland that drains its secretions only into the blood stream.

endoderm

One of three germ layers; in frog and chick embryos, it is the innermost germ layer; in mouse embryos, it is the outermost germ layer, which forms the external surface of the embryonic part of the egg cylinder.

endoderm, extraembryonic

The endoderm that contributes to the extraembryonic membranes (namely, the allantois and yolk sac).

endoderm, prospective

The layer of the embryo that forms the endoderm during gastrulation.

endolymphatic ducts

The ducts formed by the evagination of the dorsal part of the auditory vesicles.

endothelium (endothelial cells)

The inner lining of the blood vessels; in 33-hour chick embryos, it forms from the peripheral cells of each blood island.

enteric ganglia

See *terminal ganglia.*

eparterial bronchus

In 10-mm pig embryos, a distinct asymmetrical evagination of the trachea toward the right side (apparent left side in transverse sections); it forms the upper lobe of the right lung.

epiblast

In chick embryos, the surface layer of the area pellucida; in mouse embryos, the layer of the blastoderm that lines the amniotic cavity during the gastrula stage.

epiboly

The spreading of a sheet of cells that occurs as cells involute, ingress, and invaginate during gastrulation.

epibranchial placodes

The thickenings of the ectodermal layer that contribute to the semilunar ganglia of the trigeminal (V) cranial nerves, the geniculate ganglia of the facial (VII) cranial nerves, the petrosal ganglia of the glossopharyngeal (XI) nerves, and the nodose ganglia of the vagus (X) cranial nerves.

epicardium (visceral epicardium)

The outer layer of the wall of the heart; it grows downward as a cellular sheet from the region of the sinus venosus, near the dorsal mesocardium, to cover the outer surface of the entire myocardium.

epidermis

The outermost layer of the embryo and the outmost layer of the skin; it is derived from ectoderm.

epididymis

A portion of the male reproductive system; it is formed from the mesonephric duct.

epiglottis

A midline swelling in the floor of the pharynx just cranial to the glottis; it is derived from the third and fourth branchial arches.

epiphysis

See *pineal gland.*

epiploic foramen

The opening of the omental bursa into the peritoneal cavity.

epithalamus

An adult brain component derived from the diencephalon of the neural tube.

epithelial-mesenchymal interaction or induction

An interaction or induction occurring between the epithelial and mesenchymal components of a developing rudiment; it has been demonstrated experimentally that neither component can develop in the absence of the other.

epithelial to mesenchymal transformations

The conversion of an epithelium to a mesenchyme.

epitheliochorialis-type placenta

See *diffuse-type placenta.*

esophagus

A portion of the foregut; it is contained within a thick mesentery (the mesoesophagus) composed of splanchnic mesoderm.

estrus cycle

The reproductive cycle of female mice.

eustachian tubes

Also known as *pharyngotympanic tubes;* these tubes are formed by the portions of the first pharyngeal pouches that are connected to the pharynx.

external ears

See *pinnae.*

exogastrulation

A process of abnormal gastrulation; in sea urchin embryos, it causes the archenteron to evaginate rather than invaginate; in frog embryos, surface cells move, but fail to involute over the blastoporal lips—thus normal morphogenetic movements are inhibited and the formation of the archenteron is prevented.

external auditory meati

Also known as *external ear canals*; they are formed from the first branchial grooves.

external ear canals

See *external auditory meati.*

external genitalia

The external urogenital structures formed from the genital eminence.

eye muscle rudiments

Dense masses of mesodermal cells lying lateral to Rathke's pouch; they form the extrinsic eye muscles.

eye muscles

The muscles that move the eyes; they consist of the superior and inferior oblique eye muscles, and the inferior, superior, medial (or internal), and lateral (or external) rectus eye muscles.

eye muscles, inferior oblique; inferior rectus; lateral or external rectus; medial or internal rectus; superior oblique; or superior rectus

See *eye muscles.*

eyes

The organs of vision; they are formed by evagination of the diencephalon of the neural tube.

F

facial (VII) cranial nerves

Mixed (sensory and motor) cranial nerves associated with the geniculate cranial ganglia; they innervate derivatives of the second branchial arches.

falciform ligament

A mesentery connecting the liver to the ventral body wall.

fertilization

The joining of egg and sperm to form the zygote and the subsequent fusion of their pronuclei; fertilization activates development.

fertilization cone

A cone-like projection of the egg plasmalemma produced at the point of fusion with the sperm; it is involved in engulfing the sperm (nucleus, mitochondria, centrioles, and possibly tail microtubules) into the interior of the egg.

fertilization envelope (or membrane)

A layer formed when some of the expelled substances of the cortical granules harden and fuse with the elevating vitelline envelope.

Feulgen procedure

A histological procedure used to identify transplanted quail cells and their descendants in quail/chick chimeras; the Feulgen procedure stains DNA.

fiber tracts

Nerve processes in the marginal zone of the neural tube.

filopodia

Thin cellular processes that form at the broad inner ends of the bottle cells; in sea urchin embryos, secondary mesenchyme cells use filopodia to feel the basal side of the ectoderm to search for the target region, where they bind and pull the tip of the archenteron relative to it; in chick and mouse embryos, ingressing cells use their filopodia to migrate away from the primitive streak.

fimbria

The fingerlike processes along the margin of the infundibulum of the oviduct.

fin, dorsal

In 4-mm frog embryos, a dorsal flap of epidermis located above the tail spinal cord and containing a core of neural crest cells.

fin, ventral

In 4-mm frog embryos, a ventral flap of epidermis located below the tail and containing a core of neural crest cells.

flask cells

See *bottle cells.*

flexure, cervical

A bend in the embryo located at the level of the myelencephalon; due to this flexure, the most cranial regions of embryos in serial transverse sections are not seen in the most anterior sections.

flexure, cranial or cephalic

A sharp bend in the embryo located at the level of the mesencephalon; it is formed due to the prosencephalon bending toward the rhombencephalon.

flexure, tail

A bend in the embryo located at the caudal end of the embryo; in 72-hour chick embryos, the spinal cord and the notochord are cut frontally in serial transverse sections due to this flexure.

floor plate

The thin ventral, midline wall of the spinal cord.

fluorescein-labeled dextran

A fluorescent marker injected into a single cell or group of cells at the blastula stage of frog embryos, which can be demonstrated with a fluorescence microscope and the proper wavelength of light; this technique is used to construct prospective fate maps.

follicle cells

A single layer of small, flattened cells that encloses each primary oocyte in the ovary, thereby forming an ovarian follicle; follicle cells are derived from the surface layer of the ovary.

follicle stimulating hormone (FSH)

A hormone from the anterior pituitary gland that stimulates yolk accumulation by the primary oocytes within the ovary.

follicles

A cluster of spherical structures contained in each ovary; they are formed when primary oocytes become surrounded by follicle cells.

follicles, collapsed

The follicles in the ovary from which ova have escaped though the ruptured stigmas.

follicles, growing

The larger rapidly growing follicles in the ovary.

follicles, primary

The smallest follicles in the ovary.

foramen ovale

An opening that persists between the septum secundum and endocardial cushion; it is staggered with respect to the opening in the septum primum.

foramen primum (I)

The first opening in the septum primum.

foramen secundum (II)

A second opening in the septum primum that forms as the foramen primum is closing.

forebrain

The most rostral level of the incipient neural tube; the level of the early neural tube that forms the diencephalon and telencephalon.

foregut

The rudiment of the cranial end of the the gastrointestinal system; it also gives rise to the respiratory system; the foregut opens caudally via the cranial intestinal portal.

foreleg buds

Paired outgrowths from the body wall; each consists of an outer layer of ectoderm, part of which is thickened as the apical ectodermal ridge, and a core of somatic mesoderm; at the level of the foreleg buds, the ventral rami of adjacent spinal nerves are interconnected to form the branchial plexus.

Froriep's ganglion

A small but prominent ganglion on each side of the neural tube just caudal to the roots of the spinal accessory nerve; it is found at the level of the myelencephalon, and it contributes nerve fibers to the hypoglossal nerve.

FSH

See *follicle stimulating hormone.*

G

gallbladder

An expansion of the cystic duct; it stores the bile, which is produced by the liver.

gametes

The egg and sperm; they are produced, respectively, within the ovaries and testes.

gametogenesis

Development of the egg and sperm within the gonads of the adult female and male, respectively; gametogenesis is specifically called oogenesis in the female, and spermatogenesis in the male.

ganglion cells

Young neurons within the sensory retinas of the eyes.

gastrohepatic ligament

See *hepatogastric rudiment.*

gastrula

The embryo during the stage of gastrulation.

gastrulation

The process of germ layer formation; typically, gastrulation involves the formation of the primitive gut.

gene targeting

See *homologous recombination.*

geniculate ganglia

The ganglia of the facial (VII) cranial nerves.

genital eminence

A ventral swelling at the end of the trunk; it lies between the hindleg buds and the tapering tail; it is the rudiment of the external genitalia.

germ cell crescent

In chick embryos, a crescent-shaped area at the craniolateral margin of the area pellucida; it contains the primordial germ cells of hypoblast origin (the displaced hypoblast becomes condensed to form this structure).

germ layers

Also known as the *primary germ layers*; the ectoderm, mesoderm, and endoderm.

germinal epithelium

A localized thickening of visceral peritoneum in each gonad rudiment.

germinal vesicle

The enlarged nucleus of the primary oocyte; one nucleolus is present in the germinal vesicle.

glioblasts

Also known as *spongioblasts*; primitive non-nervous cells (supporting cells) that remain within the wall of the brain.

glomerular capsules

Also known as *Bowman's capsules*; the expanded spaces bounded by a very flat epithelium and filled with vascular cells, located at the medial side of each mesonephric kidney.

glomeruli

The capillaries within the glomerular capsules.

glossopharyngeal (IX) cranial nerves

The cranial nerves associated with the superior and petrosal ganglia; they innervate the third branchial arches.

glottis

The opening of the larynx into the pharynx.

gonad rudiments

Thickenings on the medial side of each mesonephric kidney; they contain primordial germ cells, and each consists of a localized thickening of visceral peritoneum and a subjacent region of condensed mesenchyme.

gonadotropin

See *chorionic gonadotropin.*

gonads

The ovaries or testes of the adult females and males, respectively; the organs where gametes develop.

gonoducts

In sea urchins, a short duct in each gonad that opens into the aboral surface near the anus; during spawning, gametes flow through these ducts into the surrounding water.

gonopores

In sea urchins, the opening of the gonoducts where gametes are shed.

gray crescent

In frog embryos, the crescent-shaped area between the heavily pigmented cortex above and the non-pigmented cortex below; its broadest region will become the dorsal surface of the embryo; in sections, it lies either to the left or right side of the blastocoel, and slightly ventral to it; as cells initiate involution, the blastopore begins to form below this area.

gut

See also specific divisions of; the gastrointestinal tract.

gut, primitive

Also known as *archenteron*; it is formed from endoderm during gastrulation

---

H

hatching enzyme

In sea urchin embryos, an enzyme produced by the blastula at the late blastula stage that weakens the fertilization envelope; the fertilization envelope ruptures due to the movements of the blastula, allowing the ciliated blastula to "hatch."

head

The rostral end of the embryo.

head blood vessels, plexus of

The capillary network of the head; it receives blood from the internal carotid arteries, and is drained by the precardinal veins.

head end of embryo

The cranial or rostral end of the embryo.

head fold of the body

A distinct curved line just cranial to the tip of the head; it is a ventrally directed fold of ectoderm and endoderm that establishes the cranial boundary of the head; also, the foregut forms due to the action of the head fold of the body.

head of sperm

The portion of the sperm that contains the acrosome and nucleus.

head process

In chick and mouse embryos, the mesodermal tongue of cells that subsequently forms the cranial part of the notochord; it is formed when prospective notochordal cells ingress through the cranial end of the primitive streak.

heart

The tube that pumps blood; it initially consists of two layers: the inner endocardium and the outer myocardium.

heart rudiments

See *cardiac primordia.*

hematopoietic organ

An organ that produces blood cells; for example, the liver temporarily produces blood cells, functioning as a hematopoetic organ.

Hensen's node

See *primitive knot.*

hepatic cords

The irregularly shaped cords of cells that constitute the lobes of the liver; they are separated by the hepatic sinusoids.

hepatic duct

The duct that connects the liver to the cystic duct and the common bile duct; it is endodermal in origin

hepatic portal vein

See *portal vein.*

hepatic sinusoids

The irregular vascular spaces or small channels that subdivide the hepatic cords of the liver.

hepatoduodenal (duodenohepatic) ligament

The mesentery that lies ventral to the duodenum; it connects the duodenum to the liver.

hepatogastric (gastrohepatic) ligament (ventral mesogaster)

The mesentery that lies ventral to the stomach; it connects the stomach to the liver.

heterochromatin

Condensed DNA within the nucleus.

hindbrain

The level of the incipient neural tube just caudal to the midbrain; the level of the early neural tube that forms the metencephalon and myelencephalon.

hindgut

The portion of the gut located caudal to the midgut.

hindleg buds

Paired outgrowths from the body wall; each consists of an outer layer of ectoderm, part of which is thickened as the apical ectodermal ridge, and a core of somatic mesoderm; at the level of the hindleg buds, the ventral rami of adjacent spinal nerves are interconnected to form the lumbosacral plexus.

homologous recombination

A molecular genetic event used to incorporate a piece of DNA into an organism to block (knock out) the function of a gene.

horseradish peroxidase

An enzyme used as a cell marker; it can be injected into a single cell or group of cells at the blastula stage to construct a prospective fate map.

hyaline layer

In sea urchin eggs, a clear layer formed by hyaline protein; it immediately encloses the surface of the zygote, and it holds the blastomeres together during subsequent cleavage.

hyaline protein

In sea urchin eggs, one of the expelled substances of the cortical granules; it forms the hyaline layer.

hypoblast

In chick and mouse embryos, the initial lower layer of the blastoderm.

hypochord

See *subnotochordal rod*

hypoglossal (XII) cranial nerves

Cranial nerves that grow out from the myelencephalon; they are motor nerves that innervate the tongue muscles.

hypothalamus

An adult brain component formed by the diencephalon of the neural tube.

I

ileum

A region of the adult small intestine; it is formed mostly by the cranial limb of the intestinal loop.

iliac arteries, external

Small lateral branches from each umbilical artery.

iliac arteries, internal

Arteries formed in part by the portion of the umbilical arteries that lie at the base of the hindleg buds.

implantation

In mouse and pig development, the process by which the blastocyst attaches to the uterine wall to initiate the formation of the placenta.

incus

One of three ear ossicles; it is a derivative of the first branchial arches.

infundibulum of brain

An evagination from the floor of the diencephalon; it forms the posterior pituitary gland.

infundibulum of oviduct

In hens, the first subdivision of the oviduct; it is expanded at its open end as a funnel-shaped structure; the ovum enters its opening 15 minutes after ovulation.

ingression

The inward migration of epiblast cells through the primitive streak; it plays a major role in chick and mouse gastrulation.

inner cell mass

The portion of the mouse blastocyst enclosed by the trophoblast; it forms the embryo and some of its extraembryonic membranes.

inner ears

The portion of the ears that develop from the auditory vesicles; each consists of the endolymphatic duct, anterior, posterior, and lateral semicircular canals, sacculus, and cochlea.

intermediate zone

Also called *mantle zone*; the middle region of the brain wall; it is formed by cells that migrate from the ventricular zone.

intersegmental arteries

A series of small arteries continuous with the descending aorta dorsally.

intersegmental veins

A series of small veins continuous dorsally with the postcardinal veins.

intersomitic furrows

The spaces that separate adjacent pairs of somites.

interventricular foramen of heart

The connection between the right and left ventricles.

interventricular foramina of brain

The narrowed connections between the lateral and third ventricles of the brain.

interventricular septum

The prominent structure that partially separates the cavity of the ventricle into right and left sides.

intestinal loop

A portion of the intestine (midgut) that herniates into the extraembryonic coelom of the umbilical cord, forming the temporary umbilical hernia.

intestinal loop, caudal limb of the

The portion of the intestinal loop that forms principally the ascending colon and most of the transverse colon of the adult large intestine; it rotates somewhat toward the left (apparent right in transverse sections) and ventrally; it is suspended between the two large mesonephric kidneys by a long, thin mesentery.

intestinal loop, cranial limb of the

The portion of the intestinal loop that forms mainly the jejunum and most of the ileum of the adult small intestine; it rotates somewhat toward the right (apparent left in transverse sections) to lie alongside the caudal limb of the intestinal loop.

intestinal portal, caudal or posterior

In 48-hour chick embryos, the cranial opening of the allantois rudiment into the subgerminal cavity; in 72-hour embryos, the cranial opening of the hindgut into the subgerminal cavity.

intestinal portal, cranial or anterior

In chick embryos, the caudal opening of the foregut into the subgerminal cavity.

intestine, large; small

Portions of the adult gut.

invagination

One of the morphogenetic movements that occurs during gastrulation; it plays a major role in sea urchin gastrulation, where cells of the vegetal plate bend inward (invaginate) to form the archenteron.

invagination, primary

In sea urchin embryos, the first phase of gastrulation; the vegetal plate initiates invagination into the blastocoel to form the beginnings of the endodermal archenteron.

invagination, secondary

In sea urchin embryos, the second phase of invagination during which the archenteron undergoes rapid elongation towards the animal pole; studies show that this phase involves rapid cell rearrangement.

involution

A morphogenetic process that plays a major role in frog gastrulation; it occurs when some cells originally located at the surface of the blastula turn inward over the blastopore lips to move into the interior.

isthmus of brain

A prominent constriction of the neural tube between the mesencephalon and metencephalon.

isthmus of oviduct

In hens, one of five regions of the oviduct; it secretes the inner and outer shell membranes; the ovum remains in the isthmus for about an hour; it is then passed to the shell gland.

J

jaw, lower

The portion of the face that is formed by the two mandibular processes.

jaw, upper

The portion of the face that is formed when each medial nasal process fuses laterally with the maxillary process, and medially with the other medial nasal process.

jejunum

A portion of the adult gut; it is formed mainly by the cranial limb of the intestinal loop.

jelly layer

In frog eggs, a protective coating that needs to be penetrated by the sperm for fertilization to occur; penetration is aided by enzymes released during the acrosomal reaction.

jugular ganglia

Ganglia of the vagus (X) cranial nerves; they are formed from accumulations of neural crest cells.

jugular veins, internal

The veins derived from the cranial ends of the paired precardinal veins.

K

knocked-out genes

See *homologous recombination.*

Koller's sickle

In chick embryos, a thickening of the epiblast (often poorly defined) at the caudal area pellucida/area opaca interface; some cells migrate from this area and join the cells that delaminated from the epiblast to form the sheetlike hypoblast.

L

lamina terminalis

The portion of the telencephalon wall lying between the two cerebral hemispheres.

larvae

The stage of some developing organisms following the stage of the embryo; for example, the prism and pluteus larvae develop in sea urchins; unlike embryos, larvae can freely swim and feed.

larvae, pluteus

A larval stage of the sea urchin during which projections of the larva begin to form—two anal arms and one oral lobe or hood.

larvae, prism

A larval stage of the sea urchin during which the embryo flattens at its ventral (oral) surface.

laryngotracheal groove

An elongation of the ventral portion of the foregut at the level where lung bud formation is initiated; it is continuous, without a distinct boundary, with the pharynx.

larynx

An adult component of the respiratory system that is formed from the laryngotracheal groove.

lateral arteries

See *mesonephric arteries.*

lateral plates

See *mesoderm, lateral plate.*

leg buds

See also foreleg buds; hindleg buds; paired structures that consist of a core of somatic mesoderm covered by skin ectoderm.

lens epithelium

The outer, thin region of each lens vesicle.

lens fibers

The inner, thick region of each lens vesicle.

lens placodes

The thickened layer of ectoderm overlying the optic vesicles; they form the lenses of the eyes.

lens vesicles

The structures formed by the invagination and thickening of the ectoderm (lens placodes) overlying the optic vesicles.

lenses

The structures formed from the lens vesicles.

LH

See *luteinizing hormone.*

ligamentum arteriosum

The ductus arteriosus is converted into this fibrous ligament after birth.

limb buds

See also *foreleg buds; hindleg buds; leg buds; wing buds; a* more general name for the foreleg and hindleg buds of frogs and mammals, or the wing and leg buds of birds.

lingual swellings, lateral

Paired structures derived from the first branchial arches; they contribute to the tongue.

liver

The large structure connected to the stomach by the hepatogastric ligament; it is subdivided into paired dorsal and ventral lobes, and it acts as a hematopoietic organ during development.

liver lobes

The subdivisions of the liver into paired dorsal and ventral units.

liver rudiments, cranial; caudal

Two prominent ventral evaginations of the foregut that contribute to the liver.

looping of heart

The process whereby the heart folds upon itself prior to forming distinct chambers.

lumbosacral plexus

The ventral rami of adjacent spinal nerves that interconnect at the level of the hindlimb buds to form a nerve plexus.

lung buds

Paired evaginations at the ventrolateral portions of the foregut; each lung bud forms a lung; each of the two lung buds bifurcates to form two lobes of each lung.

lungs

A component of the adult respiratory system; the lungs are derived from the paired lung buds; two lobes of each lung are formed when each of the two lung buds bifurcates; in the adult, the right lung contains three lobes, and the left lung contains two lobes; the upper lobe of the right lung is formed from the eparterial bronchus.

luteinizing hormone (LH)

A hormone released from the anterior pituitary gland (adenohypophysis) that triggers ovulation.

M

macromeres

In sea urchin embryos, the four larger vegetal hemisphere cells that divide longitudinally to form one tier of eight blastomeres (veg$_1$ blastomeres); they also divide latitudinally during the 6th cleavage to form two tiers of cells, for a total of 16 cells (veg$_2$ blastomeres); in frog embryos, the blastomeres at the vegetal pole.

madreporite

In sea urchins, a buttonlike structure on the aboral surface of the gonad where the signal that indicates the presence of gametes in the seawater enters and passes through the water-vascular system.

magnum

In hens, the longest subdivision of the oviduct; albumen is added to the ovum in the magnum; it takes about three hours for the ovum to traverse the magnum.

malleus

One of three ear ossicles; it is a derivative of the first branchial arches.

mammary glands

In 10-mm pig embryos, the glands forming within the mammary ridges.

mammary ridges

In 10-mm pig embryos, a pair of small ridges that lie parallel to the ventral edges of the somites, between foreleg and hindleg buds.

mandibular branches of the trigeminal (V) cranial nerves

One of three pairs of branches of the trigeminal (V) cranial nerves; they consist of mixed nerves (sensory and motor); they innervate the mandibular processes of the first branchial arches.

mandibular processes of the first branchial arches

Components of the first branchial arches; the stomodeum splits the first branchial arches into the mandibular and maxillary processes; the mandibular processes join to form the lower jaw.

mantle zone

See *intermediate zone.*

marginal zone

The outermost region of the neural tube wall; it is free of nuclei and it is the lightest staining of any region of the neural tube; it consists mainly of nerve fibers.

mature eggs or ova

See *ova.*

mature sperm

See *sperm.*

maxillary branches of the trigeminal (V) cranial nerves

One of three pairs of branches of the trigeminal (V) cranial nerves; they consist of mixed nerves (sensory and motor); they innervate the maxillary processes of the first branchial arches.

maxillary processes of the first branchial arches

Components of the first branchial arches; the stomodeum splits the first branchial arches into the mandibular and maxillary processes; the maxillary processes form the lateral portions of the upper jaw and most of the cheeks.

medulla

An adult brain component formed by the mesencephalon of the neural tube.

meiotic division, first

Primary oocytes enter the prophase stage of the first meiotic division before ovulation occurs; during this division, each primary oocyte produces a secondary oocyte and a first polar body; in males, each primary spermatocyte produces two secondary spermatocytes.

meiotic division, second

The second meiotic division follows the first meiotic division; it is initiated by each secondary oocyte as it enters the oviduct, but it then arrests in the metaphase stage; it is completed as a sperm contacts and penetrates each secondary oocyte, resulting in the formation of a second polar body and a mature ovum; in males, two spermatids are produced by each secondary spermatocyte.

membranes, extraembryonic

See also names of specific ones; the chorion and amnion, which originate from the somatopleure, and the yolk sac and allantois, which originate from the splanchnopleure.

mesencephalon

The division of the brain caudal to the diencephalon; it forms the corpora quadrigemina (superior and inferior colliculi) and the cerebral peduncles.

mesenchymal cells, head

See *mesenchyme, head.*

mesenchyme cells, primary

In sea urchin embryos, the ingressing cells at the late blastula stage that eventually detach from the vegetal plate; from mid to late gastrulation, they become organized into a ring encircling the vegetal side of the invaginating archenteron; they cluster together and assemble into a spicule; approximately 64 primary mesenchyme cells are formed; fate mapping experiments show that these cells are derived from the large micromeres and they secrete the larval skeleton; thus they contribute to the mesoderm.

mesenchyme cells, secondary

In sea urchin embryos, the cells formed by veg_2 cells; as secondary invagination is under way, secondary mesenchyme cells detach from the tip of the archenteron and undergo an epithelial-to-mesenchymal transformation to move into the blastocoel; they extend numerous filopodia toward the overlying ectoderm.

mesenchyme, head

The loosely packed cells that fill the space between the walls of the brain and the skin ectoderm; the head mesenchyme is derived from both mesoderm and neural crest cells.

mesenchyme, primary

In sea urchin embryos, the cells formed by the large micromeres.

mesenchyme, prospective head

The cells fated to give rise to the head mesenchyme after gastrulation.

mesenteric artery, superior

A large midline artery continuous with the descending aorta at about the level of the caudal end of the liver; it forms through the fusion of a pair of vitelline arteries.

mesenteric vein, superior

A vein lying above and to the left (apparent right in transverse sections) of the duodenum at the level where the left umbilical vein enters the liver.

mesocardium, dorsal

A dorsal bridge of splanchnic mesoderm that suspends the heart within the pericardial cavity.

mesocolon

The dorsal mesentery supporting the portion of the caudal limb of the intestinal loop that will form the colon region of the adult intestine.

mesoderm

One of three germ layers; it lies between the ectoderm and the mesoderm.

mesoderm, extraembryonic

The mesoderm that contributes to the extraembryonic membranes.

mesoderm, heart

The mesoderm that contributes to the heart.

mesoderm, intermediate

See *nephrotomes.*

mesoderm, lateral plate

The mesoderm that forms the most lateral mesoderm.

mesoderm, presomitic

In mouse embryos, the area of the embryo that undergoes gastrulation movements to form the somitic mesoderm (that is, the mesoderm equivalent to the prospective segmental plate mesoderm of chick embryos).

mesoderm, prospective

The area of the embryo that undergoes gastrulation movements to form the mesoderm.

mesoderm, prospective embryonic

The mesoderm that forms the embryo proper, rather than to its extraembryonic membranes.

mesoderm, prospective extraembryonic

The cells fated to give rise to the extraembyonic mesoderm.

mesoderm, prospective head

The area of the embryo that undergoes gastrulation movements to form the head mesoderm.

mesoderm, prospective heart

The cells fated to give rise to the heart mesoderm.

mesoderm, prospective lateral plate

The cells fated to give rise to the most lateral mesoderm.

mesoderm, prospective segmental plate

The cells fated to give rise to the mesoderm of the somites.

mesoderm, somitic

In mouse embryos, the mesoderm that forms the somites (that is the meosderm equivalent to the segmental plate mesoderm of chick embryos)

mesoderm, segmental plate

The mesoderm that forms the mesoderm of the somites.

mesoderm, somatic

The mesoderm that forms the dorsal layer (adjacent to the ectoderm) of the most lateral mesoderm.

mesoderm, splanchnic

The mesoderm that forms the ventral layer (adjacent to the endoderm) of the most lateral mesoderm.

mesodermal cells, ingressing prospective

In chick and mouse embryos, the cells fated to give rise to the mesoderm, at the time when they are moving through the primitive streak.

mesoduodenum

The dorsal mesentery supporting the duodenum region of the gut.

mesoesophagus

The dorsal mesentery supporting the esophagus region of the gut.

mesogaster, dorsal

The dorsal mesentery supporting the stomach region of the gut; it enlarges as the adult greater omentum; mesenchymal cells aggregate within the dorsal mesogaster and form the spleen.

mesogaster, ventral

See *hepatogastric ligament*.

mesomeres

In sea urchin embryos, they consist of the tier nearest the animal pole consisting of eight blastomeres formed during the fourth cleavage; during the fifth cleavage, they divide to form two tiers of eight cells each (an$_1$ and an$_2$ blastomeres).

mesonephric arteries

Also known as *lateral arteries*; small ventrolateral branches of the descending aorta that carry blood to the glomeruli of the mesonephric kidneys.

mesonephric duct rudiments

Also called *pronephric duct rudiments*; solid longitudinal structures formed by the elongation of the pronephric cords on each side of the pronephric kidneys after their separation from the underlying nephrotome; in 48-hour chick embryos, they are located lateral to the descending aorta or dorsal aortae, and they extend caudally and undergo cavitation.

mesonephric ducts

Tiny ducts formed by the cavitation of the mesonephric duct rudiments; they form the epididymis, vas deferens, and ejaculatory duct of the adult male reproductive system, and they also give rise to an evagination that forms the seminal vesicle; they degenerate in females.

mesonephric tubule rudiments

The rudiments of the tubules of the mesonephric kidneys.

mesonephric tubules

The tubules of the mesonephric kidneys.

mesonephric kidneys

The structures formed collectively by the mesonephric tubules.

metanephric kidneys

In birds and mammals, the third or definitive kidneys; they begin to function in late prenatal and early postnatal life.

metaphase

A phase of mitosis in which chromosomes are arranged on the metaphase plate.

metencephalon

The division of the brain just caudal to the mesencephalon; along with the myelencephalon, it is derived from the rhombencephalon.

micromeres

In sea urchin embryos, they consist of the tier nearest the vegetal pole consisting of four smaller blastomeres, which later divide to form the large and small micromeres; in frog embryos, the blastomeres in the animal pole.

micromeres, large

In sea urchin embryos, the upper tier of cells formed by the original four micromeres; fate mapping studies have shown that primary mesenchyme cells are derived from the large micromeres.

micromeres, small

In sea urchin embryos, the lower tier of cells nearest the vegetal pole formed by the original four micromeres.

microtubules of sperm tail

Cytoskeletal elements arranged as nine outer doublets and two inner singlets.

midbrain

The level of the incipient neural tube just caudal to the forebrain; the level of the early neural tube that forms the mesencephalon.

middle ear cavities

See *tympanic cavities*.

midgut

The portion of the gut between the foregut and hindgut.

midpiece of sperm

The portion of the sperm that contains mitochondria and a pair of centrioles.

mitochondria of sperm

Energy-producing organelles located in the midpiece of the sperm.

mitral valve

See *bicuspid valve*.

morphogenetic movements

The form-shaping movements of tissue, especially prevelant during gastrulation and neurulation.

morula

In sea urchin and mouse embryos, a solid ball of blastomeres formed during late cleavage.

motor horns

Paired enlargements of the ventral part of the intermediate zone of the spinal cord.

mouth or mouth opening

In invertebrate embryos, the structure formed from the blastopore (in protostomes) or as a new opening (in deuterostomes); in vertebrate embryos, the structure formed by the rupture of the oral membrane; the cavity of the gut opens into the stomodeum through this opening.

muscle layers of gonads

Muscles within the gonads that undergo contraction to shed the eggs and sperm from the ovaries and testes, respectively, during the breeding season.

myelencephalon

The most caudal region of the brain; it is characterized by the presence of the neuromeres; along with the metencephalon, it is derived from the rhombencephalon.

myocardium

The outer, thicker layer of the heart that surrounds the endocardium; it is derived from the splanchnic mesoderm, and it is widely separated from the endocardium by the cardiac jelly.

myotomes

Plates of lightly stained cells lying just medial to the dermatome of the somites; they are the source of the skeletal muscles of the trunk and tail.

N

nasal cavities

The structures lined by the invaginated nasal placodes.

nasal pits

Also known as *olfactory pits;* the structures formed through the invagination of the nasal placodes; most of the cells within their walls consist of young neurons.

nasal placodes

Also known as *olfactory placodes;* paired regions of thickened skin ectoderm located ventral to the prosencephalon; they eventually form the linings of the nasal cavities; they are induced to form by the adjacent regions of the telencephalon.

nasal processes, lateral

Elevated regions of head mesenchyme covered by skin ectoderm on the lateral side of each nasal pit; the lateral nasal processes eventually form the sides of the nose.

nasal processes, medial

Elevated regions of the head mesenchyme covered by skin ectoderm on the medial side of each nasal pit; each medial nasal process will eventually fuse laterally with a maxillary process and medially with the other medial nasal process to form the upper jaw.

neck muscles

Muscles innervated by the spinal accessory nerves.

nephrogenic tissue

A dark mass of cells just outside each renal pelvis; it forms the secretory tubules of the metanephric kidneys.

nephrotomes

Also called *intermediate mesoderm; a* slender area of mesodermal cells lateral to each somite pair; the nephrotome on each side produces the pronephric cords.

nerve fibers, mixed

Nerves that have both sensory and motor fibers; for example, the trigeminal (V) cranial nerve and the facial (VII) cranial nerve.

nerve fibers, motor

Nerves that have motor fibers only; for example, the oculomotor (III) cranial nerves and the trochlear (IV) cranial nerves; motor nerves grow out from motor nuclei toward the areas they innervate.

nerve fibers, sensory

Nerves that have sensory fibers only; for example, the olfactory (I) cranial nerves and the auditory (VIII) cranial nerves; sensory nerves grow out from ganglia toward the areas they innervate.

neural crest cells

Ectodermal cells derived from the roof of the neural tube shortly after neural groove closure; they give rise to a multitude of structures including pigment cells, nerve cells located outside of the neural tube, and mesenchymal cells in the head; they can be identified as a straggling line of cells wedged between the skin ectoderm and neural tube.

neural folds

The paired folds that form at the lateral margins of the neural groove; each fold consists of an inner layer of neural ectoderm and an outer layer of skin ectoderm; the neural folds fuse to establish the neural tube and an overlying layer of skin ectoderm.

neural groove

The space bounded by the neural ectoderm and the neural folds; when the neural folds approach one another and fuse, the neural groove is closed.

neural plate

The initial rudiment of the central nervous system; it is formed by thickening of the ectoderm overlying the head mesenchyme and segmental plate mesoderm; formation of the neural plate is induced in part by the mesoderm (perhaps also the endoderm).

neural plate, mesencephalon level of

The portion of the neural plate that forms the mesencephalon region of the neural tube.

neural plate, prosencephalon level of

The portion of the neural plate that forms the prosencephalon region of the neural tube.

neural plate, prospective

The portion of the ectoderm that is fated to form the neural plate through induction.

neural plate, rhombencephalon level

The portion of the neural plate that forms the rhombencephalon region of the neural tube.

neural plate, spinal cord level

The portion of the neural plate that forms the spinal cord region of the neural tube.

neural tube

The tubular rudiment of the central nervous system; it is formed during neurulation by the folding of the neural plate into a neural groove and tube.

neural tube defects

Birth defects in which abnormal development of the neural tube occurs.

neurohypophysis

See *pituitary gland, posterior.*

neuromeres

A series of enlargements of the neural tube that characterize the rhombencephalon region.

neurons, young

Precursor cells that give rise to nerve cells (neurons).

neurons, young motor

Precursor cells that give rise to motor nerve cells (neurons).

neurons, young sensory

Precursor cells that give rise to sensory nerve cells (neurons).

neuropore, caudal or posterior

The caudal opening of the neural tube, prior to closure of the neural groove at the caudal end of the embryo.

neuropore, cranial or anterior

The cranial opening of the neural tube, prior to closure of the neural groove at the cranial end of the embryo.

neurula

The embryo during the stage of neurulation.

neurulation

The process of neural tube formation.

node (organizer)

The structure of the mouse embryo that is located at the rostral end of the primitive streak; it has the ability to organize the body axis and the tube-within-a-tube body plan.

nodose ganglia

Ganglia of the vagus (X) cranial nerves; they are formed from epibranchial placodes in the area of the fourth branchial arches.

nose

A portion of the face; the sides of the nose are formed by the lateral nasal processes.

notochord

A midline rod of cells flanked by the head mesenchyme and segmental plate mesoderm; the head process contributes to its cranial end; its cranial tip induces the formation of the infundibulum; the notochord induces the formation of the floor plate of the neural tube.

notochord, prospective

The cells that are fated to form the notochord.

nuclei of blastomeres

The part of blastomeres (cells of the gastrula during cleavage) containing the chromosomes.

nuclei of cells

The part of cells containing the chromosomes.

nuclei, motor

Accumulations of neural ectodermal cells that lie within the ventral part of the brain and spinal cord; they consist of the cell bodies of young neurons within the intermediate zone of the neural tube; their axons grow out of the neural tube to form motor fibers.

nucleolus

An organelle contained within the nucleus; it is rich in RNA.

nucleus of egg

An organelle of the egg containing DNA.

nucleus of sperm

An organelle of the sperm containing DNA.

nucleus of zygote

An organelle of the zygote containing DNA derived from the male and female pronuclei.

nuptial pads

Swellings located on the forelegs of mature male frogs.

O

oblique vein of the left atrium

A vein of the heart formed from the left common cardinal vein.

oculomotor (III) cranial nerves

Motor nerve fibers that innervate four pairs of extrinsic eye muscles—inferior oblique, and inferior, superior, and medial rectus eye muscles.

olfactory (I) cranial nerves

Sensory nerve fibers that innervate the lining of the nasal cavities; their axons are produced by the young neurons that originate within the nasal placodes.

olfactory lobes

An adult brain component formed from the telencephalon of the neural tube.

olfactory placodes

See *nasal placodes.*

olfactory receptors

The dendrites that function as receptors for the sense of smell; they are produced by the young neurons that originate within the nasal placodes.

omental bursa

A closed cavity to the right (apparent left in sections) of the stomach.

omentum, greater

A mesentery formed by the enlargement of the dorsal mesogaster.

omentum, lesser

A mesentery consisting of the hepatogastric and hepatoduodenal ligaments.

oocytes, primary

Cells formed when oogonia enlarge and become surrounded by a layer of follicle cells; they enter the prophase stage of the first meiotic division, and each undergoes the first meiotic division, producing a secondary oocyte and first polar body; the first meiotic division is completed during ovulation.

oocytes, secondary

Cells formed after primary oocytes complete the first meiotic division during ovulation.

oogenesis

Development of the ovum; it occurs in the ovaries of the mature female; one mature egg and the first and second polar bodies are generated during oogenesis.

oogonia

Cells derived from the primordial germ cells after they enter the embryonic ovaries; oogonia undergo rapid mitotic divisions.

oolemma

Also called *plasmalemma*; it encloses the female pronucleus and its surrounding cytoplasm.

ophthalmic branches of the trigeminal (V) cranial nerves

One of three pairs of branches of the trigeminal (V) cranial nerves; they consist of mixed nerves (sensory and motor); they innervate the regions of the eyes.

optic (II) cranial nerves

See also optic nerve fibers; these are cranial nerves that have axons that remain within the marginal zone throughout their extent; they are not actually true nerves, but are fiber tracts located within derivatives of the diencephalon wall; they are sensory nerve fibers that innervate the retinas.

optic cup, dorsal lip of the; ventral lip of the

The dorsal and ventral margins, respectively, of the optic cup (the rudiment of the eye).

optic cups

The rudiments of the eyes; each cup is derived from a lateral evagination from the diencephalon, which secondarily invaginates at its blind end to form a double layered optic cup— a thick layer and a thin layer.

optic fissures

Ventral gaps in the optic cups formed by the ventral invagination of the optic cups and stalks; they are important in development because optic nerve fibers from the retina grow back to their proper connections within the diencephalon through the wall of the fissure; also, the cavity of the fissure provides a pathway for blood vessels to enter the optic cup; they later close.

optic nerve fibers

See also optic (II) cranial nerves; nerve fibers that arise from the retinas and grow back to their proper connections within the diencephalon, through the wall of the optic fissures.

optic stalks

The structures that connect the optic cups to the diencephalon of the neural tube.

optic sulci (cups)

In mouse embryos, the rudiments of the eyes.

optic vesicles

The rudiments of the eyes in frog and chick embryos; they are the two largest lateral evaginations of the neural tube; they evaginate from the diencephalon, and they are in close contact laterally with the overlying skin ectoderm.

opticoels

The spaces between the two layers of the optic cups; the opticoels are continuous with the cavity of the diencephalon.

oral arms

In sea urchin pluteus larvae, two projections that grow out from the oral lobe.

oral lobe or hood

In sea urchin embryos, a large unpaired projection of the body of the pluteus larva.

oral membrane

A double-layered membrane consisting of the stomodeal ectoderm and the foregut endoderm; it separates the foregut from the stomodeum, and it ruptures to form the mouth opening.

oral surface of sea urchin

See *ventral surface of sea urchin*.

organizer

See *node; dorsal lip of blastopore; primitive knot*

organogenesis

The phase of development during which organs and organ systems develop.

organs of Corti

See *spiral organs of Corti*.

ostium of oviduct

The opening of the oviduct.

otic placodes

See *auditory placodes*.

otocysts

See *auditory vesicles*.

ova

The cells produced as the sperm contacts and penetrates secondary oocytes.

ovaries

A component of the functional reproductive system of the adult female in which oogenesis occurs.

oviducts

A component of the functional reproductive system of the adult female; in the hen, it is subdivided into 5 regions— infundibulum, magnum, isthmus, shell gland, and vagina.

ovoposition

The process of laying of the egg (for example, laying of the hen's egg).

ovulation

The rupture of the ovarian follicle and the release of its contained oocyte; fully grown primary oocytes undergo this process in response to hormones secreted by the anterior pituitary gland.

P

pacemaker of heart

The sinus venosus is absorbed by the wall of the right atrium to become the pacemaker of the adult heart.

palatine tonsils

Tonsils derived from the second pharyngeal pouches.

pancreas

A component of the gastrointestinal system formed by the fusion of the dorsal and ventral pancreatic rudiments.

pancreatic duct, dorsal

Also known as the *duct of Santorini*; it connects the dorsal pancreatic rudiment to the duodenum, and it persists in the adult pig for transport of pancreatic secretions involved in digestion; this structure degenerates in humans.

pancreatic duct, ventral

Also called *duct of Wirsung*; it connects the ventral pancreatic rudiment to the common bile duct; it degenerates in the pig and persists in humans.

pancreatic rudiment, dorsal

A dorsal outgrowth from the duodenum; it is connected to the duodenum by the dorsal pancreatic duct; it fuses with the ventral pancreatic rudiment to form the pancreas.

pancreatic rudiment, ventral

A ventral outgrowth of the common bile duct; it is connected to the common bile duct by the ventral pancreatic duct; it fuses with the dorsal pancreatic rudiment to form the pancreas.

parafollicular cells

See *C cells*.

parasympathetic division of the autonomic nervous system

A portion of the peripheral nervous system associated with the cranial and sacral regions of the embryo.

parasympathetic terminal ganglia

See *terminal ganglia*.

parathyroid glands, inferior

Glands derived from the dorsal portions of the third pharyngeal pouches; they later migrate caudal to the fourth and fifth pharyngeal pouches; in the adult they lie caudal to the superior parathyroid glands.

parathyroid glands, superior

Glands derived from the fourth pharyngeal pouches.

pericardial cavity

The space between the heart and the parietal pericardium; it is the portion of the coelom surrounding the heart.

pericardial cavity, rudiments of

In early chick embryos, two large coelomic spaces that project toward the midline at the level of the lateral body folds; they later fuse to form a single cavity surrounding the heart.

pericardium, parietal

The thin layer of cells surrounding the heart that is derived from the somatic mesoderm.

pericardium, visceral

See *epicardium*.

peritoneal cavity

The cavity formed by the fusion of the paired rudiments of the peritoneal cavity.

peritoneal cavity, rudiments of

The paired portions of the coelom lateral to the gut; they fuse to form a single peritoneal cavity surrounding the gut.

peritoneum, parietal

The thin layer of somatic mesoderm on the internal surface of the ventral and lateral body walls.

peritoneum, visceral

The thin layer of splanchnic mesoderm on the surfaces of the abdominal organs.

perivitelline space

The space that forms between the oolemma and the fertilization membrane following fertilization.

petrosal ganglia

Cranial ganglia of the glossopharyngeal (VIII) cranial nerve; they are derived from epibranchial placodes.

pharyngeal pouches

Lateral evaginations of the pharynx.

pharyngeal pouches, first

Lateral evaginations of the pharynx just caudal to the first branchial arches; each pouch consists of an upper portion connecting to the pharynx that forms the Eustachian tube, and a lower portion forming the middle ear cavity.

pharyngeal pouches, second

Lateral evaginations of the pharynx just caudal to the second branchial arches; they are the source of the adult palatine tonsils.

pharyngeal pouches, third

Lateral evaginations of the pharynx just caudal to the third branchial arches; the ventral portions of the third pharyngeal pouches fuse in the midline to form the thymus gland; the dorsal portions of the third pharyngeal pouches form the inferior parathyroid glands.

pharyngeal pouches, fourth

Lateral evaginations of the pharynx just caudal to the fourth branchial arches; they are the sources of the adult superior parathyroid glands.

pharyngeal pouches, fifth

Each pouch is derived from a small dorsolateral evagination of each fourth pharyngeal pouch; the fifth pharyngeal pouches form the ultimobranchial bodies, and they are incorporated in the thyroid gland, where they give rise to c cells.

pharyngotympanic tubes

See *eustachian tubes*.

pharynx

The region of the foregut that gives rise to the pharyngeal pouches.

pia mater

The outer vascular layer of mesenchymal cells just outside the brain wall, which together with the thin roof plate of the myelencephalon, constitute the rudiments of the choroid plexus.

pigment cells

Cells derived from neural crest cells; they are responsible for pigmentation of the skin.

pigment granules

Granules contained within the retinas; they are responsible for the dark coloration of the eyes.

pineal gland

Also known as *epiphysis*; a midline dorsal evagination of the diencephalon.

pinnae

Also known as *auricle of the external ear*; the portion of the ear formed by the fusion of hillocks from the first and second branchial arches on either side of the first branchial grooves.

pituitary gland, anterior

Also known as *adenohypophysis*; a gland that secretes hormones, such as luteinizing hormone that trigger the fully grown oocytes to undergo ovulation; it also stimulates yolk accumulation by release of follicle stimulating hormone; it is derived from Rathke's pouch.

pituitary gland, posterior

Also known as *neurohypophysis*; it is derived from the infundibulum.

placenta

The organ of mammals necessary for the survival of the embryo as it develops in utero.

placental membrane or barrier

The membrane of the placenta that separates maternal and fetal blood.

plasmalemma

The external membrane of the cell.

pleura, parietal

The portion of the lining of the pleural cavities surrounding the lung buds, but not covering the splanchnic mesoderm constituting the surfaces of the developing lungs.

pleura, visceral

The portion of the lining of the pleural cavities surrounding the lungs, and covering the surfaces of the developing lungs but not the adjacent walls of the cavity; it is derived from splanchnic mesoderm.

pleural cavities

The paired portions of the coelom lateral to the developing lung buds.

plexus of head blood vessels

See *head blood vessels, plexus of*.

polar body, first

The non-functional cell formed when the primary oocyte undergoes the first meiotic division; it is pinched off between the vitelline membrane/zona pellucida and the large secondary oocyte.

polar body, second
The non-functional cell formed during the second meiotic division of the oocyte, as the sperm contacts and penetrates the secondary oocyte.

polyingress
See *delaminate*.

polyspermy
The entrance of more than one sperm into the egg.

pons
An adult brain component derived from the metencephalon.

portal vein
A large vein connecting the superior mesenteric vein to the ductus venosus within the liver.

postcardinal veins
Veins that lie dorsolateral to the dorsal aortae in the trunk and tail; they drain blood into the common cardinal veins.

precardinal veins
Veins that lie dorsolateral to the dorsal aortae in the region of the aortic arches; each vein extends caudad from the plexus of the head blood vessels and joins the postcardinal and common cardinal veins; they give rise to the superior vena cava and internal jugular veins.

prechordal plate
The mesoderm and endoderm rostral to the notochord, beneath the midline of the forebrain level of the neural plate and tube.

preoral gut (Seessel's pouch)
The part of the foregut that lies cranial to the oral membrane; it later degenerates.

primitive folds
See *primitive ridges*.

primitive groove
The portion of the primitive streak that is flanked by the primitive ridges; it is continuous with the primitive pit cranially.

primitive knot
Also known as *Hensen's node*; in chick embryos, the thickened cranial end of the primitive streak that partially surrounds the primitive pit; it regresses during gastrulation.

primitive pit
A depression partially surrounded by the primitive knot; it is continuous caudally with the primitive groove.

primitive ridges
Ridges that mark the lateral margins of the primitive streak; they flank the primitive groove and are continuous caudally with the neural folds; the transition between the primitive ridges and the neural folds is gradual.

primitive streak
A midline thickening of the epiblast; cells ingress through the primitive streak during gastrulation to form the endoderm and mesoderm.

primordial germ cells
In chick embryos, cells derived from hypoblast found in the germ cell crescent; they undergo extensive migration and enter the embryonic gonads, where they form oogonia in the female and spermatogonia in the male; in frog embryos, they originate from the yolky endoderm in the vegetal pole of the blastula; in mouse embryos, they originate from the yolk sac endoderm.

proamnion
In early chick embryos, the region of the blastoderm that still lacks mesoderm; after formation of the head fold of the body, it lies beneath the head, separated from the head by the subcephalic pocket; it consists of an upper layer of ectoderm and a lower layer of endoderm; its name is a misnomer because the proamnion is not the source of the amnion .

proctodeum
A ventral invagination of skin ectoderm toward the hindgut.

pronephric cords
Solid masses of cells produced by the nephrotome; they constitute the paired rudimentary pronephric kidneys; they later separate from the underlying nephrotome on each side, and they elongate to form the pronephric ducts/mesonephric duct rudiments.

pronephric duct rudiments
See *mesonephric duct rudiments*.

pronephric ducts
Along with the pronephric tubules, they constitute the pronephric kidneys.

pronephric kidneys
Kidneys that consist of pronephric ducts and tubules; they are functional in frog larvae.

pronephric ridges
In frog embryos, lateral bulges toward the coelom that are formed by the pronephric kidneys.

pronephric tubules
Along with the pronephric ducts, they constitute the pronephric kidneys.

pronucleus, female or egg
The nucleus of the mature egg; it fuses with the male pronucleus to form the zygote nucleus, completing the process of fertilization.

pronucleus, male or sperm
The nucleus of the penetrating sperm after it enlarges within the ovum; it unites with the female pronucleus to form the zygote nucleus, completing the process of fertilization.

prophase
The first stage of meiosis and mitosis.

prosencephalon
The most rostral portion of the neural tube; it gives rise to the telencephalon and diencephalon of the brain.

prospective fate
What an area of cells in the blastula/gastrula stage is destined to become.

prospective fate map
A map at the blastula/gastrula stage indicating the location of specific groups of cells prior to and during gastrulation.

protostomes
The type of organism in which the blastopore forms the mouth.

pulmonary arteries
Arteries that originate as caudal outgrowths from the sixth aortic arches; they supply the lungs.

pulmonary trunk
One of two arteries formed by the bulbar septum within the conotruncus; it is continuous with the two sixth aortic arches.

pulmonary veins

The veins formed when the single pulmonary vein continuous with the left atrium branches to form four veins; they drain blood from the lungs and open directly into the left atrium in the adult.

R

radial symmetry

A type of body symmetry characteristic of echinoderms such as the sea urchin.

radialization

An effect observed when early stages of sea urchin embryos are treated with solutions of LiCl; after such treatment, larvae retain their radial symmetry.

rami communicantes

See *spiral nerves, visceral rami of.*

rami of spinal nerves

See *spinal nerves, dorsal rami of; ventral rami of; visceral rami of.*

Rathke's pouch

A small vesicle between the infundibulum and foregut; it is the rudiment of the anterior pituitary gland; it develops as an outgrowth from the stomodeum, due to induction by the infundibulum.

rectum

The level of the gut that forms by partitioning the cloaca by the cloacal septum.

renal pelvis

The region of the metanephric kidneys formed by the expansion of the ureters.

retinae, pigmented

The thin outer layers of the optic cups that contain pigment granules.

retinae, sensory

The thickened inner layers of the optic cups, next to the lens vesicles.

rhodamine-labeled dextran

A type of cell marker (which can be demonstrated with a fluorescence microscope after illumination with the proper wavelength of light) that can be injected into a single cell or group of cells at the blastula stage to construct a prospective fate map.

rhombencephalon

The portion of the neural tube that gives rise to the metencephalon and myelencephalon.

rhombomeres

The neuromeres of the hindbrain.

ribs

Skeletal elements derived from the sclerotome in the thorax region.

roof plate

The dorsal midline wall of the brain and spinal cord; the myelencephalon has a thin roof plate, which together with the pia mater, forms the rudiment of the choroid plexus.

roots of cranial nerves

See specific names of cranial nerves.

roots of spinal nerves

See *spinal nerves, dorsal roots of; ventral roots of.*

S

sacculi

The ventral portions of the auditory vesicles.

sclera

A portion of the eye derived from neural crest cells and formed when the head mesenchyme surrounding each pigmented retina condenses.

sclerotomes

A portion of the somites; they are the source of the vertebral column and ribs.

secretory tubules

The part of the metanephric kidneys formed by the condensed nephrogenic tissue.

Seessel's pouch

See *preoral gut.*

segmental plates

See *mesoderm, segmental plate.*

semicircular canals, anterior; lateral; or posterior

Three canals derived from the utriculus portion of each auditory vesicle.

semilunar ganglia

The ganglia of the trigeminal (V) cranial nerves; epibranchial placodes and neural crest cells contribute to the formation of these ganglia.

seminal vesicles

A portion of the male reproductive system; they are derived from evaginations from the mesonephric ducts.

septum membranaceum

A down growth from the endocardial cushion and part of the bulbar septum that closes the interventricular foramen.

septum primum (I)

The first partition that separates the cavities of the right and left atria.

septum secundum (II)

The second partition that separates the cavities of the right and left atria; it lies to the right (apparent left in transverse sections) of the septum primum.

septum transversum

A major source of the adult diaphragm.

sexual maturity

The age at which an organism has acquired the ability to reproduce.

shell

In hens, the calcified structure that is formed by the shell gland in a process that requires 20 hours.

shell gland

A portion of the hen's oviduct; the ovum is carried to this gland after it remains within the isthmus for about an hour; its major function is to form the calcified shell.

shell membranes, inner; outer

In hens, membranes secreted by the isthmus of the oviduct.

shoulder muscles

Muscles that are innervated by the spinal accessory (XI) cranial nerves.

sinoatrial region

A portion of the early heart; it forms the sinus venosus and atrium.

sinoatrial valve

A projection into the cavity of the right atrium at the entrance of the sinus venosus; it consists of right and left valve flaps.

sinus terminalis

In chick embryos, a circular blood vessel separating the area vasculosa from the area vitellina; it may appear dark or light depending on whether it contains blood cells.

sinus venosus

A region of the heart derived from the sinoatrial region.

skeletal muscles

The type of muscles derived from somitic myotomes.

skin

A layer at the surface of the body formed in part from the ectoderm.

somatopleure

Collectively, the ectoderm and the somatic mesoderm.

somites

Paired blocks of mesoderm lying ventrolateral to the spinal cord.

somitogenesis

The process of somite formation.

sperm

The gamete produced by the adult male; each sperm contains a head, midpiece, and tail.

spermatids

A type of cell formed during spermatogenesis; four are generated from each secondary spermatocyte during spermatogenesis; they remain in the testes for some time to undergo a maturation process, after which they become mature sperm.

spermatocyte, primary

A type of cell formed during spermatogenesis; each is formed by the enlargement of a spermatogonium; each primary spermatocyte undergo the first meiotic division, producing two secondary spermatocytes.

spermatocytes, secondary

A type of cell formed during spermatogenesis when a primary spermatocyte undergoes the first meiotic division; each undergoes the second meiotic division to produce two spermatids.

spermatogenesis

The process of sperm formation in the testes.

spermatogonia

A precursor cell formed when the primordial germ cells undergo migration and enter the embryonic testes.

spicules

In sea urchins, a triradiate pair of structures that consititutes the larval skeleton; formed by assembly of primary mesenchyme cells into a triradiate shape, they are crystalline in nature and composed of calcium carbonate and magnesium carbonate deposited in an organic matrix; they are deposited within membrane-bound compartments inside the syncytial cables.

spinal accessory (XI) cranial nerves

Cranial nerves consisting of only motor nerve fibers; they innervate derivatives of the fourth branchial arches and certain neck and shoulder muscles.

spinal cord

The portion of the neural tube caudal to the brain; it has a thin roof plate and floor plate, and thick lateral walls.

spinal ganglia

Aggregations of neural crest cells alongside the neural tube, with a segmentation corresponding to the segmentation of the somites; they contain the cell bodies of sensory neurons.

spinal nerves

The nerves that innervate the trunk; they are formed from nerve fibers whose cells bodies lie both within the ventrolateral walls of the spinal cord and the spinal ganglia.

spinal nerves, dorsal rami of

One of the branches of the spinal nerves.

spinal nerves, dorsal roots of

The portion of the spinal nerves consisting of axons entering the spinal cord, spinal ganglia, and dendrites entering the spinal ganglia.

spinal nerves, ventral rami of

One of the branches of the spinal nerves.

spinal nerves, ventral roots of

The portion of spinal nerves consisting of axons leaving the spinal cord.

spinal nerves, visceral rami of

Also known as *ramus communicantes*; one of the branches of the spinal nerve that extends ventromedially on each side toward an accumulation of neural crest cells.

spiral ganglia

One of the subdivisions of the acoustic ganglia of the auditory (VIII) cranial nerves; dendrites from cells bodies within these ganglia terminate in the spiral organ of Corti within each cochlea.

spiral organs of Corti

The receptor organs for sound formed by the cochleas of the inner ears.

splanchnopleure

Collectively, the endoderm and splanchnic mesoderm.

spleen

An adult organ formed by the aggregation of mesenchymal cells within the dorsal mesogaster.

spongioblasts

See *glioblasts*.

stapes

One of three ear ossicles; it is derived from the second branchial arches.

sternocleidomastoid muscles

Shoulder muscles innervated by the spinal accessory (XI) cranial nerves.

stigma of follicle

The area where the ovarian follicle ruptures during ovulation.

stomach

A region of the adult gastrointestinal tract; it undergoes a rotation of 45 degrees along its longitudinal axis.

stomach, greater curvature of

The side of the stomach formed by the original dorsal surface of the stomach.

stomach, lesser curvature of

The side of the stomach formed by the original ventral surface of the stomach.

stomodeum

An ectodermal invagination that partially splits the first branchial arches into two processes.

subcardinal anastomosis

A midline fusion of right and left subcardinal veins.

subcardinal veins

Small veins lying ventral to each mesonephric kidney; they fuse to form the subcardinal anastomosis.

subcaudal pocket

In chick embryos, an ectoderm-lined space that separates the caudal end of the embryo from the underlying blastoderm.

subcephalic pocket

In chick embryos, an ectoderm-lined space that separates the head from the blastoderm.

subclavian arteries

Arteries that carry blood to the forelegs; the left subclavian artery is derived exclusively from a left intersegmental artery at the level of the foreleg buds; the right subclavian artery is derived from a right intersegmental artery at the level of the foreleg buds.

subclavian veins

Irregularly shaped veins that develop at the base of the foreleg buds.

subgerminal cavity

In chick embryos, a space that separates the center of the blastoderm from the underlying yolk; this space allows cells to move into the interior during formation of the hypoblast.

subnotochordal rod

Also known as *hypochord*; in frog embryos, a small cluster of mesodermal cells lying beneath the notochord in caudal regions.

suckers, ventral

See *adhesive glands.*

sulcus limitans

A component of the spinal cord that appears as a slight depression on each side partially separating the alar and basal plates.

superior ganglia

The ganglia of the glossopharyngeal (IX) cranial nerves; they originate from neural crest cells.

sympathetic chain ganglia

See *chain ganglia.*

sympathetic collateral ganglia

See *collateral ganglia.*

syncytial cables

In sea urchins, the membrane-bound compartments where spicules are deposited; they are formed from fused filopodia.

syngamy

The process whereby male and female pronuclei fuse to form the zygote nucleus.

T

tail

The caudal portion of the embryo that extends caudal to the hindleg buds.

tail bud

A mass of cells derived from the cranial part of the primitive streak; it contributes cells to the neural tube, segmental plates, and mesenchyme of tail.

tail end of embryo

The caudal end of the embryo.

tail fold of the body

The body fold that establishes the caudal boundary of the embryo.

tail gut

A portion of the gut located caudal to the cloacal membrane; it eventually degenerates.

tail of sperm

A portion of the sperm; it contains microtubules arranged as nine outer doublets and two inner singlets; it is about 50 micrometers long in sea urchins.

target region for invaginating archenteron

In sea urchins, a patch of ectoderm where filopodia firmly attach and pull the tip of the archenteron subjacent to it.

telencephalon

The level of the brain cranial to the diencephalon and derived from the prosencephalon; it forms lateral oval-shaped expansions called the cerebral hemispheres.

terminal ganglia

Parasympathetic ganglia of the autonomic nervous system that form from accumulations of neural crest cells.

test

The external skeleton or "shell" of the sea urchin.

testes

The organs where sperm develop.

thalamus

An adult brain component derived from the diencephalon.

theca folliculi externa

A thin sheath of connective tissue that forms the surface layer of the ovary.

theca folliculi interna

A sheath that partially surrounds each ovarian follicle.

thymus gland

An adult organ formed from the fusion of the ventral portions of the two third pharyngeal pouches in the midline.

thyroglossal duct

The original connection of the thyroid gland to the floor of the pharynx; it later degenerates.

thyroid gland

An endocrine gland lying beneath the floor of the pharynx.

thyroid rudiment

The rudiment of the thyroid gland.

tongue muscles

Muscles innervated by the hypoglossal (XII) cranial nerves.

tongue rudiments

Rudiments that form the tongue; they consist of a pair of lateral lingual swellings, a midline tuberculum impar, and a midline copula; they protrude into the pharynx.

torsion

Or twisting; a process whereby a change in the axis of the body of the embryo occurs; due to this, in the 48-hour chick embryo, the cranial end of the body lies on its left side, and the yolk sac lies between the amnion and the yolk.

trachea

One of two structures derived from the laryngotracheal groove.

transgenic mice

Mice genetically engineered to carry a foreign gene.

trapezius muscles

A shoulder muscle innervated by the spinal accessory (XI) cranial nerves.

tricuspid valve

The valve that develops within the right atrioventricular canal.

trigeminal (V) cranial nerves

Cranial nerves associated with the semilunar ganglia; they innervate derivatives of the first branchial arches.

tritiated thymidine

A cell marker that is incorporated specifically into DNA, thus radioactively labeling cells.

trochlear (IV) cranial nerves

Cranial nerves that innervate the superior oblique eye muscles.

trophoblast

The outer layer of the mouse blastocyst.

trunk

The portion of the body between the head and tail.

tuberculum impar

One of the midline tongue rudiments; it is derived from the first branchial arches.

tube-within-a-tube body plan

See *body plan*.

turning

In mouse embryos, the process in which the "inverted" embryo becomes c-shaped.

tympanic cavities

The middle ear cavities; they are formed by the lower portion of the first pharyngeal pouches; they eventually surround the ear ossicles.

tympanic membranes

The ear drums; they are derived from the first closing plates.

U

ultimobranchial bodies

See *pharyngeal pouches, fifth*.

umbilical arteries

Also called *allantoic arteries*; two large arteries within the umbilical cord and continuous with the descending aorta.

umbilical cord

The connection between the embryo and its placenta; it contains Wharton's jelly.

umbilical hernia

A temporary condition formed when the intestinal loop herniates into the extraembryonic coelom of the umbilical cord.

umbilical veins

Also called *allantoic veins*; right or left umbilical veins within the umbilical cord; the right umbilical vein is smaller than the left umbilical vein because it degenerates; after degeneration, all oxygenated blood flows from the placenta to the embryo via the left umbilical vein.

ureters

The small ducts that emerge from the dorsal side of each mesonephric duct; they connect the metanephric kidneys to the bladder.

urogenital sinus

One of two regions formed from the cloaca with formation of the cloacal septum.

uterine cavity

The cavity of the uterus, a portion of the female reproductive system in mammals.

uterine cervix

The opening to the uterus of mammals, adjacent to the vagina.

uterine horns

The paired structures that constitute the uterus in mice.

uterine wall

The wall of the uterus, a portion of the female reproductive system in mammals.

uterus

A portion of the female reproductive system; in mice, the uterus is subdivided into two uterine horns; the uterus joins the vagina at the uterine cervix.

utriculi

The dorsal portion of the auditory vesicles.

V

vagina

A portion of the female reproductive system.

vaginal plug

In mice, coagulated semen within the vagina that provides evidence of mating.

vagus (X) cranial nerves

Cranial nerves associated with the jugular and nodose ganglia; they innervate mainly derivatives of the fourth branchial arches.

valve flaps

Components of the sinoatrial valve; the left valve flap forms part of the septum secundum.

vas deferens

A portion of the male reproductive system formed by the mesonephric ducts.

vasa efferentia

See *efferent ductules*.

vegetal hemisphere

In frog eggs, the hemisphere opposite the animal hemisphere.

vegetalization

One of the main effects that occurs when sea urchin embryos are treated with solutions of LiCl; the development of the vegetal half of the embryo predominates over the development of the animal half, and embryos develop with expanded vegetal structures, such as gut.

vegetal plate

In sea urchin embryos, the thickened vegetal pole of the blastula; its cells detach to form the primary mesenchyme cells.

vegetal pole
The pole of the egg that lies directly opposite the animal pole; it corresponds to the caudal end of the future embryo; it is located in the vegetal hemisphere; its cells form the endoderm.

vena cava, inferior
A large vein that joins the ductus venosus within the liver; it opens into the sinus venosus.

vena cava, superior
A large vein derived in part from a portion of the right precardinal vein and the right common cardinal vein.

ventral surface of sea urchin
The flattened surface of the prism larva, opposite the dorsal surface.

ventricle of heart
The major component of early heart; it subdivides into right and left ventricles with the formation of the interventricular septum.

ventricles, first and second
See *ventricles, lateral.*

ventricles, lateral
Or first and second ventricles; the cavities of the cerebral hemispheres; they are broadly continuous with the cavity of the middle portion of the telencephalon of the neural tube.

ventricle, third
The cavity of the diencephalon of the neural tube.

ventricle, fourth
The cavity of the rhombencephalon where the choroid plexus is formed; it is continuous with the cerebral aqueduct.

ventricular zone
The innermost zone of the neural tube, next to the lumen; it is rich in nuclei; cells migrate from this zone to the middle region of the neural tube to form the intermediate zone.

vertebrae
Skeletal elements formed by the fusion of the caudal half of one sclerotome with the cranial half of the next caudal sclerotome.

vertebral arteries
Irregularly shaped arteries that join the basilar artery; each of these arteries receives blood directly from a subclavian artery.

vertebral column
The skeleton surrounding the spinal cord; it is derived from mainly the sclerotome.

vestibular ganglia
One of the two subdivisions of each acoustic ganglion; dendrites from cell bodies of young neurons within these ganglia terminate in equilibrium receptors in the semicircular canals, utriculi, and sacculi.

vital dyes
Dyes used to stain areas of the blastula to determine the prospective fate of each area for prospective fate mapping.

vitelline arteries
Paired arteries continuous with the caudal dorsal aortae; they supply the yolk sac.

vitelline blood vessels
The blood vessels of the yolk sac.

vitelline envelope (or membrane)
In sea urchin embryos, the extracellular layer surrounding the plas-

malemma; in frog embryos, the extracellular layer deposited in the ovary between the follicle cells and plasmalemma of the primary oocyte.

vitelline membranes, inner and outer, of chicken eggs
The membranes surrounding the yolk of chicken eggs.

vitelline vein, common
A prominent vein that joins the superior mesenteric vein; it is derived from an anastomosis between the distal end of the right and left vitelline veins.

vitelline veins
Two large veins that enter the heart; they drain the yolk sac.

W

water-vascular system
In sea urchins, the system connected to the madreporite where chemoreceptor cells trigger the release of gametes in the seawater.

Wharton's jelly
The extracellular matrix of the umbilical cord.

wing buds
In chick embryos, paired buds that consist of a core of somatic mesoderm covered by skin ectoderm.

Y

yolk
The material of the egg that provides a food source for the development of the embryo.

yolk plug
In frog embryos, a circular plug that consists of a protruding mass of yolk-filled endodermal cells located between the very prominent dorsal blastoporal lip above and the less prominent ventral blastoporal lip below.

yolk sac
The extraembryonic membrane that overgrows the yolk; it is formed from splanchnopleure.

Z

zona pellucida
In mouse embryos, the extracellular layer surrounding the egg plasmalemma; it is equivalent to the vitelline envelope/membrane of sea urchin eggs, and the vitelline membrane of frog eggs.

zygote
The egg after fertilization; it undergoes cleavage.

Use the space below for notes.

Index